Astrophysics and Space Science Proceedings
Volume 42

More information about this series at http://www.springer.com/series/7395

Nicola R. Napolitano • Giuseppe Longo •
Marcella Marconi • Maurizio Paolillo •
Enrichetta Iodice
Editors

The Universe of Digital Sky Surveys

A Meeting to Honour the 70th Birthday of Massimo Capaccioli

Springer

Editors
Nicola R. Napolitano
INAF - Observatory of Capodimonte
Napoli, Italy

Giuseppe Longo
Department of Physics "Ettore Pancini"
University "Federico II"
Naples, Italy

Marcella Marconi
INAF - Observatory of Capodimonte
Napoli, Italy

Maurizio Paolillo
Department of Physics "Ettore Pancini"
University "Federico II"
Naples, Italy

Enrichetta Iodice
INAF - Observatory of Capodimonte
Naples, Italy

ISSN 1570-6591 ISSN 1570-6605 (electronic)
Astrophysics and Space Science Proceedings
ISBN 978-3-319-19329-8 ISBN 978-3-319-19330-4 (eBook)
DOI 10.1007/978-3-319-19330-4

Library of Congress Control Number: 2016933675

Springer Cham Heidelberg New York Dordrecht London

Printed on acid-free paper

Springer International Publishing AG Switzerland is part of Springer Science+Business Media (www.springer.com)

Foreword

Over the last decades, digital sky surveys performed with dedicated telescopes and finely tuned wide field cameras have revolutionised astronomy. They have become the main tool to investigate the nearby and far away universe, thus providing new insights in the understanding of the galaxy structure and assembly across time and the dark components of the universe, as well as of the history of our own galaxy. They also have opened the time domain leading to a new understanding of the transient phenomena in the universe.

The astronomical community is pursuing ambitious programmes using new and planned facilities both ground based (e.g. SDSS, VST, VISTA, Skymapper, DEC-@BLANCO and LSST) and from the space (cf. EUCLID). These technologically challenging efforts are producing an unprecedented synoptic (multiwavelength and multi-epoch) view of the universe. Needless to say, the ever-increasing amount, quality and complexity of the data have also led to a stronger connection with other fields of the human endeavour such as databases technology, computer sciences, advanced statistics and machine learning.

By providing public access to top quality data, digital surveys have also changed the everyday practice of the astronomical community who do not need as much as in the past to have direct access to large observing facilities. The full scientific exploitation of these surveys has also triggered significant advances in technology (both from the space and from the ground) and in the field of multiobject spectroscopy.

This conference aims at summarising the results from the main current and past digital sky survey projects and at discussing how these can be used to inspire ongoing projects and better plan the future ones. For this reason the conference will be articulated in long review talks and a small number of shorter contributed talks.

This conference takes place 30 years after the 78th IAU Colloquium *Astronomy with Schmidt Telescopes*, organised at the Asiago Observatory by Massimo Capaccioli who has just turned 70. The conference will therefore be the best venue where to celebrate his life-long involvement in the advancement of science. The conference was held at the Observatory of Capodimonte, located on top of a hill dominating the vibrant bay of Naples with a wonderful view of the Vesuvio and the pearls of the gulf, Sorrento and Capri.

Naples, Italy
June 2015

Enrica Iodice
Giuseppe Longo
Marcella Marconi
Nicola Napolitano
Maurizio Paolillo

Introduction

The challenge of discovery, for astronomers, consists in understanding the physical laws of the Universe. Nowadays this research has to address a major problem regarding the constituents that make up all its density budget and the processes that have shaped the stars and galaxies that we see today. If one looks at the last four decades, we can see that our understanding of the Universe has dramatically changed – we have discovered that the Universe is made up of more than 95 % of unordinary matter and energy that we still don't understand. Dark matter, proposed in the first half of last century and definitely confirmed in 1970s, represents about 25 % of this unknown fraction of the Universe. It is an elusive form of matter that does not shine and that interacts with regular matter (i.e. baryons) only through its gravitational pull. Dark energy, discovered at the end of the 1990s, is filling up the other 70 % of the density budget and is yet an unveiled form of energy acting like a negative pressure, causing the acceleration of the expansion of the Universe, which started with the Big Bang.

Despite our ignorance on the nature of these constituents, the great success of modern cosmology is that it can nicely predict their effects on the evolution of the Universe. The attempt to understand its nature is a challenge so important in modern physics that it is driving not only astronomical research but also particle physics. Today's largest experiments including the word-class particle accelerators (e.g. the Large Hadrons Collider at CERN), have as one of their primary missions the one of finding "the dark matter particle", and a new generation of detectors is being built specifically to address this challenge. If unsuccessful, the time will come to start thinking of alternative theories, including new physics, which will require us to build a new paradigm for the evolution of the Universe. On the other hand, astronomers are building the observational and theoretical tools to understand the other dark ingredient: dark energy. Current and future large sky surveys are all designed to yield significant insights into the puzzle of the nature of this component. In particular, they are trying to constrain its equation of state, which has left its imprint in the large-scale structure of galaxies. This particular challenge is driving the efforts of the astronomical community worldwide and it is based on the observations of the current and next generation of survey telescopes from the ground (e.g. the large synoptic survey telescope, LSST) and from space (e.g. the ESA/NASA Euclid mission).

Astronomy was born as a science based on surveys since Hipparchus, in the second century B.C., observed about 850 stars, motivated by the appearance, in 134 B.C., of a new star in Scorpio, which led him to doubt the Aristotelian claim of the invariability of the sky. After more than 2000 years, astronomy still remains a survey-driven science, which is expected to provide new unexpected discoveries and inspire new generations of researchers. In the conference "The Universe of Digital Sky Surveys", we made an attempt to summarize what has been the role of the astronomical surveys in the digital era. We have started from one of the most successful photometric and spectroscopic surveys, the Sloan Digital Sky Survey, and gathered oral contributions from ongoing state-of-art sky surveys like the Dark Energy Survey and the ESO public surveys VST-ATLAS and the Kilo-Degree Survey (KiDS), to arrive at the future missions from space, like Euclid and WFIRST. Several contributions have emphasized how critical the role of current and future spectroscopic surveys will be: current surveys like e-BOSS, Manga, GAMA and future surveys with the ESO 4MOST and MUSE spectrographs will put more stringent constraints on the dark energy equation of state via the study of the Baryonic oscillation and the so-called redshift-space distortions as well as on galaxy formation and evolution by mapping their internal chemodynamics. Finally, they will provide the census of groups and cluster to compare against cold dark matter (CDM) simulation prediction and test the CDM scenario. ESO public surveys will play a pivotal role in the universe quest, thanks to the latest facilities on Paranal – the near-infrared 4 m telescope VISTA and the optical 2.6 m VST. A special place has been given to the latter in the conference program as the VST was conceived, in the late 1990s, at the Observatory of Capodimonte, where this conference has been held. Since then a long and troubled, but successful, effort has succeeded to place this powerful machine right aside the giant VLT on Cerro Paranal, in 2011. A conference section has been devoted to the ESO commitment to map the sky through large astronomical surveys, and here the VST and VISTA public surveys have illustrated the wealth of discoveries expected in the very near future. Finally, a special section was dedicated to the first results from VST observations with the Italian guaranteed time of observations (GTO), all led by Italian groups with a large participation of international partners. The ongoing programs, started in Fall 2011, are investigating the formation and structure of the Milky Way as well as the spatially resolved star formation history of the Small Magellanic Cloud, the formation of galaxies from the local group to the Virgo and Fornax clusters, up to superclusters, like Shapley. VST-GTO programs have also explored deep regions of space to look for new supernovae, AGNs and to study galaxy formation out to redshift z 1. These have been among the first scientific results based on the VST to be delivered to the community, while the ones based on the ESO public surveys (ATLAS, KiDS and VPHAS+) are about to be published in the near future, including the observations of weak lensing from galaxy groups, galaxy formation and evolution, and the discoveries of new satellites of our Galaxy and high-redshift quasars.

All these achievements have a putative father who, after all, has been the motivation of the whole conference itself and the key factor in bringing together so many astronomers from different fields – Massimo Capaccioli. As the "father" of VST, this facility represents his most recent contribution and long-standing legacy to astronomy and to the exploration of the Universe. After more than 40 years of teaching and researching, after his key contributions to the understanding of galaxy structure and formation, to the study of globular cluster, supernovae and finally dark matter, the VST was with no doubt his greatest effort to provide new generations of scientists with a tool to make forefront science and allow the European community to compete with other world-class facilities. Massimo turned 70 on April and this conference was meant to celebrate, together with his birthday, the VST dream come true, thanks to Massimo's perseverance, despite overwhelming difficulties.
To build a world-class telescope is the dream of most astronomers, the achievement of just a few of them.

Naples, Italy
June 2015

Nicola R. Napolitano
Maurizio Paolillo

Contents

Part II News from VST

Chapter 1
Massimo Capaccioli: A Life for Astronomy

G. Longo and The Organizing Committee

1.1 The Early Years

Massimo Capaccioli was born in 1944 in Castel del Piano, a small village in the "Maremma", a region in the heart of Tuscany. At the age of eight he moved to Mestre, near Venice, maintaining however a deep affection for his roots, often defining himself as a "maremmano". In Mestre he attended schools and in 1964 obtained the diploma at the Scientific Lyceum Giovan Battista Benedetti. He remained in Mestre other three years, commuting every day to Padua to follow the courses of the Laurea in Physics. In 1967 his parents moved to Trieste and he decided to live in Padua, which would become his home for the next 25 years. There, in 1969, he graduated in Physics discussing a thesis on the Hubble law under the supervision of Leonida Rosino and Francesco Bertola with whom he also published his first paper [1]. The same year he was appointed assistant professor in astronomy at the University of Padua, a position he maintained until 1976 when he was promoted adjunct professor of celestial mechanics. From 1972 to 1976, he also thought Mechanics and Cosmology at the University of Lecce, and General Physics and Introductory Analysis at the Faculty of Architecture of the University of Venice. These were years of intense scientific and academic work during which Massimo built a rich network of international collaborations. First among all, his long lasting collaboration with Gérard Henr de Vaucouleurs, among the fathers of modern observational cosmology who at the time was professor of Astronomy at the University of Texas in Austin. Capaccioli visited Austin several times and in 1985 was also appointed visiting professor. Quickly moving up in the academic career, Capaccioli became first associate professor (1980) then, in 1986, Director of Research at the Astronomical Observatory of Padua, a non-academic

---

G. Longo (✉)
Department of Physics "Ettore Pancini", University "Federico II" of Naples, Italy

N.R. Napolitano et al. (eds.), *The Universe of Digital Sky Surveys*, Astrophysics and Space Science Proceedings 42, DOI 10.1007/978-3-319-19330-4_1

position which he held until 1990, when he was finally appointed full professor of celestial mechanics at the University of Padua. In the meanwhile, however, the scientific and personal baricenter of his life was slowly shifting to Naples where in 1992 Capaccioli was appointed Director of the Astronomical Observatory of Capodimonte. A professional and personal transition which became definitive in 1995, when he was called to cover the chair of astronomy at the University Federico II in Napoli.

During his Padova years, Massimo published some of his most significant contributions to astronomy. First of all, in a seminal series of papers, he pioneered the field of galaxy photometry and kinematics, paying special attention to the properties of early type galaxies (see for instance [2–4]). His most striking discovery took place in 1975 when, in collaboration with Francesco Bertola, he demonstrated that, against all current beliefs, elliptical galaxies were not flattened by rotation [5]. A result immediately explained by James Binney in terms of an anisotropy of the velocity dispersion tensor. This discovery was labeled as a "Copernican Revolution" in the field of elliptical galaxies by Beatrice Tinsley, and triggered a whole new field of research which is still very active. Most of his work on the kinematical properties of galaxies was driven by a deep interest in the properties of dark matter. An interest which Massimo still cultivates. Other significant contributions regarded the calibration of the cosmic distance scale [cf. 6, 7], the discovery of flaring stellar disks in edge on lenticulars [8]. In 1978 he also was among the first [9] to use the International Ultraviolet Explorer (IUE) to study the ultraviolet emission from galaxies. A line of research which eventually led to the discovery of the so called Ultraviolet Rising Branch: a sharp rise of the spectral energy distribution in the far ultraviolet which was completely unexpected for early type galaxies [10].

In the Padua years, another thread of his scientific activity regarded the study of the collective properties of early type galaxies [11–13] which led to the discovery that there exist two families of ellipticals separated by a well defined cutoff in absolute magnitude and effective radius . Last but not least, in collaboration with his former student Giampaolo Piotto, Massimo obtained significant results on the properties of globular clusters [cf. 14, 15]. It is worth stressing that most of these works were made in collaboration with his students, who nowadays hold permanent and prestigious positions in many institutes around the world.

1.2 The Napoli Years

Massimo Capaccioli's years in Naples have been difficult and full of events. When the Astronomical Research Council (CRA) appointed him as director of the local observatory, the situation he found was far from optimal. In spite of its historical and beautiful infrastructures, the Observatory of Capodimonte had little scientific momentum and was far from the main stream of Italian and international research. Funds were scarce, and the few staff researchers were scattered in too many research lines. Furthermore, the astrophysics group at the University Federico II was small and, due to a series of events which would be too long to recapitulate, it had little if any interaction with the astronomers at the Observatory. Last but not least, Capaccioli inherited from the previous director Mario Rigutti, the task of completing the construction of the observing station in Castelgrande built around a telescope of only 1.5 m of diameter; a telescope which already at its inception was far from the mainstream of national research. Making a long and troublesome story short, I shall just mention some of the problems that Massimo faced. His arrival in Naples had coincided with the well-known "Mani Pulite" (clean hands) investigation, that in a few years, would decapitate the Italian political class. In those years of turmoil, the fear of being prosecuted paralyzed the public administration, making almost impossible to spend public money. In spite of this unfriendly atmosphere and of many other difficulties, among which also the bankruptcy of the first company in charge of the construction, Massimo and the administrative council of the Observatory took several non trivial administrative risks but in 2001 they succeeded in completing both the telescope and the observing station. But let us return to what was happening at the Observatory in Naples. The beginning was slow and, while gaining visibility in the academic and social life of his new town, Capaccioli steadily guided the Observatory out of the hole. In those years he was appointed member of the scientific council of the Italian Institute of Physical Studies establishing a long lasting friendship with its founder and president Gerardo Marotta. Marotta helped Massimo in connecting to the local and national authorities and in obtaining the first grants which were needed to modernize the observatory. He also became a consultant for the science park "Città della Scienza" for which he cured the first astronomical exhibits, and a member of the "Accademia Pontaniana" (the oldest scientific academy in the world) and of the National Society for Science Letters and Arts of which he was also elected general president for the period 2000–2002. Both academies were hosted in a magnificent but quite disrupted building which Massimo was capable to refurbish by obtaining a 1.8 million euro grant.

After moving to the University of Naples in 1995, in order to modernize the curricula and to increase the number of researchers, Massimo obtained a perpetual grant from the Italian Ministry for University Research Science and Technology (MURST) aimed at funding new permanent positions (at all levels) for astrophysics. Due to the increased personnel and to the improving relationship with the astronomers at the OAC, in 2005, the University Federico II started a new master course in Astrophysics and Space Sciences which, in spite of the initial success, was closed after only four years due to an unexpected and myopic reform of the University. In the meanwhile, after obtaining funds from the new National Consortium for Astronomy and Astrophysics (CNAA) and from the Local Government, he started an intense program of external activities which included, among the others, the refurbishing of the Museum of Astronomical Instruments, a program of Senior Visiting Scientists, as well as a series of public outreach initiatives and events. In order to modernize the Observatory, Capaccioli used his scientific prestige to introduce the observatory staff in several international project and to start new ones. To list just a few: the construction of the wide field imager for the 2.2 m ESO telescope, the design and construction of the VIMOS spectrograph for VLT, the Planetary Nebulae Spectrograph to be installed at the William Herschel Telescope in Canary Islands. Last but not least the VLT Survey Telescope or VST, which will be discussed in the final paragraph. In the meanwhile Capaccioli succeeded in convincing first the CNAA and then the new National Institute for Astrophysics (INAF, established in 2002), to potentiate the scientific staff of the observatory which, in less than 15 years, raised from 7 to 32 permanent members and saw more than 70 people employed on temporary positions. Most of these young collaborators were selected among his former students, but he also tried to incentivate other Italian and foreign capable researchers to move to Naples in order to improve and internationalize the scientific environment . In the meanwhile Capaccioli continued his intense scientific activity and without even trying to summarise it, I shall just mention his contributions to the study of dark matter in early type galaxies using planetary nebulae as kinematical tracers [cf. 16–19], and to the properties of early type galaxies [20, 21].

1.3 The VST Project

In 1995 Capaccioli had obtained two large grants totaling about 18 million euro. Instead of dispersing it in a myriad of minor projects, Capaccioli decided to invest the bulk of the grants in the construction of a modern wide field telescope devoted only to survey work. This decision was formalized in the fall of 1996 with the start of the VST project. The year after, a scientific committee chaired by G.P. Vettolani was appointed to prepare a report on the scientific value of the proposed telescope and to start the interaction with ESO. The project was officially presented in 1998 and had its formal green light in October 1998 when ESO and the Observatory of Capodimonte signed a Memorandum of Understanding (MOU). According to the MOU the OAC committed to the construction of the telescope and ESO to build the in situ infrastructures and to sustain the operational costs for a ten years period. In order to understand how brave was this decision it is worth to quote a few lines from the speech given by Capaccioli in 1998, at the official presentation of the project.

> We are a small Observatory, if compared to the 'giants' of Italian astronomy, all located in the Northern part of the Country. We are placed in a splendid corner of the world, full of history beauty and intelligence, but heavily penalized by a dark past and a difficult present. In the last few years both locally and centrally it has been understood that one of the instruments for the rescue of Naples and of the whole of Southern Italy is 'culture'. This has been the rationale leading to the allocation of a significant amount of special funds to our Observatory, mainly by the Ministry of Education and Research. Suddenly we became a small institute with a lot of money: a welcome fortune, but extremely difficult to invest effectively in view of its 'una tantum' nature, of the bureaucratic boundary conditions and lack of manpower and traditions.

Indeed, the construction of the VST proved to be a complex and arduous task. In 1999 the design was completed, and already the next year construction begun. In 2000 the call for tenders for the construction of the primary and secondary

mirrors was won by the Russian Lytkarino Optiks (LZOS) which delivered the first mirrors in 2002. These mirrors were shipped by boat to Chile but, unfortunately and for reasons that are still largely unknown, during a stopover in a South American harbor, they were destroyed. ESO, who was in charge of the shipping, took full responsibility of the accident and a second set of mirrors was immediately ordered. Due to some problems related to the reorganization of LZOS, the new mirrors – both of excellent quality-reached Paranal only in 2007. In the meanwhile all the mechanical and electronic subsystems had been realized and as early as 2002, the assembly of the telescope had begun in the Mecsud Factory of Scafati, a small town South of Napoli. In parallel, Massimo also launched an international collaboration finalized to the construction of OMEGACAM, at the time the largest CCD camera ever produced (298 Mpix). The camera was completed in 2006. In 2005 and in spite of the enormous progress made by the observatory, with an unjustifiable political decision, INAF decided to change the OAC director and Capaccioli was appointed director of the newly established VST-Center. This unwise decision not only slowed down, almost to a stop, the growth of the local astronomical community, but also created non trivial problems during the final delicate phases of the VST construction. However, in spite of many problems related to a sudden shortage of funds and to some delays in the completion of the infrastructures as well to some engineering problems which required a partial revision of the original design, Massimo struggled to complete the telescope and the VST saw its first light on the 8-th of June 2011. Already the first images proved that the telescope and the camera had excellent imaging capabilities, well beyond the expectations. The VST was offered to the community early in 2012 [22] and is currently engaged in a large number of outstanding surveys such as the Kilo Degree Survey (KiDS) [23], VST Early Type Galaxies Survey [VEGAS 24], VOICE [25], to quote just a few. I shall stop my short biography here but, in closing, I wish to emphasize that Massimo has left a profound mark in the history of Neapolitan astronomy. Under his guidance the Observatory of Capodimonte and the University have entered a new era and gained international reputation. Furthermore, he has formed a new generation of young and motivated scientists, who are currently active not only in Naples but also in prestigious institutes all over the world.
His legacy will last forever.

References

1. Bertola, F., Capaccioli, M.: A new determination of the mass of M87. Mem. SAIt **41**, 57 (1970)
2. Barbon, L., Benacchio, L., Capaccioli, M.: Photometric study by a numerical mapping technique of the trio of galaxies in Leo. A&A **51**, 25 (1976)
3. de Vaucouleurs, G., Capaccioli, M.: Luminosity distribution in galaxies. I – The elliptical galaxy NGC 3379 as a luminosity distribution standard. ApJ **40**, 699 (1979)
4. Capaccioli, M., de Vaucouleurs, G.: Luminosity distribution in galaxies. II – a study of accidental and systematic errors with application to NGC 3379. APJSS **52**, 465 (1983)

5. Bertola, F., Capaccioli, M.: Dynamics of early type galaxies. I. The rotation curve of the elliptical galaxy NGC 4697. ApJ **200**, 439 (1975)
6. Capaccioli, M., Fasano, G.: On the distance of the giant spiral galaxy M101. A&A **83**, 345 (1980)
7. Capaccioli, M., della Valle, M., D'Onofrio, M., Rosino, L.: Properties of the novae population in M31. AJ **97**, 1622 (1989)
8. Capaccioli, M., Vietri, M., Held, E.V.: The z-structure of the disk of NGC 3115. MNRAS **234**, 335 (1988)
9. Bocksemberg, A., et al.: IUE observations of extragalactic objects. Nature **275**, 34 (1978)
10. Bertola, F., Capaccioli, M., Holm, A.V., Oke, J.B.: International ultraviolet explorer observations of M87. ApJ **237**, L65 (1980)
11. Capaccioli, M., Caon, N., Rampazzo, R.: Further evidence of continuity between elliptical and disk galaxies. MNRAS **242**, 24 (1990)
12. Capaccioli, M., Caon, N.: On the correlation between and for early-type galaxies. MNRAS, **248**, 523 (1991)
13. Capaccioli, M., Caon, N., D'Onofrio, M.: Families of galaxies in the mu(e)-R(e) plane. MNRAS **259**, 323 (1992)
14. Piotto, G.P., Capaccioli, M., Ortolani, S., Rosino, L.: C-M diagram and luminosity function of the galactic globular cluster NGC 7099. I – photographic photometry. AJ **94**, 360 (1987)
15. Djorgovski, G.S., Piotto, G.P., Capaccioli, M.: What determines the stellar mass functions in globular clusters? ApJ **105**, 2148 (1993)
16. Arnaboldi, M., Capaccioli, M., et al.: Studies of narrow polar rings around E galaxies. I. Observations and model of AM2020-504. A&A **267**, 21 (1993)
17. Capaccioli, M., et al.: Deep kinematics and dynamics of edge-on S0's. I. A&A **274**, 69 (1993)
18. Romanowsky, A.J., et al.: A dearth of dark matter in ordinary elliptical galaxies. Science **301**, 1696 (2003)
19. Cortesi, A., et al.: Planetary Nebula Spectrograph survey of S0 galaxy kinematics – II. Clues to the origins of S0 galaxies. MNRAS **432**, 1010 (2013)
20. Tortora, C., et al.: Colour and stellar population gradients in galaxies: correlation with mass. MNRAS **407**, 144 (2010)
21. Romeo, A.D., et al.: The evolution of the galaxy red sequence in simulated clusters and groups. MNRAS **389**, 13 (2008)
22. Schipani, P., et al.: VST: from commissioning to science. SPIE **8444**, article id. 84441C (2012)
23. De Jong, J., et al.: The first and second data releases of the Kilo-Degree survey. A&A **582**, 62 (2015)
24. Capaccioli, M., et al.: VEGAS: a VST Early-type GAlaxy Survey. I. Presentation, wide-field surface photometry, and substructures in NGC 4472. A&A **581**, 10 (2015)
25. De Cicco, D., et al.: SUDARE-VOICE variability-selection of active galaxies in the Chandra Deep Field South and the SERVS/SWIRE region. A&A **574**, 112 (2015)

Part I
The Universe of Imaging Surveys

Chapter 2
Digital Sky Surveys from the Ground: Status and Perspectives

T. Shanks

Abstract I first review the status of Digital Sky Surveys. The focus will be on extragalactic surveys with an area of more than 100 deg^2. The Sloan Digital Sky Survey is the archetype of such imaging surveys and it is its great success that has prompted great activity in this field. The latest surveys explore wider, fainter and higher resolution and also a longer wavelength range than SDSS. Many of these surveys overlap particularly in the S Hemisphere where we now have Pan-STARRS, DES and the ESO VST surveys, and our aim here is to compare their properties. Since there is no dedicated article on the VST ATLAS in this symposium, we shall especially review the properties of this particular survey. This easily fits onto our other main focus which is to compare overlapping Southern Surveys and see how they best fit with the available NIR imaging data. We conclude that the Southern Hemisphere will soon overtake the North in terms of multiwavelength imaging. However we note that the South has more limited opportunities for spectroscopic follow-up and this weakness will persist during the LSST era. Some new perspectives are offered on this and other aspects of survey astronomy.

2.1 Introduction

Digital imaging sky surveys are one of the most powerful tools for the modern astronomer. Their scientific heritage includes the twentieth Century Schmidt photographic surveys that surveyed the sky from Palomar, Siding Springs and La Silla to $g \approx 21$m in blue and red bands (e.g. Cannon [2], Woltjer [17]). It took some time for CCD imaging cameras to catch up in terms of field-of-view/area with the photographic surveys but by the beginning of the twenty-first Century they had. The success of the Sloan Digital Sky Survey (Stoughton et al. [11]) has been enormous and the new generation of sky surveys can only hope to emulate its success.

Table 2.1 shows the list of extragalactic imaging surveys of area >100 deg^2 that have been done until now from their starting point around the year 2000. They start

T. Shanks (✉)
Department of Physics, Durham University, Durham, UK
e-mail: tom.shanks@durham.ac.uk

N.R. Napolitano et al. (eds.), *The Universe of Digital Sky Surveys*, Astrophysics and Space Science Proceedings 42, DOI 10.1007/978-3-319-19330-4_2

Table 2.1 Recent optical and NIR extragalactic imaging sky surveys with an area of >100 deg^2. Magnitude limits are quoted in r_{AB} and K_{Vega}

Survey	Type	Epoch	Bands	Lim. mag.	deg^2	N/S	Seeing (arcsec)
DENIS	NIR	1997–2003	iJK	$K \approx 12$	20,000	South	3
SDSS	Visible	2000–2005	*ugriz*	$r \approx 22.7$	14,500	North	1.2
CFHT RCS2	Visible	2002–2009	*grz*	$r \approx 24.8$	830	N+S	0.9
CFHTLS Wide	Visible	2003–2012	*ugriz*	$r \approx 25$	157	North	0.9
2MASS	NIR	1997–2001	*JHK*	$K \approx 14.3$	All sky	N+S	1.5
UKIDSS	NIR	2005–2012	*YJHK*	$K \approx 18.4$	7,500	North	0.9
WISE	Mid-IR	2010–2012	$3.4 - 22\,\mu$m	$W1 \approx 17$	All Sky	N+S	6
Pan-Starrs 3π	Visible	2010–2014	*grizy*	$r \approx 22.8$	30,000	N+S	1.1
SkyMapper	Visible	2009–	*uvgriz*	$r \approx 22.0$	20,000	South	2.5
VST ATLAS	Visible	2011–	*ugriz*	$r \approx 22.7$	4,700	South	0.9
VST KiDS	Visible	2011–	*ugri*	$r \approx 24.6$	1,500	South	0.7
VISTA VHS	NIR	2010–	*YJKs*	$Ks \approx 18.2$	18,000	South	0.7
VIKING	NIR	2010–	*zYJHKs*	$Ks \approx 19.5$	1,500	South	0.9
DES	Visible	2013–	*grizy*	$r \approx 25.0$	5,000	South	0.9
DECaLS	Visible	2015–	*grz*	$r \approx 23.6$	9,000	North	1.2
HSC Wide	Visible	2015–	*grizy*	$r \approx 26.0$	1,400	South	0.7

with the DENIS NIR Survey from a 1-m telescope covering the S Hemisphere to $K \approx 12$ and SDSS covering the N Hemisphere in the optical to $r \approx 22.7$ from a 2.5-m telescope at Apache Point, New Mexico. The CFHT MegaCAM instrument also featured powerfully in the early digital surveys with its 1 deg field and high UV sensitivity producing the Red Cluster Sequence surveys to $r \approx 24.8$ over 900 deg^2 and then the CFHT Legacy Survey with its Wide component extending over some 157 deg^2. NIR detectors were increasing in size, leading then to the 2MASS survey in JHK over the full sky and UKIDSS much deeper in YJHK but with only partial coverage of the northern sky. Also completed by 2012 was the WISE survey in the 3.6, 4.5 12 and 22 μm bands. This was, of course, a satellite survey but as we shall see this meshes in with the ground-based optical surveys that it is hard to leave out of this list. This leads to the more modern surveys that we are going to feature specifically in this review. These include PanSTARRS 3π survey which covers 30,000 deg^2 in *grizy* predominantly in the North but also reaches as far South as Dec $= -30$ in the Southern Hemisphere. Then there are the ESO Southern Hemisphere surveys using the VLT Survey Telescope in the optical for KiDS and ATLAS and using the VISTA telescope in the NIR for the VISTA Hemisphere Survey (VHS) and for VIKING. Finally there is the DES and DECaLS surveys imaging in respectively *grizy* to $r \approx 25.2$ over 5,000 deg^2 in the Southern Hemisphere and *grz* to $r \approx 23.6$ in the Southern half of the SDSS footprint.

Currently it is becoming apparent that the greatest digital imaging strides are currently being taken in the Southern Hemisphere due to the availability of

VISTA, DECam and VST OmegaCam. The focus of this review is to assess the complementarity of the Southern optical and NIR imaging surveys. The case that will be made is that DES is a powerful optical survey but can still be matched by combinations of the other surveys such as KiDS, Pan-STARRS and ATLAS. Also DES may be less well matched than the others in terms of the depth of its accompanying NIR surveys.

2.2 ESO Surveys

Here we call the 'ESO surveys' the surveys by the ESO VST (Schipani et al. [8], see Fig. 2.1) and VISTA (Sutherland et al. [13]) telescopes. We shall discuss the ATLAS (Shanks et al. [9, 10]) and KiDS (de Jong et al. [4]) surveys from VST and the VHS (McMahon et al. [6]) and VIKING (Sutherland [12]) surveys from VISTA. We shall discuss the VST ATLAS survey at greater length since there is no dedicated ATLAS article elsewhere in this volume. The sky footprints of the ESO ATLAS, KiDS, VHS, VIKING and PanSTARRS 3π, DES and PanSTARRS surveys are shown in Fig. 2.2.

Fig. 2.1 The 2.61-m ESO VLT survey telescope (VST) in its dome at Paranal observatory

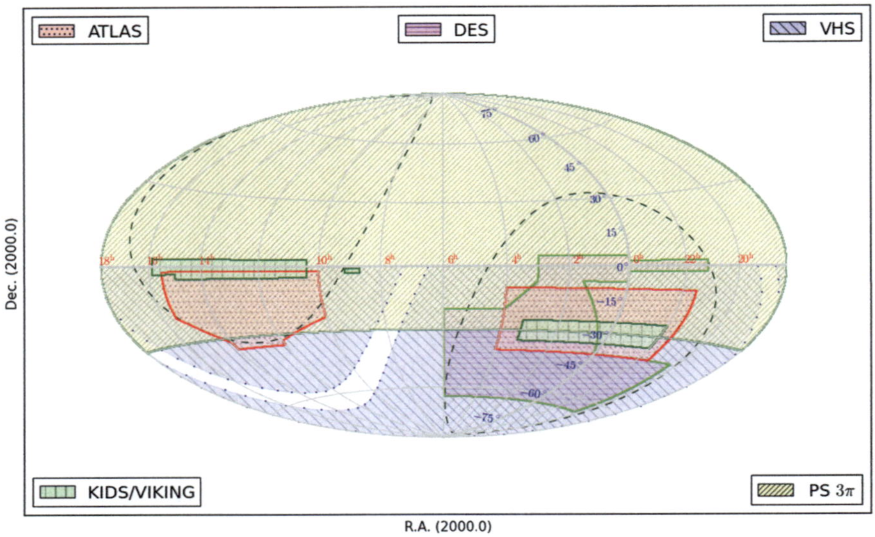

Fig. 2.2 The sky areas covered by the ESO VST ATLAS, KiDS, ESO VISTA VHS, VIKING, DES and PanSTARRS 3π surveys

2.2.1 *VST ATLAS*

The ESO VST ATLAS targets 4,700 deg^2 of the Southern Hemisphere in the *ugriz* bands to similar depths to the SDSS survey in the North. Shanks et al. [10] report that the throughput for VST + OmegaCam is usually similar to SDSS but the seeing for VST in all bands is significantly better than for SDSS. For example, ATLAS achieves a median of $\approx 0.''8$ FWHM in the *i* band and $\approx 1.''0$ in the *u* band; the equivalent numbers for SDSS are $1.''2$ and $\approx 1.''5$ so in general a 50 % improvement. This means there is generally an advantage for ATLAS over SDSS in terms of the 5σ magnitude limit for stellar sources amounting to ≈ 0.2 mag in *ugr* and rising to ≈ 0.3 mag in *i* and ≈ 0.7 mag in *z* where the throughput is also 50 % higher than for SDSS. For resolved galaxies the ATLAS and SDSS magnitude limits are more comparable. All of the ATLAS *ugriz* bands have $\approx 2\times$ longer exposure times than the 54*s* of SDSS to accommodate the smaller $0.''2$ pixel size of OmegaCam to account for the increased readout noise and for the use of grey time in the *iz* bands. There is also an ongoing 'Chilean ATLAS *u*-band extension' (PI L. Infante) where the ATLAS *u*-band exposure time is doubled to 4 min.

The original science aims of ATLAS were first to exploit the *u*-band coverage of ATLAS to make UVX quasar searches. Now one such survey has already been done, the 2QDES pilot survey, comprising $\approx$10,000 quasar redshifts to $g \approx 22.5$ (Chehade et al. 2015 in prep.). We found that *u* selection still reaches deeper than cuts that use the WISE W1, W2, such as $g-i : i-W1$, although these appear to show less contamination by stars (see Fig. 2.3). Another top priority for VST ATLAS is to search for the Integrated Sachs-Wolfe effect in the Southern Hemisphere by

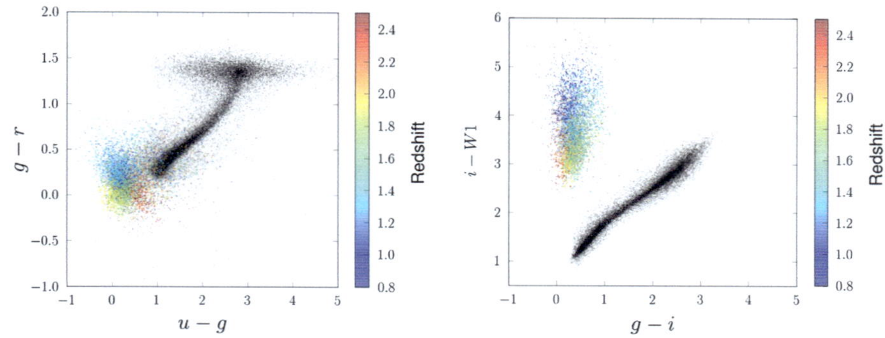

Fig. 2.3 *Left*: *ugr* quasar selection. *Right*: *giW*1 quasar selection. In an ATLAS-based survey of ≈10,000 quasars, Chehade et al. (2015 in prep). find that although the *giW*1 separates quasars from stars more cleanly, *ugr* selection still reaches fainter limits, $g \approx 22.5$ mag

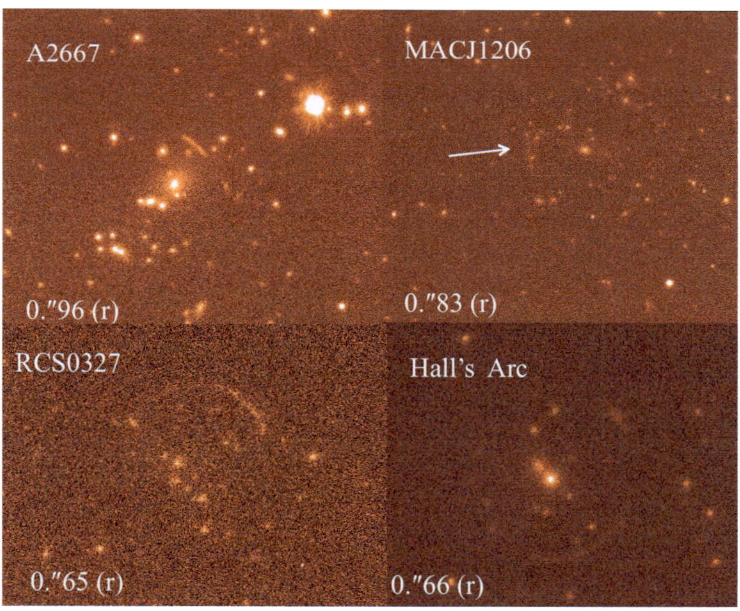

Fig. 2.4 Previously known galaxy clusters with gravitationally lensed arcs detected in subarcsecond ATLAS images

cross-correlating the positions of Luminous Red Galaxies (LRGs) and microwave background fluctuations. Here again we shall be selecting LRGs using traditional *griz* colour cuts complemented by WISE magnitudes at higher redshifts. Photo-z catalogues for ATLAS are being prepared and we find the best results using ANNz neural networks that incorporate both ATLAS *ugriz* and WISE *W*1, *W*2 magnitudes.

The improved seeing also opens up new avenues for ATLAS projects in terms of both finding multiple quasar lenses and lensed arcs in galaxy clusters (see Fig. 2.4).

Most ATLAS tiles have at least one band with significantly sub-arcsecond seeing. P.L. Schechter has therefore initiated a programme to find quadruple lenses in ATLAS images. Meanwhile Carnall et al. [3] have combined ATLAS and WISE photometry to confirm spectroscopically confirm 3 bright $z > 6$ quasars, the first from ATLAS.

2.2.2 VST KiDS

KiDS (PI K Kuijken) is the premier VST survey, aimed at covering 1,500 deg^2 in *ugri* split between an NGC field at the equator and an SGC field at Dec $= -30$. These field comprise the original areas of the 2dF Galaxy Redshift Survey. The survey aims to reach $r \approx 25.2$ with the best seeing on the VST reserved for the KiDS r band where the median seeing is currently $\approx 0.''7$. The main purpose is to do weak lensing tomography and, with the remarkably distortion-free field of VST and 9-band photometry (including VISTA VIKING *zYJHK*) for photo-z, KiDS seems ideally positioned. KiDS DR2 released 172 deg^2 with another year's worth of data already taken (de Jong et al. [4]). The DR2 coverage means KiDS' area is already larger than the CFHTLS Wide survey, the previously largest lensing survey. The main problem for KIDS is that currently it is estimated to take until 2020 to complete. The i band is already reasonably complete but *ugr* are further behind. Figure 2.5 shows weak lensing results from KiDS that measure the mass correlation function at $z \approx 0.4$ (K. Kuijken, priv. comm.).

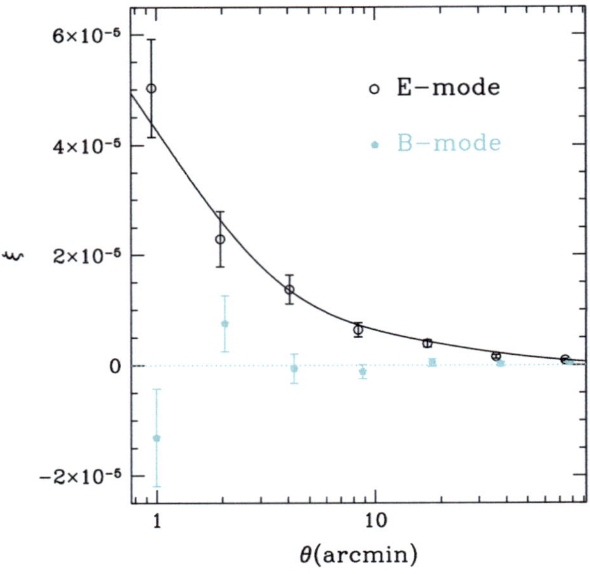

Fig. 2.5 Weak lensing shear results from KiDS, measuring the mass correlation function at $z \approx 0.4$. The low amplitude of the B-modes provides upper limits on how systematics affect the lensing E-modes

2.2.3 *VISTA VHS*

VISTA VHS (McMahon et al. [6]) is the sister survey to VST ATLAS and also to DES. VISTA is a 4.1-m telescope sited near Cerro Paranal and dedicated to NIR imaging. Its 1.5 deg^2 tile 'images' are composed of 6 individual pawprints. VHS-ATLAS covers 5,000 deg^2 mainly overlapping the VST ATLAS survey with exposures of Y(120s), J(60s), Ks(60s). VHS-DES covers 4,500 deg^2 (excluding the 500 deg^2 of the SGC VIKING footprint) and complements the DES survey with exposures of J(120s) and Ks(120s). At lower latitudes there is also VHS-Galactic Plane survey covering another 8,200 deg^2 with exposures of J(60s), K(60s). Thus in the ATLAS area VHS reaches $Ks = 18.2$(Vega), similar to the UKIDSS limit, and in the DES area it reaches $Ks = 18.5$. Here again VHS takes the poorer seeing VISTA time with the better seeing going to VIKING and other projects such as UltraVISTA. One of the prime aims of VHS is to look for high redshift quasars and therefore that is why there is so much emphasis on Y, J. (In the DES area Y is done by DECam). Other aims are to detect the faintest brown dwarf stars and to provide NIR magnitudes to help measure photo-z for the ATLAS and DES surveys. Currently VHS ATLAS is $\approx$50 % complete and VHS-DES is $\approx$70 % complete.

2.2.4 *VISTA VIKING*

The VIKING survey (Sutherland [12]) aims to cover the same 1,500 deg^2 area as KiDS in $zYJHKs$ to a 5σ detection limit of $Ks \approx 19.5$ (Vega). VIKING includes the 2dF and GAMA galaxy redshift survey areas, the Herschel ATLAS, WALLABY-ASKAP HI survey and the 2dFLens survey. VIKING is now back on the sky again after a gap of about a year due to management changes. VIKING has now covered $\approx$1,000 deg^2 with at least one of ZYJ or $JHKs$ visits. This is a good match to the KiDS coverage in the i-band.

Scientifically, the survey has already found several high redshift $z > 6$ quasars (Venemans et al. 2013 [16] – see Fig. 2.6). At the opposite extreme of redshifts,

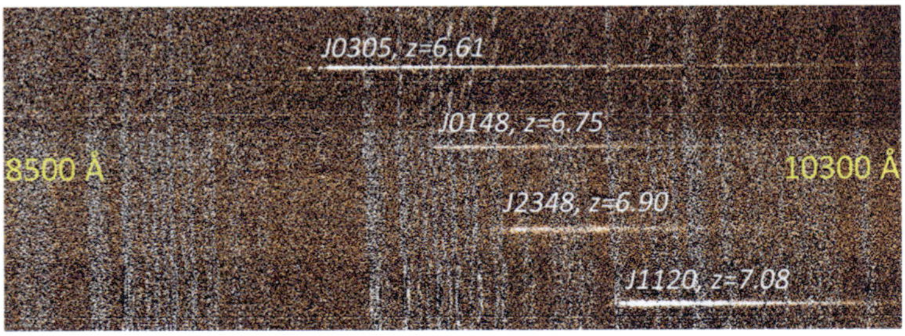

Fig. 2.6 3 VIKING $z > 6$ quasars compared to the $z \approx 7$ quasar from UKIDSS

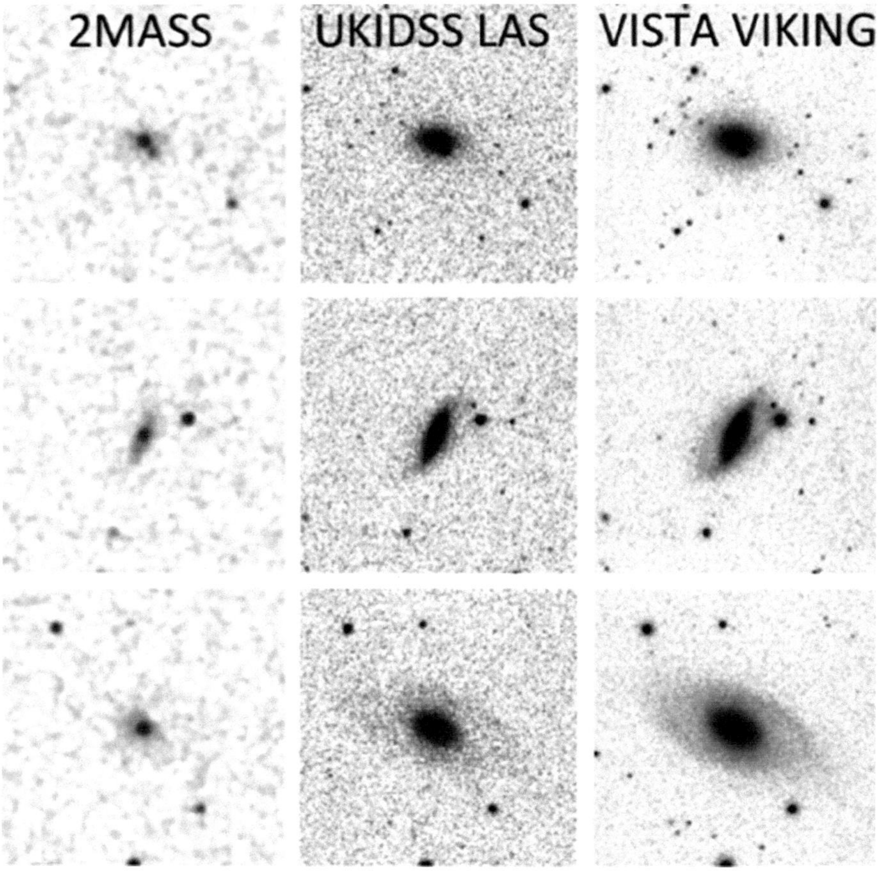

Fig. 2.7 *Left*: 2MASS *Middle*: UKIDSS *Right*: VIKING. These K-band images of 3 galaxies shows the significantly fainter surface brightness detection limits of VIKING as compared to 2MASS and UKIDSS

the additional depth of VIKING over UKIDSS LAS or the VHS means that galaxy surface brightness sensitivity is significantly better (see Fig. 2.7). So for brighter galaxies at $z < 0.2$, NIR morphological parameters can be extracted and this is a primary focus of the GAMA survey.

The parameters extracted from the imaging (total stellar mass, bulge-to-disk radio, etc.) are a key component of the GAMA project (PIs Driver and Hopkins) that covers a significant part of the VIKING/KiDS area with AAOmega to recover spectra for a much denser sampling than traditional redshift surveys. GAMA will be complemented with the WALLABY radio survey that will recover the 21 cm HI emission of $z < 0.25$ galaxies.

The largest Open Time Key Projects with Herschel was H-ATLAS (PI Eales). It covers $\approx$550 deg^2 of which $\approx$400 deg^2 is in the VIKING/KiDS footprint. The wide redshift range of the Herschel sources, accentuated by lensing, means that NIR imaging is crucial to the identification of these sources.

The VIKING/KiDS NGC area will also be covered by the HyperSuprimeCam (Miyazaki et al. [7]) Wide survey area (2015–2020) which will image *grizY* to 26.5, 26.1, 25.9, 25.1 and 24.4, $\approx$2.5 mag fainter than the KiDS limits and $\approx$1 mag fainter than the DES 5 Year limits (see Table 2.2). In total HSC Wide will cover 1,400 deg^2 at Dec $\approx$ 0 deg.

2.3 Pan-STARRS 3π Survey

The Pan-STARRS 3π survey (Tonry et al. [14]) is an ambitious survey aimed at covering 3/4 of the sky in the *grizy* optical-NIR bands (see Fig. 2.8). It uses a 1.8-m, $f/4.4$ telescope at Haleakala Observatory on Maui, Hawaii with an $\approx$7 deg^2 field-of-view. The 3π survey was done with a cadence of 6 epochs with each epoch consisting of a pair of exposures taken $\approx$25 min apart. Individual exposure times vary between 30–45s skewed towards the bluer bands. The detectors are OTA arrays which have high red sensitivity and pixel widths of $0.''26$. Seeing is typically $1.''1 - 1.''3$. It is interesting to compare light grasp + throughput with VST ATLAS. ATLAS has a 2.1$\times$ bigger mirror area and better throughput in *gri*. We estimate that PS1[1] has $\approx$1.5$\times$ higher throughput in the z band and 2$\times$ less throughput in the g band. We make a comparison of ATLAS exposure times, median seeing and magnitude limits with Pan-STARRS 3π (and KiDS and DES) in Table 2.2. The

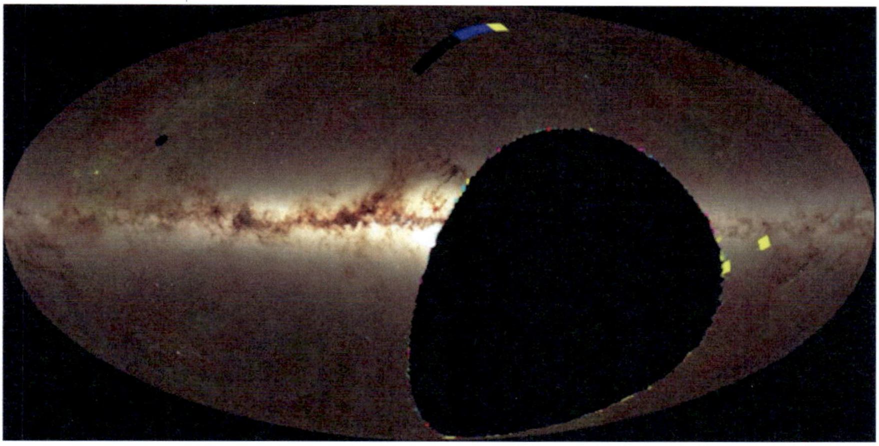

Fig. 2.8 The Pan-STARRS PS1 3π survey. The unsurveyed area has Dec < -30 deg

[1] PS1 is the first of 4 planned Pan-STARRS telescopes.

Table 2.2 ATLAS-PanSTARRS PS1 3π-KiDS-DES comparison. ATLAS median seeing for ESO A,B classified tiles. ATLAS Mag Lim corresponds to the median 5σ magnitude detection limit for stars as measured in a $1''$ radius aperture. PS1 Mag Lim is the measured 5σ magnitude limit inside a $3''$ diameter aperture. KiDS and DES Mag Lims are 5σ limits for stars aperture corrected in a $2''$ diameter aperture. Sky brightness is measured in ABmag/arcsec^2. All magnitudes are quoted in the AB system. DES Y1 and Y5 refers to the DES Year 1 and Year 5 exposures and magnitude limits

Survey/Band	g	r	i	z
ATLAS exposure	2×50 s	2×45 s	2×45 s	2×45 s
Mean PS1 exposure	8.4×43 s	8.4×40 s	8.8×45 s	9.7×30 s
DES Y1 exposure	4×90 s	4×90 s	4×90 s	4×90 s
KiDS exposure	900 s	1,800 s	1,080 s	500 s
DES Y5 exposure	10×90 s	10×90 s	10×90 s	10×90 s
ATLAS seeing	$0.''95$	$0.''90$	$0.''81$	$0.''84$
PS1 seeing	$1.''33$	$1.''19$	$1.''13$	$1.''08$
KiDS seeing	$0.''8$	$0.''7$	$0.''75$	$0.''8$
DES seeing	$1.''0$	$0.''94$	$0.''94$	$0.''94$
ATLAS Mag Lim	23.14	22.67	21.99	20.87
PS1 Mag Lim	23.05	22.85	22.45	21.85
DES Y1 Mag Lim	25.0	24.5	24.1	23.3
KiDS Mag Lim	25.0	24.8	23.8	(23.1)
DES Y5 Mag Lim	25.5	25.0	24.6	23.8

total exposure time of PS1 is 5.2× longer in g and 4× longer in i. However, the seeing is 20 % worse in all bands and sky brightnesses are approximately the same. Point source magnitude limits should therefore be similar in gr, ≈0.5 mag fainter in i and ≈1 mag fainter in z (see Table 2.2). Similar conclusions apply to gri for the SDSS-PanSTARRS comparison but the PanSTARRS advantage over SDSS in z is now ≈1.7 mag for stars (Shanks et al. [10]).

2.4 Dark Energy Survey

The Dark Energy Survey (DES) is aimed at KiDS depths over ATLAS areas. It uses the Dark Energy Camera (Flaugher [5]) installed at the prime focus of the CTIO 4-m Blanco Telescope. It is aimed at covering 5,000 deg^2 over a period of 5 years starting in August 2013. It aims to do a pass with 4×90 s exposures in $grizy$ for the first 2 years increasing this to 10×90 s by the end of Year 5. DECam has high throughput especially in i and z due to its deep depletion CCDs. The DECam pixel size is $0.''263$. The quality of its imaging can be seen in Fig. 2.9 which compares DES and HST images of gravitationally lensed arcs in a galaxy cluster.

The aims of DES are mostly cosmological to determine the dark energy equation of state by a variety of probes including cluster counts, weak lensing and large-scale structure. They are also covering 30 deg^2 in time domain mode to search

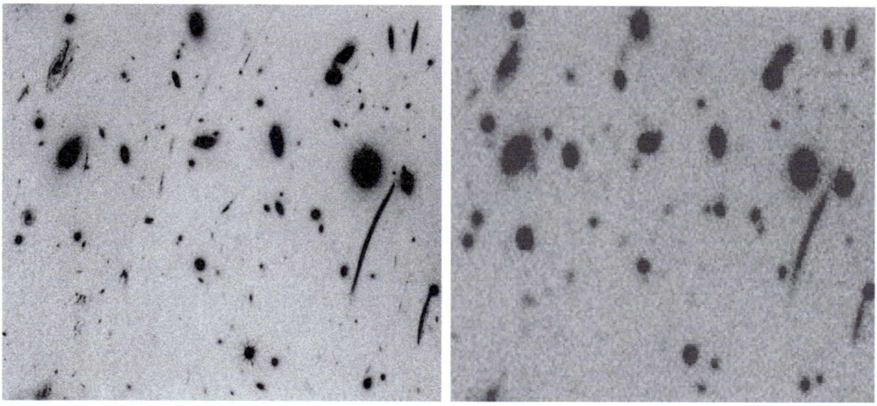

Fig. 2.9 *Left*: HST F435W image of galaxy cluster RXJ2248 from the CLASH collaboration. *Right*: DES *riz* image of the same cluster

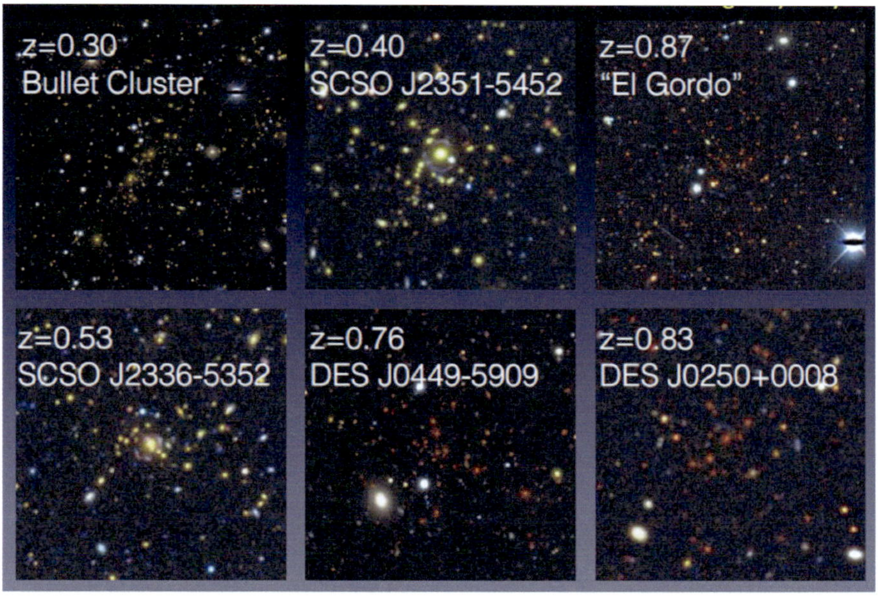

Fig. 2.10 DES images of lensing clusters out to redshift, $z \approx 0.9$ (E. Rykoff et al. priv. comm.)

for supernovae. But it is the wide field aspect of DES that is our focus here. The cluster counts include cross-correlation with the South Pole Telescope CMB SZ measurements (see Fig. 2.10). Much will be learned about the astrophysics of cluster gas as well as cosmology. The large-scale structure aims include galaxy and quasar power spectra at the largest scales partly to constrain neutrino masses. Here there will be a need for accurate photo-z but these will only be available out to $z \approx 1$ given the lack of deeper NIR photometry beyond Y. Clearly the VHS *JHKs* photometry

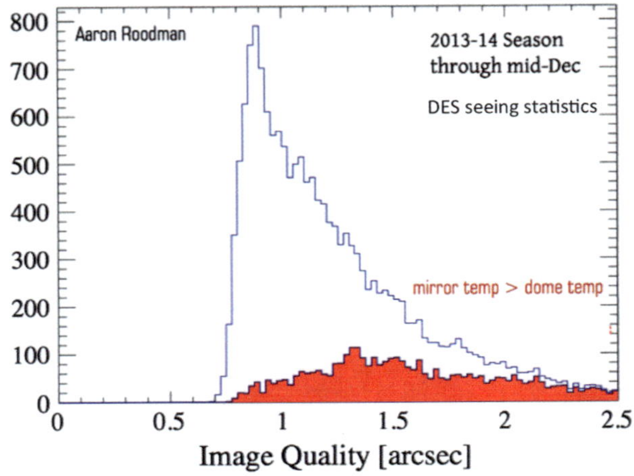

Fig. 2.11 DES seeing statistics from Year 1 (A. Roodman priv. comm.). The median (sub-selected) seeing in the r-band is $0.''94$ FWHM compared to $0.''9$ specification. The *red histogram* demonstrates the importance of maintaining the mirror temperature below the dome temperature

will provide opportunities for very high redshift quasar z-dropouts at $z > 7$ but there may be a future need for VISTA to provide *JHKs* imaging to VIKING depths to fully exploit DES *grizy* over the full survey area.

For DES, the main target must be weak-lensing shear. DES certainly provides enough r-band exposure to reach $r \approx 25$ at 5σ after 5 years. The other two necessary components are the consistency of the PSF over the full focal plane and the delivered seeing. Figure 2.11 shows that it is possible for DES to provide median $0.''94$ seeing in the r band, close to its $0.''9$ specification for weak lensing. However, this is mainly produced by sub-selecting the best seeing for r. There also seems to be a lower limit to DECam seeing of $0.''75$ in the r band. On the issue of the uniformity of the PSF there is less information available, although weak lensing maps have already been produced using DES data.

2.5 DES Versus Other Surveys

Clearly other surveys are now playing David to the Goliath survey that is DES. In terms of DES versus ATLAS, DES reaches further South on the SGC side of the sky than ATLAS whereas ATLAS also covers the NGC side. The 5th band for ATLAS is the u band to $u = 22.4$ (including Chilean ATLAS u extension) whereas the 5th band for DES is y with the 5σ limit $y = 22.7$ in Year 5. As can be seen from Table 2.2 ATLAS median seeing is slightly better in all bands than DES. Pan-STARRs 3π also has its 5th band in y rather than u, reaching $y = 20.85$ at 5σ. Again its seeing is worse than both DES and ATLAS. KIDS is most competitive with DES in the r band where KiDS' seeing is significantly better (see Fig. 2.12). DES Y5 r magnitude limit is 0.2 mag fainter. In the other bands, DES Y5 will be significantly (0.5–0.8 mag) fainter than the KiDS limits. VIKING, however, is well matched to KiDS and VHS is not so well matched to DES except for the purpose

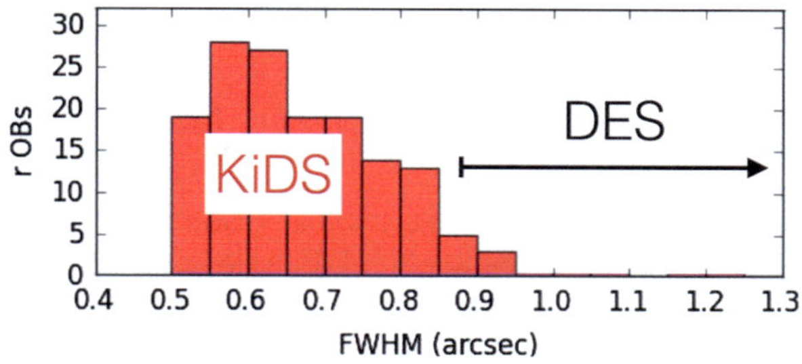

Fig. 2.12 KiDS seeing statistics compared to the DES Year 1 range from Fig. 2.10 (K. Kuijken, priv. comm.)

of high redshift quasar searches. For many other purposes VHS is a better match to ATLAS. The same could be said for WISE – it is a better match to SDSS depth surveys such as ATLAS than it is to KiDS or DES.

SDSS has also been a great success because of its access to Luminous Red Galaxies extending out to $z \approx 0.75$. This can now be extended to $z \approx 1$ by incorporating WISE W1 and W2 data into ATLAS. By $z \approx 1.3$ the 4,000 Å break has redshifted out of the optical bands and $z > 1.3$ LRGs are likely too faint even for UKIDSS LAS and VHS. Emission line galaxies out to $z \approx 0.7$ can be selected at ATLAS depth (Bielby et al. [1]) but it is undeniable that deeper surveys such as DES can in principle probe out to $z \approx 2$. But the lack of matched NIR data will make it difficult to measure photo-z past the VHS limit. However, the lack of a deep u band makes it hard to look for $2 < z < 3.5$ dropouts. The $3.5 < z < 4.5$ g dropouts are a possibility but even the DES 5-year limits will make it tough to detect these.

The basic advantage of SDSS depth in the optical is that the galaxy number counts turn over within the SDSS range and most of the fainter galaxies are also intrinsically faint, so the majority of galaxies are probing the same volume as those at brighter limits. Of course, the brightest of the more distant galaxies are there but until g dropouts become available at $z \approx 4$, these will remain expensive to identify even for a survey like DES. The effective volume of the lower redshift is optimised where the galaxies reach the knee in the luminosity function at M*. At this point correlation functions and power spectra produce their smallest errors in the most cost-effective exposure time.

The great overpowering advantage that the depth of DES allows is of course for cosmological weak-lensing studies. The image quality is $0.''94$ and this is judged adequate for this purpose. The main competition here is from KiDS where the median seeing is $0.''7$ a 25 % improvement on DES. The KiDS magnitude limit in r is 24.8 mag which is lies between the DES 1–2 year and 5 year limits. Of course, weak-lensing demands uniform PSF's across the field-of-view and this has already been demonstrated for KiDS and this is still under scrutiny for DES. Speed of survey is also slower for KiDS than DES given its $\approx 3\times$ bigger field and $\approx 2.4\times$ bigger mirror area. The KiDS SGC area is half covered by DES and so this should make for interesting lensing comparisons with DES. In this area at least DES will ultimately be able to incorporate the VIKING JK data for photo-z.

2.6 Conclusions

2.6.1 No Access to DES? Stay Calm and Don't Panic!

So what can be done if you are at an institute with no access to DES? The message is to stay calm – in the Southern sky at least the ESO Public Surveys can help. Much of the low lying fruit can be accessed with SDSS depth surveys such as ATLAS and then there is KiDS, VIKING and VHS too. Over wider areas the PanSTARRS 3π survey will soon become available and its $\approx 70\,\text{deg}^2$ of Medium Deep Survey fields will also be available. For ATLAS there is the pleasing match of the WISE Mid-IR surveys to its depth, now being exploited for LRG and high-z quasar studies. The prime science area for DES is weak lensing but even here KiDS can compete with its combination of similar *r*-band depth and $\approx 25\,\%$ better seeing.

In the future, the 8-m LSST (Tyson [15]) will reach an AB depth of 26.1–27.5 in *ugrizy* over the whole Southern sky. For synoptic surveys, LSST will have an unparalleled reach but meanwhile there is much work that can be done now in PS1 MDS and the overlap areas between PS1 3π, ATLAS, SDSS, KiDS and DES, not to mention VHS and VIKING.

2.6.2 But No Access to Spectroscopic Follow-Up?

For LSST photometric redshifts are hugely important. Indeed, the LSST 'black book' barely makes mention of spectroscopic follow-up. Generally, the Southern sky looks weakest compared to the Northern Hemisphere in terms of spectroscopic follow-up. There is AAT 2dF + HERMES and there will be MOONS but the latter has a small $\approx 25'$ field similar to Gemini FMOS but with more fibres ($\approx$1,000). VISTA 4MOST with $\approx$1,000 fibres over $3\,\text{deg}^2$ should be excellent for ATLAS follow-up but perhaps needs more fibre numbers to match the high sky densities of galaxies and stars available from the deeper surveys. There have been more ambitious designs made such as NG2dF and VXMS to make multislit spectrographs to cover $> 3\,\text{deg}^2$ but these have generally fallen foul of the current need for high resolution ($R > 10{,}000$) spectroscopy for Galactic Archaeology (coupled to GAIA astrometry), which is done better with fibres. But there remains possibilities at 2dF and at DECam to build extreme multislit spectrographs where $\approx$10,000 galaxy redshifts can be measured simultaneously. Combining photo-z information from the deep imaging surveys with low S/N spectroscopic redshifts, reduce the redshift error from $\pm 10{,}000$ to $\pm 100\,\text{km s}^{-1}$ is also a possibility that is worth investigating. The main aim of these surveys would be to measure redshift-space distortions combined with lensing tomography.

There may be a contribution to spectroscopic surveys that VST can make once its current imaging surveys are completed. Even extreme multislit spectrographs may miss interesting objects because of the need to target previously imaged galaxies and stars to manufacture the slit mask. However slitless spectroscopy could offer a potentially exciting route for a telescope like VST. An objective prism is infeasible because of the large mirror diameter. Instead, it has been suggested that a grens for VST that might be easily accommodated in the filter wheel could offer low dispersion spectroscopy over the full 1 deg field (R. Content, 2014, priv. comm.). There would be a price to pay in terms of sky noise but previous experience at UKST and CFHT suggest that 30 min exposures could produce good emission line spectra for ≈ 100 $g < 22$ mag quasars in a typical VST field. It may also be possible simultaneously to measure redshifts for at least several hundred strong emission-line and absorption-line galaxies. But for slitless the main advantage is that we get to observe ALL the spectra with no need for pre-selection for fibres or masks, with immediate advantage for future searches for gravitational wave source counterparts etc.

2.6.3 Ground-Based Image Competition for EUCLID?

Given the excellent seeing that VST has experienced, for the final perspective, it may also be worth thinking about improving its optics to see if even better seeing could be accessed. It might not take too much of an improvement to make VST competitive with EUCLID, for example, at least over small areas of a few hundred square degrees with 'sub-selected' ground-based seeing. Even if this proved impossible with VST, it would make a good proving ground for a more sophisticated telescope of similar size to see what are the limits of what can be achieved by a ground-based telescope on an excellent site such as Paranal. Recall that this goal of using active *and* adaptive optics over a wide field was already the ambitious original aim of the PanSTARRS project. Maybe this idea came too early for the technology and perhaps its time will come again?

Acknowledgements I should like to thank N. Metcalfe and B. Chehade (Durham Univ.), M.J.Irwin and E. Gonzalez-Solares (CASU Cambridge) and R.G. Mann, M.A. Read and colleagues from WFAU, Edinburgh for all their help with the VST ATLAS survey. We are also deeply grateful to the OmegaCAM and VST teams and the ESO VST observers whose dedication makes the ESO VST surveys possible. I should also like to thank K Kuijken (Leiden), A.C. Edge (Durham), O. Lahav (UCL) and S. Smartt (QUB) for allowing me to present their images illustrating highlights from the KiDS, VIKING, DES and PanSTARRS surveys. Finally I should like to thank the Scientific Organising Committee for inviting me to give this review at the Universe of Digital Sky Surveys meeting in honour of the 70th birthday of Prof. Massimo Capaccioli.

References

1. Bielby, R., Shanks, T., Sawangwit, U., et al.: Photometric selection of emission-line galaxies, clustering analysis and a search for the integrated Sachs-Wolfe effect. MNRAS **403**, 1261 (2010)
2. Cannon, R.D.: Sky surveys with the UK 1.2-metre Schmidt Telescope. In: Capaccioli, M. (ed.) Astronomy with Schmidt-Type Telescopes, IAU Colloquium 78, vol. 110, p. 25. Reidel, Dordrecht (1984)
3. Carnall, A.C., Shanks, T., Chehade, B., et al.: Two bright $z > 6$ quasars from VST ATLAS and a new method of optical plus mid-infrared colour selection. MNRAS **451**, L16 (2015)
4. de Jong, J.T.A., Kuijken, K., Applegate, D., et al.: The Kilo-degree survey. The Messenger **154**, 44 (2013)
5. Flaugher, B., Diehl, H.T., Honscheid, K., et al.: The dark energy camera (2015). arXiv:1504.02900
6. McMahon, R.G., Banerji, M., Gonzalez, E., et al.: First scientific results from the VISTA hemisphere survey (VHS). The Messenger **154**, 35 (2013)
7. Miyazaki, S., Komiyama, Y., Nakaya, H., et al.: Hyper Suprime-Cam. In: Proceedings of SPIE, Amsterdam, vol. 8446, p. 84460Z (2012)
8. Schipani, P., Capaccioli, M., Arcidiacono, C., et al.: VST: from commissioning to science. In: Proceedings of SPIE, Amsterdam, vol. 8444, p. 84441C (2012)
9. Shanks, T., Belokurov, V., Chehade, B., et al.: VST ATLAS first science results. The Messenger **154**, 38 (2013)
10. Shanks, T., Metcalfe, N., Chehade, B., et al.: The VLT survey telescope ATLAS (2015). arXiv:1502.05432
11. Stoughton, C., Lupton, R.H., Bernardi, M., et al.: Sloan digital sky survey: early data release. Astr. J **123**, 485 (2002)
12. Sutherland, W.: VIKING: the VISTA Kilo-degree INfrared Galaxy survey. Science from the Next Generation Imaging and Spectroscopic Surveys, ESO, vol. 40 (2012)
13. Sutherland, W., Emerson, J., Dalton, G., et al.: The visible and infrared survey telescope for astronomy (VISTA): design, technical overview, and performance. Astron. Astrophys. **575**, A25 (2015)
14. Tonry, J.L., Stubbs, C.W., Lykke, K.R., et al.: The Pan-STARRS1 photometric system. ApJ **750**, 99 (2012)
15. Tyson, A.: LSST and Dark Energy, Observing Dark Energy. ASP Conference Series, vol. 339, p. 95. ASP, San Francisco (2005)
16. Venemans, B.P., Findlay, J.R., Sutherland, W.J., et al.: Discovery of three $z > 6.5$ quasars in the VISTA kilo-degree infrared galaxy (VIKING) survey. ApJ **779**, 24 (2013)
17. Woltjer, L.: Schmidt telescopes as discovery instruments. In: Capaccioli, M. (ed.) Astronomy with Schmidt-Type Telescopes, IAU Colloquium 78, vol. 110, p. 3. Reidel, Dordrecht (1984)

Chapter 3
Public Surveys at ESO

M. Arnaboldi, N. Delmotte, S. Geier, M. Hilker, G. Hussain, L. Mascetti, A. Micol, M. Petr-Gotzens, M. Rejkuba, and J. Retzlaff

Abstract We provide an overview of the ESO public survey projects being carried out at the ESO survey telescopes, VISTA and VST, on the VLT instruments, UVES, FLAMES and VIMOS, and on the NTT, with SOFI and EFOSC, at the La Silla-Paranal Observatory. We outline the motivations behind these extensive projects, the policies and the facilities supporting these programmes. We illustrate how their data products are validated and published through the ESO Science Archive Facility, in order to make them accessible to the astronomical community. We conclude with an outlook on the future of public surveys at ESO.

3.1 Motivation

Observational astronomy is in an era of surveys. Projects like SDSS, Pan-STARRS, SkyMapper and LSST are characterised by large investments in "survey-systems", which include dedicated telescopes and instruments, a large community of astronomers involved in the science projects and large networks for the data distribution. The goals of these "survey-systems" are to enable new science in a variety of fields and serve broad communities. ESO has a strong background in survey projects ESO/SERC Southern Sky survey (1974–1987), and the EIS survey (1997–2004): this avenue has now received more impetus. ESO currently operates two dedicated survey telescopes, VISTA [5] and VST [1, 3], and organises survey projects that include the use of the available instrumentations at the La Silla-Paranal Observatory. The legacy value and the scientific excellence of the survey program are reviewed periodically during peer reviews organised by ESO. The current implementation of the ESO policies for public surveys is such that the community carries out all the activities that go beyond those enlisted in the ESO mission statement. Because the ESO public surveys are managed within the framework of the VLT/VLTI science policies (ESO council meeting 104, 17–18

M. Arnaboldi (✉) • N. Delmotte • S. Geier • M. Hilker • G. Hussain • L. Mascetti • A. Micol • M. Petr-Gotzens • M. Rejkuba • J. Retzlaff
ESO, K. Schwarzschild str. 2, 85748 Garching, Germany
e-mail: marnabol@eso.org

N.R. Napolitano et al. (eds.), *The Universe of Digital Sky Surveys*, Astrophysics and Space Science Proceedings 42, DOI 10.1007/978-3-319-19330-4_3

December 2004), the primary point of publication of the reduced products from the ESO public surveys is the ESO Science Archive Facility (SAF).

3.2 What the ESO Public Surveys Projects Are and What Their Status Is

3.2.1 Public Imaging Surveys

The ESO public surveys projects are very large programmes that require observing time at the telescope that last longer than 2 years. Their observing strategies include pencil beam deep surveys on cosmological fields, to wide area surveys covering from few hundreds to thousands of squared degrees, and up to the whole Southern Hemisphere. They have a legacy value for the astronomical community at large because they cover a broad range of research topics: from the study of the Milky Way and the Local group, stellar astrophysics, to galaxy evolution, cosmology and high redshift universe. They support synergies among each other and are complementary either in wavelengths (optical vs near-infrared) or in terms of observing modes, for example spectroscopic follow-up of interesting candidates identified via their colors or morphological properties. Important rules that apply to all these projects are that public imaging surveys are carried out in service mode, all raw data become immediately public and the survey teams commit to deliver reduced data, images and catalogues, within yearly release.

We summarize the general observational parameters of the nine imaging surveys, currently being carried out at the survey telescopes, VISTA and VST, in Tables 3.1 and 3.2 respectively. The six VISTA surveys started in April 2010 and have now completed their fifth year of science operations. The three VST public surveys started in October 2011 and are in their fourth year of science operations. A more detailed description of the survey goals of the imaging public surveys is available from the Public Surveys web pages: http://www.eso.org/sci/observing/PublicSurveys.html.

All the nine imaging surveys are either approaching or nearing completion of the initial allocated time defined on the basis of their approved survey management plans. The fraction of completion ranges from more than 80 % (VHS) to 60 % (UltraVISTA), with values in between for the other VISTA imaging surveys. For the VST surveys, the fraction of completeness varies from more than 72 % (ATLAS), 50 % (VPHAS+) and down to 30 % (KIDs). Two of the VISTA surveys (VVV, VIKING) and two of the VST surveys (ATLAS and VPHAS+) will complete the execution of their allocated time by 2017.

Table 3.1 General observational parameters for the six VISTA public surveys. The columns illustrate the public survey (col.1), the main scientific classification of the surveys scientific goal (col.2.), the targeted total area (col. 3), the filters (col.4), the magnitude limits (10σ AB for VMC, otherwise 5σ AB) in the different filters (col.5) and the observing hours completed up to April 1st, 2015 (col.6)

Survey ID	Science topic	Area (deg^2)	Filters	Magnitude limits (mag)	Obs. time (h)
Ultra-VISTA	Deep high-z	1.7 deep	YJ HKs	25.7 25.5 25.1 24.5	1,315
Ultra-VISTA	Deep high-z	0.73 ultra deep	NB118	26.7 26.6 26.1 25.6 26.0	1,315
VHS	Whole sky	17,800	YJ HKs	21.2 21.1 20.6 20.0	3,174
VIDEO	Deep high-z	12	ZYJ HKs	25.7 24.6 24.5 24.0 23.5	1,298
VVV	Galactic MW	560	ZYJ HKs	21.9 21.1 20.2 18.2 18.1	1,815
VIKING	Extragalactic	1,500	ZYJ HKs	23.1 22.3 22.1 21.5 21.2	1,748
VMC	Resolved SFH	180	YJ Ks	21.9 21.4 20.3	1,284

Table 3.2 General observational parameters for the three VST public surveys. The columns illustrates the public survey (col.1), the main scientific classification of the surveys main scientific goal (col.2.), the targeted total area (col. 3), the filters (col.4), the magnitude limits (10σ AB) in the different filters (col.5) and the observing hours completed up to April 1st, 2015 (col.6)

Survey ID	Science topic	Area (deg^2)	Filters	Magnitude limits (mag)	Obs. time (h)
KIDs	Extragalactic	1,500	u g r I	24.1 24.6 24.4 23.4	1,213
ATLAS	Wide area	4,000	u g r I z	22.0 22.2 22.2 21.3 20.5	1,134
VPHAS+	Stellar astrophysics	2,000	u g H$\|$*alpha* r I z	21.8 22.5 21.6 22.5 21.8	540

3.2.2 Public Spectroscopic Surveys

There are four public spectroscopic surveys being carried out at the ESO facilities now. The first two projects, Gaia ESO and PESSTO, started in January 2012 and are now in their third year of visitor mode operations. Following the call for VIMOS public surveys in 2014, two additional surveys were approved: they are LEGA-C and VANDELS. Their observations started in October 2014. Hence ESO is currently managing 13 Public Surveys on ESO telescopes of the La Silla-Paranal Observatory. Here we describe concisely the science goals of the Public Spectroscopic Surveys. Gaia ESO: this survey targets 100000 stars distributed among the major component of the Milky Way (MW) galaxy and in 100 open clusters. It has a strong synergy with the Gaia survey and will provide the phase space structure and abundances for the stellar population in the MW. Spectra are obtained with FLAMES and FLAMES/UVES spectrographs on VLT UT2. Target selection comes from the imaging surveys VHS, VVV etc. The current allocation entails 240 nights over 4 years, with a possible extension to a fifth year pending a review of the project.

PESSTO: the goal of this survey is to carry out the spectroscopic follow-up of about 150 transient objects in an unbiased sample of nearby galaxies to understand the physics of supernovae explosion, and achieve a statistical sample of SN progenitors in the nearby universe. Three hundred and sixty nights are allocated to this survey over 4 years at the NTT, with EFOSC2 and SOFI. As for Gaia ESO, this survey may be extended for another year after a successful review.

VANDELS: this spectroscopic survey with VIMOS targets star forming galaxies in the redshift range $2.5 < z < 7.0$ and passive galaxies in the redshift range $1.5 < z < 2.5$, in the CANDELS and CDF South ultra-deep fields. The goals are to measure metallicities and kinematics of the ionized gas in these systems. A total of 914 h are allocated at VIMOS on VLT UT3. The survey is expected to be completed in 4 years.

LEGA-C: this spectroscopic survey with VIMOS targets 3,000 early-type galaxies in the COSMOS field in the redshift range $0.6 < z < 1.0$. The goals are to understand how galaxies grow in mass through measurements of their dynamical masses, stellar ages and metallicities. A total of 1,010 h are allocated to this project with VIMOS on VLT UT3, with expected completion in 4 years. More information on ESO public surveys is available at http://www.eso.org/sci/observing/PublicSurveys.html. Similarly to the imaging surveys, all raw data from the spectroscopic surveys become immediately public and the survey teams commit to deliver reduced data, extracted spectra and catalogues, within yearly release.

3.3 Management of Public Surveys at ESO

Public surveys at ESO are projects executed within the framework of the VLT/VLTI policies for science operations. In this scheme, ESO supports the public survey teams for standard telescope operations (service mode and support for visitor mode observations), the delivery of the raw data, the management of the archive and of the reduced products, plus the organization of the high level peer reviews to monitor the surveys progress and ensure the excellence and coordination of the overall program. The public survey teams are responsible for the definition of the observing strategy, the final quality control of scientific data and for the data reduction. They deliver the reduced data for publication through the ESO SAF (http://archive.eso.org/cms.html). These reduced data are delivered in yearly incremental releases, and a final release including publication of the catalogs is planned at the end of each survey project. Observations for the ESO public surveys are carried out at the survey telescopes, VISTA and VST, at the spectrographs UVES/FLAMES, VIMOS, and at the NTT with SOFI and EFOSC. In Table 3.3 we report technical data and efficiency parameters for survey telescopes at different observatories. VISTA and VST have singularly an "on-average" efficiency, when compared with other wide field telescopes. Together they make ESO the only Observatory that provides wide field capabilities with a wavelength coverage from the UV (0.33 μm) to the Ks band (2.15 μm). The ESO spectrographs supporting the spectroscopic surveys provide multiplexing capabilities over FoV of tens arcmin diameter (FLAMES/VIMOS) to single slits (UVES/EFOSC/SOFI). Spectral resolution ranges from ten thousands

Table 3.3 Survey facilities with their technical parameters (telescope diameter, field of view (FoV), pixel size, wavelength range) and figure of merit (Entendue) for optical and NIR survey telescopes

Facility	Telescope diameter (m)	FoV (deg^2)	Pixel size (arcsec $pixel^{-1}$)	Wavelength range (μm)	Entendue ($m^2\ deg^2$)
SkyMapper	1.4	5.7	0.5	0.32–0.95	6.6
Pan-STARRS1	1.8	9.0	0.3	0.4–1.15	16.3
SDSS	2.5	2.4	0.4	0.33–1.0	6.0
MegaCAM	3.6	1.0	0.2	0.34–0.95	7.6
CTIO(Deacm)	4.0	4.0	0.2	0.6–1.0	16.0
VISTA	4.0	1.65	0.34	0.88–2.15	5.2
SuprimeCAM	8.0	0.25	0.2	0.36–1.0	13.5
VLT HAWKI	8.2	0.016	0.1	1.0–2.2	0.81
VLT VIMOS	8.2	0.06	0.2	0.36–1.0	3.2
LBC LBT	8.4	0.15	0.23	0.32–1.0	8.1
VST	2.6	1.0	0.21	0.33–1.0	5.5
Subaru Hyper-SuprimeCAM	8.0	1.5	0.17	0.36–1.0	94
LSST	8.4	3.5	0.2	0.33–1.1	319

(UVES/FLAMES) to few thousands (VIMOS/EFOSC/SOFI) depending on the science goals, with wavelength coverage from the optical to the near-infrared.

3.4 Legacy Value of ESO Public Survey Projects and Community Engagement

The entire ESO public surveys generate a very large data volume, of the order of 10 TB of raw data each year, whose science quality is ensured by the constant monitoring of the instrument stability and standard calibrations. Because these large data volumes are very homogenous and stable, a standard pipeline processing is capable of removing the instrumental signature and producing reduced data in physical units for further scientific analysis and processing. In this context, "physical units" signifies that images are astrometrically and photometrically calibrated, and one-dimensional (1D) spectra are uniquely associated with the RA,Dec position of the target object, they are wavelength calibrated and have physical flux unit. Thanks to the ESO public surveys, ESO set up a process of preparation, submission, validation and publication of reduced data through the ESO SAF, which is identified as "Phase 3" [2]. Within the Phase 3 process, the data provider, i.e. the survey PI or his/her delegates, is responsible for the data generation, their scientific quality and the documentation of the released data. ESO is responsible for the definition of the data format and the metadata information, which jointly are refereed to as the "Phase

3 standard". ESO also provides the Phase 3 tools to validate, upload and manage data submissions, the user support to the data providers, and carries out an audit process for the content validation of the released data [4]. The Phase 3 process was launched in 2011 and more than 30 TB of reduced data were ingested and published through the ESO SAF since the last year (2014). The metadata of the reduced products are key information for their publications through the query forms of the ESO SAF and a prerequisite for their further dissemination through the VO network. The metadata encode the level of calibration and support the archive exploration and queries by high level quality parameters, like for example the limiting magnitude and the FWHM of the point-spread-function for imaging data, or the signal-to-noise ratio and the spectra resolution for 1D spectra. By adopting the Phase 3 standard, the reduced products from ESO public surveys, but also large programs and ESO-in-house pipeline processed data, can be served co-jointly through the same ESO SAF query form, which is available at http://archive.eso.org/wdb/wdb/adp/phase3_main/form. Since the first data release in December 2011, more than 15 TB and more than half a million files were retrieved from the ESO SAF! In Fig. 3.1 we show the sky coverage of the survey image products available at the ESO SAF and released via the Phase 3 process.

In addition to images and spectra, the ESO SAF published the scientific catalogues at the ESO catalogue query interface (http://www.eso.org/qi/). The catalogues from ESO public surveys come into three flavors: source lists, multi epoch photometry (light curves) and high level catalogues:

1. Source-lists are fits tables containing a list of sources with their photometric parameters in a single band. A source list inherits the metadata information from

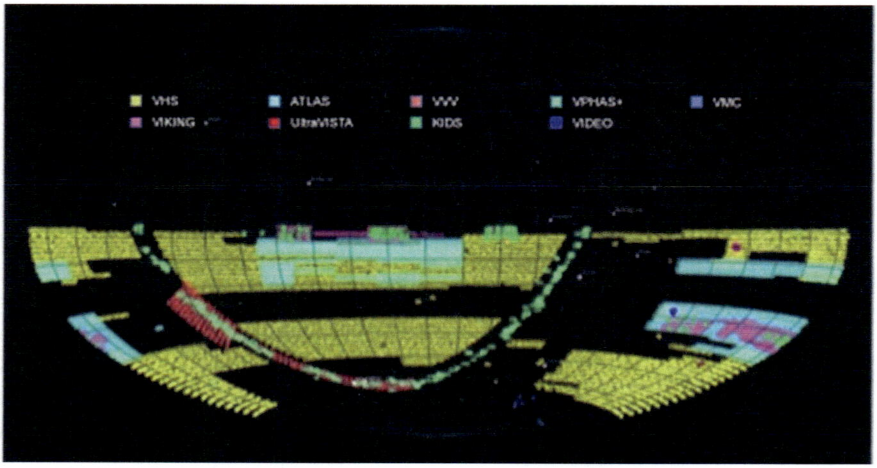

Fig. 3.1 Sky coverage of the ESO public survey products available at the ESO SAF and published through the Phase 3 process – Full sky (Hammer-Aitoff projection). The total area covered amounts to almost 10,000 deg^2 in the Southern Hemisphere. These products included all VISTA and VST data acquired before Oct 1st, 2013

the image it is extracted from and these products are delivered mostly by imaging surveys.
2. The multi epoch photometry or light curves contain the photometric parameters measured at different epochs for all sources.
3. The high-level catalogues contain either the multi-band aperture matched photometric measurements for sources in the whole area or sub-areas of the imaging surveys, or the physical parameters (i.e. EWs, abundances, redshifts, etc.) measured from spectra of individual objects, completed at a given date. One of the attribute that qualifies the science quality of a high-level catalog is the uniqueness of the source identifier. Having unique entries is very important for all those science goals addressing star/galaxy counts, the luminosity functions of stars/galaxies on large areas on the sky and galaxy evolution studies.

Further dissemination of catalog products via the Virtual Observatory is implemented by the on-going collaboration with the Centre de Donnes astronomiquées de Strasbourg (CDS). Such collaboration adds science content to the catalogs from ESO public surveys by cross-matching them with those coming from other projects, like for example X-rays and Gamma-ray experiments.

The community is informed of any new data releases via publication of dedicated announcements on the ESO science newsletter and on the ESO SAF website. The public surveys projects are also advertised by articles published on the Messenger by the ESO survey team and the PIs of the surveys (e.g. the Messenger volume n.154). ESO Press releases are also important channels to advertise the scientific results from the ESO public surveys and their data. The public surveys were the focus of an ESO workshop in 2012; the scientific results of these surveys will be discussed next at the workshop "Rainbows on the Southern Sky: legacy value of ESO Public Surveys and Large Programmes" (http://www.eso.org/sci/meetings/2015/Rainbows2015.html) in October 2015 at the ESO headquarters in Garching. Scientific workshops and the periodic reviews organised by ESO with the public survey panels are important occasions to foster scientific debate and identify the way forward to ensure the continuing scientific excellence of the survey program.

3.5 Future Outlook

The Next-Generation Transit Survey (NGTS) achieved first light at ESO's Paranal Observatory in northern Chile on January 14, 2015. This project will search for transiting exoplanets – planets that pass in front of their parent star and hence produce a slight dimming of the star's light that can be detected by sensitive instruments. The goal of this facility is to discover Neptune-sized and smaller planets, with diameters between two and eight times that of Earth. Their data products, in particular the light curves of their most likely candidates, will be ingested through Phase 3 in the ESO SAF and become available to the community. The first data release is expected 2 years after the start of science operations. In

March 2015, ESO published the announcement of a call for new surveys with the Arizona Radio Telescope. These projects will be selected and added to the currently on-going survey program in 2015. The horizon of the public surveys at ESO includes also a new call for imaging surveys with VISTA in 2016, and, on a longer timescale 4MOST (4-m Multi-Object Spectroscopic Telescope) from 2020. 4MOST is a very large field (goal $> 5\,\text{deg}^2$) multi-object spectrograph with up to 3,000 fibres and spectral resolutions of 5,000 and 20,000 for VISTA. The science cases will cover the Gaia follow-up for chemistry and kinematics of the Galaxy and the redshift surveys of targets from the eROSITA X-ray mission. The future of Public Surveys at ESO is bright!

References

1. Arnaboldi, M., Capaccioli, M., Mancini, D., et al.: VST: VLT survey telescope. Messenger **93**, 30 (1998)
2. Arnaboldi, M., Delmotte, N., Geier, S., et al.: Phase 3 status and access to science data products from ESO public surveys. Messenger **156**, 24 (2014)
3. Capaccioli, M., Schipani, P.: The VLT survey telescope opens to the sky: history of a commissioning. Messenger **146**, 2 (2011)
4. Retzlaff, J., Arnaboldi, M., Romaniello, M., et al.: Data products in the ESO science archive facility. In: Proceedings of the SPIE, Montréal, vol. 9149, 8pp., id. 914903 (2014)
5. Sutherland, W., et al.: The visible and infrared survey telescope for astronomy (VISTA): design, technical overview, and performance. A&A **575**, 25 (2015)

Chapter 4
Galaxy Evolution in the Era of Digital Surveys: A Theoretical Overview

G. De Lucia

Abstract In this review I will summarize the status of modern theories of galaxy formation and evolution. I will briefly introduce the main techniques employed and highlight recent successes and open problems.

4.1 Introduction

The explosion of digital surveys in the last decade has ushered in a new era of data richness and complexity. About 3 years ago, it was estimated that more than 1 PB (petabyte) of public data were electronically accessible, and that the data volume was growing at a rate of about 0.5 PB per year. Projections indicated that by 2020, more than 60 PB of archived data should be accessible to astronomers [1]. The growth of data in size is going to accelerate dramatically in the next years as new projects move into operation (e.g. LSST, ALMA, Euclid, SKA, just to name a few) so that it will likely turn out that the numbers given above are under-estimated. The datasets at our disposal are not just large (and will be huge in the next future!), but also multi-dimensional (i.e. for each astrophysical system, we have estimates for a number of different physical properties). Therefore, the big challenge for the future is not that of data storage, rather that of data mining (i.e. clever and efficient algorithms need to be developed in order to search within available databases, match data, and actually use them for scientific analyses).

On the theory side, we have witnessed a similar evolution: very large cosmological simulations have been completed over the last decades, by taking advantage of the rapidly growing computer performance and, at the same time, of the development of more sophisticated numerical algorithms. Theoretical models of galaxy formation are providing more and more accurate observables, and several predictions are nowadays available for completely independent models. Much of these data have been made publicly available through the development of dedicated databases [14]. This experience demonstrates that there is considerable interest from

G. De Lucia (✉)
INAF – Astronomical Observatory of Trieste, via G.B. Tiepolo 11, I-34143, Trieste, Italy
e-mail: delucia@oats.inaf.it

N.R. Napolitano et al. (eds.), *The Universe of Digital Sky Surveys*, Astrophysics and Space Science Proceedings 42, DOI 10.1007/978-3-319-19330-4_4

the astronomical community, and that this way of publishing results of theoretical research provides significant stimulus for additional scientific investigations.

4.2 Theoretical Models of Galaxy Formation

That of galaxy formation and evolution is a subject of great complexity. The very simple reason for that is that it is the result of a complex network of physical processes, most of which are quite poorly understood. Difficulties grow dramatically when one realizes that the different physical processes at play are entangled in a complex network of actions, back-reactions and self-regulations, operating on vastly different physical (from the scale of a black-hole to that of the Universe) and time scales (from the life-time of the most massive stars to the age of the Universe).

Despite this complexity, in the past years different methods have been developed to model galaxy formation and evolution in a cosmological context.

In semi-analytic models of galaxy formation, the evolution of the baryonic components of galaxies is modelled using *simple* yet *physically and/or observationally motivated prescriptions*. Modern semi-analytic models take advantage of high-resolution N-body simulations to specify the location and evolution of dark matter haloes, which are assumed to be the birth-places of luminous galaxies. Since pure N-body simulations can handle very large number of particles, this approach can access very large dynamic ranges in mass and spatial resolution. In addition, the computational costs are limited so that the method allows a fast exploration of the parameter space, and an efficient investigation of different specific physical assumptions. The main drawback of semi-analytic models is that they do not treat explicitly the gas-dynamics. This is instead done in direct hydrodynamical simulations. As a tool for studying galaxy formation, however, it is worth reminding that these methods are still limited by relatively low mass and spatial resolution, and by computational costs that are still prohibitive for simulations of galaxies throughout cosmological volumes. In addition, and more importantly, physical processes such as star formation, feedback, etc. still need to be modelled as *sub-grid*[1] *physics*, either because the resolution of the simulation becomes inadequate to treat a specific physical process or because (and this is, unfortunately, almost always true) we simply do not have a complete theory for the particular physical process under consideration.

In the following, I will discuss separately of 'central' and 'satellite' galaxies. The distinction is obvious (and convenient) from the theoretical point of view: here, central galaxies constitute a 'special' class of objects by construction, as they are the (in semi-analytic models often also the only) objects on which gas is allowed to cool. For observational data, the distinction is 'easy' (but now always, e.g. in

[1]The term is somewhat misleading as it gives the impression that the limits can be removed by increasing the resolution.

the Coma cluster there are two very bright members located in the proximity of the peak of the X-ray emission) for massive clusters, but typically very difficult, and perhaps inappropriate, for lower mass systems. With the aid of numerical simulations, however, this distinction is now routinely applied to observational data (for a discussion, see e.g. [6]).

4.3 Central Galaxies

One crucial ingredient to model the formation of massive central galaxies is the feedback from Active Galactic Nuclei (AGN). It has long been suspected that this powerful source of energy might play a crucial role in the so called 'cooling flow' problem, i.e. the observation that gas at the centre of massive galaxy clusters is apparently not condensing and turning into stars (at least not at a significant rate), although the observed X-ray emission implies a cooling time that is much shorter than the age of the system. Early galaxy formation models accounted for this form of feedback by simply assuming that cooling would be suppressed above some critical halo mass or circular velocity [e.g. 5, 13]. In the last decade, more sophisticated AGN feedback models have been included and these have confirmed that the 'radio-mode' feedback indeed represents a crucial ingredient to suppress the otherwise excessive number density of massive galaxies, and to keep their stellar populations old.

The efficiency of this feedback mechanism in suppressing the cooling flow at the centre of massive galaxy clusters has been extensively investigated using detailed hydrodynamical simulations. This work has shown that results strongly depend on a number of unknown parameters e.g. the duty cycle of the AGN, gas viscosity, and the geometry of the energy injection [see e.g. 2, 19]. Simulations have also clarified that the influence of this form of feedback, as resulting from current implementations, is not limited to the very central regions of galaxy clusters. Indeed, it results in an efficient extraction of enriched gas from star forming regions at high redshift. This affects for example the metallicity pattern of the intra-cluster medium leading to flatter metallicity gradients [7].

While current model implementations successfully reproduce the old stellar populations observed for massive galaxies, however, they all fail to reproduce their observed chemical abundances. Figure 4.1 shows the stellar mass-metallicity relation as predicted by the semi-analytic model presented in [3]. The model predicts a relatively tight metallicity-mass relation with a steep slope and a pronounced turn-over at the most massive end. The latter is in marked contrast with observational measurements. Similar results are found for completely independent semi-analytic models, as well as in hydrodynamical simulations [15, 17].

The middle and right panels of Fig. 4.1 show how the mass-weighted ages and metallicities of the central galaxies of massive haloes are affected when switching off the radio mode feedback (red histograms), and when adopting a different scheme for stellar (supernovae) feedback. As expected, when suppressing AGN feedback,

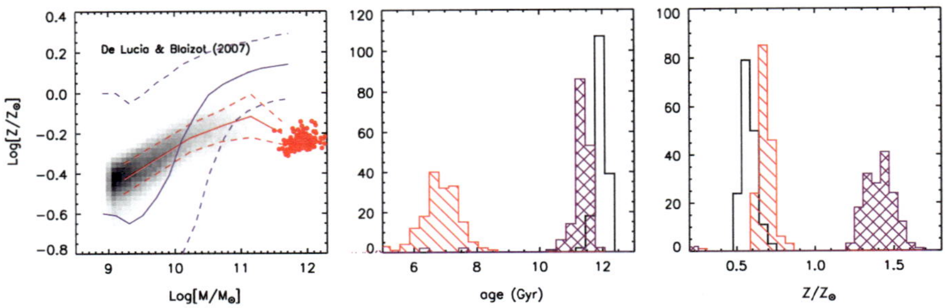

Fig. 4.1 From [4]. The *left panel* shows the metallicity-stellar mass relation predicted by the semi-analytic model presented in [3]. The *grey map* shows the distribution of model galaxies. The *red solid line* shows the median, and *dashed red lines* the 16th and 84th percentiles of the distribution. *Filled circles* show the location of the central galaxies of haloes with $M_{200} > 5 \times 10^{14}\,M_{\odot}$. *Purple lines* show the observational measurements by [10]. The *middle* and *right panels* show the distributions of mass-weighted ages and stellar metallicities for these massive central galaxies. *Black histograms* correspond to the fiducial model described in [3], while *red* ones correspond to the same model but with AGN feedback switched off. *Purple histograms* correspond to a model with AGN feedback but using a different scheme for stellar feedback

late time gas cooling allows recent star formation episodes that make the galaxies more massive and significantly younger. The stellar metallicity is however not significantly altered because the metallicity of the cooling gas is relatively low ($\sim 0.2\,Z_{\odot}$, in agreement with observational data). Modifying the scheme adopted for stellar feedback can have more drastic effects on the metallicity distributions, as shown by the purple cross hatched histograms. In this case, the model adopted results in less efficient outflows, and therefore longer star formation histories and higher stellar metallicities. Unfortunately, this model also over-predicts the number density of massive galaxies. These findings suggest that the difficulty of current models in simultaneously reproducing the observed ages and metallicities of the most massive galaxies highlights a fundamental problem with the schemes currently adopted to model star formation, feedback, and recycling of gas and metals.

The formation of the most massive galaxies is believed to be connected with the formation of the diffuse stellar component. Hydrodynamical simulations suggest that most of the stars found in the diffuse component of massive clusters come from particles that are unbound during galaxy mergers, with a minor fraction coming from tidal stripping of satellite galaxies [16]. Unfortunately, however, simulation results do not converge: increasing the resolution typically leads to increasing fractions of diffuse light. In addition, a recent work by [18] has pointed out that a significant portion of the diffuse component identified in hydrodynamical simulations forms in cold gas clouds stripped from infalling structures. Additional work is needed to clarify if (and how much of) this 'intra-cluster star formation' is due to spurious numerical effects (e.g. fluid instabilities, that are not well treated in smoothed particle hydrodynamical codes, might be able to destroy these clouds and reduce this contribution).

4.4 Satellite Galaxies

Figure 4.2 shows the observed (blue) stellar mass function for galaxies in clusters at low and intermediate redshifts. Observational data are compared with predictions from the galaxy formation model published in [3]. In this case, central galaxies have been excluded both for the data and for the models. The figure clearly shows that this particular model over-predicts the number densities of the satellite galaxies with mass just below the knee of the mass function. At low redshift, the model also over-predicts the number densities of the most massive galaxies, but the statistics here are lower and the large errors on the estimated stellar mass can affect significantly the shape at the massive end of the mass function. Recent theoretical work has pointed out that this excess of low-to-intermediate mass galaxies plagues all recent published semi-analytic models as well as hydrodynamical simulations [21, and references therein]. When including a strong stellar feedback, models are able to reproduce the observed galaxy stellar mass function in the local Universe, but they then consistently over-predict the number density of sub-M_* galaxies at higher redshift. Satellite model galaxies also tend to be less active than estimated observationally [22].

Early attempts to address these problems focused in particular on the assumption that the hot gas reservoir associated with infalling galaxies would be instantaneously stripped after accretion. Although improved, however, results from more sophisticated treatments were still not found to be in good agreement with observational data [8, 11]. Fontanot et al. [9] showed that the excess of low and intermediate mass galaxies in the models is mainly driven by an over-efficient formation at high redshift of central galaxies, with circular velocities $\sim 100 - 200\,\mathrm{km\,s^{-1}}$. In fact, this is the very same problem that is found in hydrodynamical simulations: gas cooling is very efficient at high redshift and in small and compact haloes. This leads to the formation of clumps that then merge via dynamical friction transferring angular momentum to the dark matter and resulting in disk galaxies that are more compact and rotate less rapidly than observed spirals. Once again, the likely solution to these problems reside in a better treatment of the feedback and recycling of gas. In a recent work, [12] showed that a model that includes an explicit dependence of the gas recycling time-scale on halo mass can successfully reproduce the observed evolution of the galaxy stellar mass function. The predicted passive fraction of low mass galaxies remains, however, significantly larger than observational measurements. The model also adopts some extreme values for other physical parameters (e.g. the chemical yield).

4.5 Final Remarks

As discussed above, recent theoretical work has identified a number of persistent problems that are common to independently developed models, and to different methods to model the formation of galaxies in a hierarchical context. In particular,

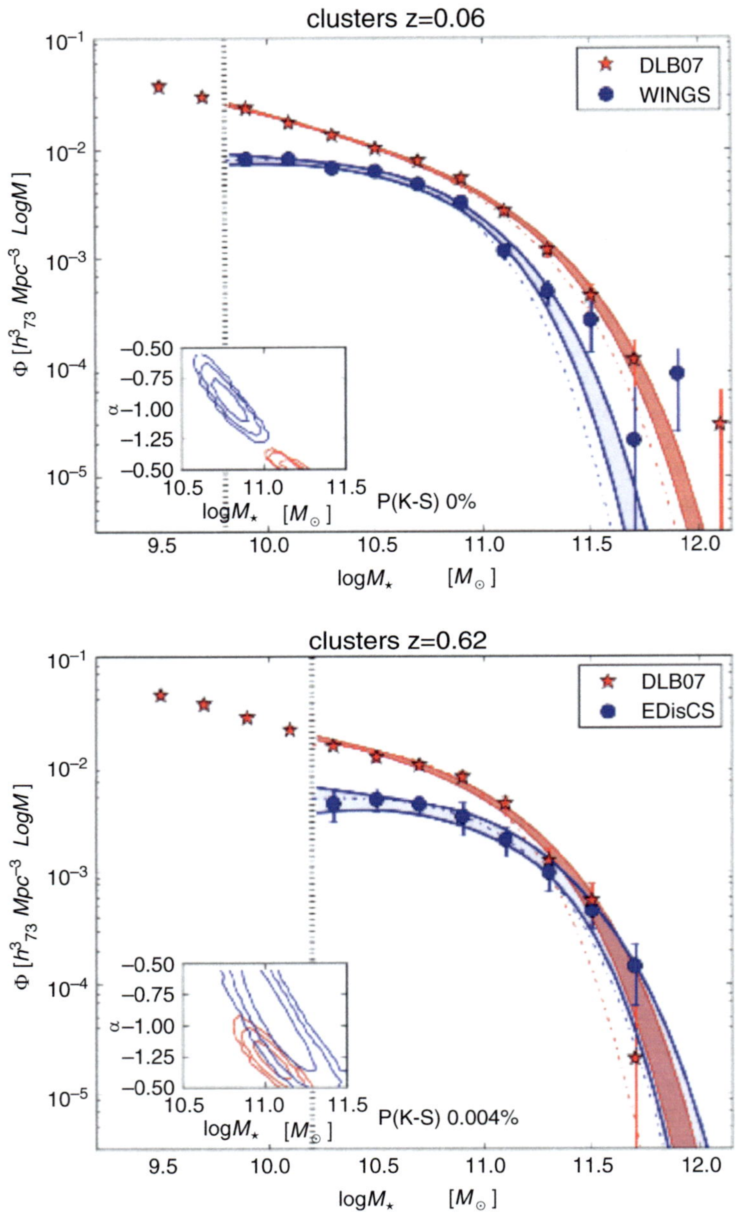

Fig. 4.2 From [20]. Observed (*blue*) and projected model (*red*) mass functions, for galaxies in clusters at low (*top*) and intermediate (*bottom*) redshift. Error bars on the y-axis are computed combining the Poissonian errors and the uncertainties due to cosmic variance or cluster-to-cluster variations. *Solid lines* and *shaded areas* represent Schechter fits with 1σ errors, *dashed lines* represent the mass function obtained from the deconvolution with the uncertainties on masses. See original paper for details

the number densities of low-to-intermediate mass galaxies are over-predicted; galaxies in this mass range are typically older than observed; the metal content of massive central galaxies is under-estimated with respect to observational determinations. I argue that these are probably different manifestations of the same problem, that likely relies in our (simplified) treatment of the self-regulation between star formation and feedback.

The era of 'data tsunami' has just started. In order to be prepared to interpret in a theoretical framework the new wave of data coming, it is necessary, on the one hand, to clarify the origin of the above mentioned problems and identify plausible and physically motivated solutions. On the other hand, further developments of the available models are required so as to extend the range of predicted observables. E.g. an explicit modelling of the transition between atomic and molecular gas and a detailed treatment of the chemical compositions of different galactic phases are necessary to take advantage of ongoing and planned programmes. Only through keeping this very close link between theoretical predictions and observable data, will it be possible to shed light on the physical processes governing galaxy formation and evolution as a function of cosmic time.

References

1. Berriman, G.B., Groom, S.L.: arXiv/1111.0075 (2011)
2. Dalla Vecchia, C., Bower, R.G., Theuns, T., Balogh, M.L., Mazzotta, P., Frenk, C.S.: MNRAS **355**, 995 (2004)
3. De Lucia, G., Blaizot, J.: MNRAS **375**, 2 (2007)
4. De Lucia, G., Borgani, S.: MNRAS **426**, L61 (2012)
5. De Lucia, G., Kauffmann, G., White, S.D.M.: MNRAS **349**, 1101 (2004)
6. De Lucia, G., Muzzin, A., Weinmann, S.: New Astron. Rev. **62**, 1 (2014)
7. Fabjan, D., Borgani S., Tornatore, L., Saro, A., Murante, G., Dolag, K.: MNRAS **401**, 1670 (2010)
8. Font, A.S., et al.: MNRAS **389**, 1619 (2008)
9. Fontanot, F., De Lucia, G., Monaco, P., Somerville, R.S., Santini, P.: MNRAS **397**, 1776 (2009)
10. Gallazzi, A., Charlot, S., Brinchmann, J., White, S.D.M., Tremonti, C.A.: MNRAS **362**, 41 (2005)
11. Guo, Q., et al.: MNRAS **413**, 101 (2011)
12. Henriques, B.M.B., White, S.D.M., Thomas, P.A., Angulo, R.E., Guo, Q., Lemson, G., Springel, V.: MNRAS **431**, 3373 (2013)
13. Kauffmann, G., Colberg, J.M., Diaferio, A., White, S.D.M.: MNRAS **303**, 188 (1999)
14. Lemson, G., Virgo Consortium, t.: astro-ph/0608019 (2006)
15. McCarthy, I.G., et al.: MNRAS **406**, 822 (2010)
16. Murante, G., Giovalli, M., Gerhard, O., Arnaboldi, M., Borgani, S., Dolag, K.: MNRAS **377**, 2 (2007)
17. Planelles, S., Borgani, S., Fabjan, D., Killedar, M., Murante, G., Granato, G.L., Ragone-Figueroa, C., Dolag, K.: MNRAS **438**, 195 (2014)
18. Puchwein, E., Springel, V., Sijacki D., Dolag K.: MNRAS **406**, 936 (2010)
19. Sijacki, D., Springel, V.: MNRAS **366**, 397 (2006)

20. Vulcani, B., De Lucia, G., Poggianti, B.M., Bundy, K., More, S., Calvi, R.: ApJ **788**, 57 (2014)
21. Weinmann, S.M., Pasquali, A., Oppenheimer, B.D., Finlator, K., Mendel, J.T., Crain, R.A., Macciò, A.V.: MNRAS **426**, 2797 (2012)
22. Weinmann, S.M., van den Bosch, F.C., Yang, X., Mo, H.J.: MNRAS **366**, 2 (2006)

Chapter 5
Photometric Surveys of the Galactic Bulge and Long Bar

O. Gerhard, C. Wegg, and M. Portail

Abstract The Galactic bar and box/peanut bulge can be studied in an unrivaled manner, star-by-star, with detailed chemical information and full 3D kinematics. Because of intervening dust this is greatly facilitated by the availability of wide field deep NIR photometric surveys. Here we summarize recent results on the three-dimensional structure of the bulge and the long bar region, based on 2MASS, UKIDSS, and particularly the ongoing VVV survey. We also summarize results from dynamical models for the Galactic bulge constructed with the Made-to-Measure method.

5.1 Photometric Surveys of the Galactic Bulge and Bar

It has been established from gas kinematics, near-Infrared (NIR) photometry, star counts, and stellar kinematics, that the Galactic bulge is barred [see 9, for an extensive recent review of the Galactic bulge]. Red clump giants (RCGs) are one of the primary tools used to establish the structure of the barred bulge. RCGs are core Helium burning stars and provide an approximate standard candle [13]. Large NIR photometric studies have now surveyed the entire bulge with sufficient depth to use RCGs as a probe. In Fig. 5.1 we show the deepest K-band surveys: the ongoing VVV survey [10] covers the central 10° and the southern galactic plane, UKIDSS [7] has surveyed the galactic plane at positive longitudes, while 2MASS remains the deepest survey away from the galactic plane. Only isolated areas remain without data of sufficient depth to use RCGs as a probe.

O. Gerhard (✉) • C. Wegg • M. Portail
Max-Planck-Institut für Extraterrestrische Physik, Giessenbachstrasse, 85748 Garching, Germany
e-mail: gerhard@mpe.mpg.de; wegg@mpe.mpg.de; portail@mpe.mpg.de

N.R. Napolitano et al. (eds.), *The Universe of Digital Sky Surveys*, Astrophysics and Space Science Proceedings 42, DOI 10.1007/978-3-319-19330-4_5

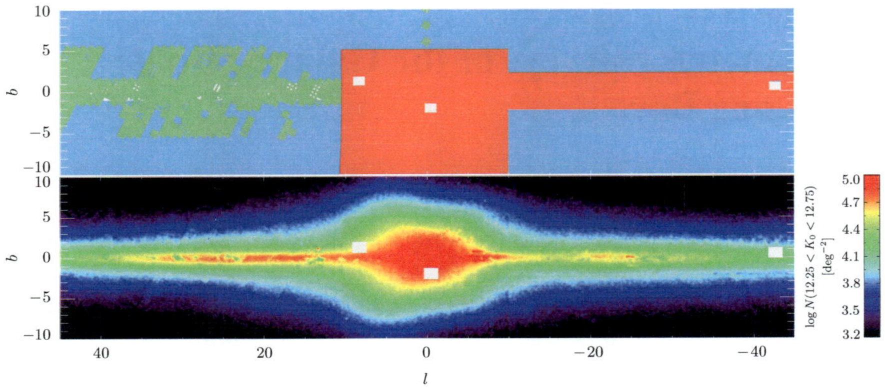

Fig. 5.1 In the *top figure* we show the deepest NIR surveys of the inner Galaxy: VVV DR2 in *red*, UKIDSS in *green*, and 2MASS in *blue*. *Grey regions* are those without data of sufficient depth to study clump stars i.e. close to the plane without VVV or UKIDSS data where 2MASS is insufficient. In the *lower panel* we show the resultant surface density of stars in the extinction corrected K-band magnitude range $12.25 < K_0 < 12.75$ (From [16])

5.2 The Structure of the Inner Milky Way

The VVV survey in particular is about 4 mag deeper and has a higher resolution than 2MASS. We exploited this in [14] to estimate the three-dimensional density using RCGs as a standard candle. We constructed extinction and completeness corrected magnitude distributions of VVV data for 338 fields covering the Galactic bulge. In these magnitude distributions RCGs were identified statistically as an excess over the otherwise smooth and nearly exponential background of non-RCG stars, which are mostly other giants in the bulge. The distribution of magnitudes of RCGs in each field arises from the convolution of the RCG luminosity function with the line-of-sight density. In each field we estimate the density by deconvolving using a variation on the Lucy-Richardson algorithm. The process of estimating the line-of-sight density is shown for one field in Fig. 5.2. These line-of-sight densities are then assembled into a non-parametric three-dimensional density assuming only eight-fold mirror symmetry about the principle axes of the bar with small symmerization errors. By varying details of the map construction process, such as the background fitting and the details of the luminosity function, we find the density seems robust at the $\sim$10 % level, while the symmetrization gives a bar angle of $(27 \pm 2)^\circ$.

The resultant 3D density is shown in Fig. 5.3. From above the bulge is barred with axis ratio (2.1:1) while from the side a strong peanut or X-shape is seen similar to some external galaxies such as the prototypical peanut bulge in NGC128 [15].

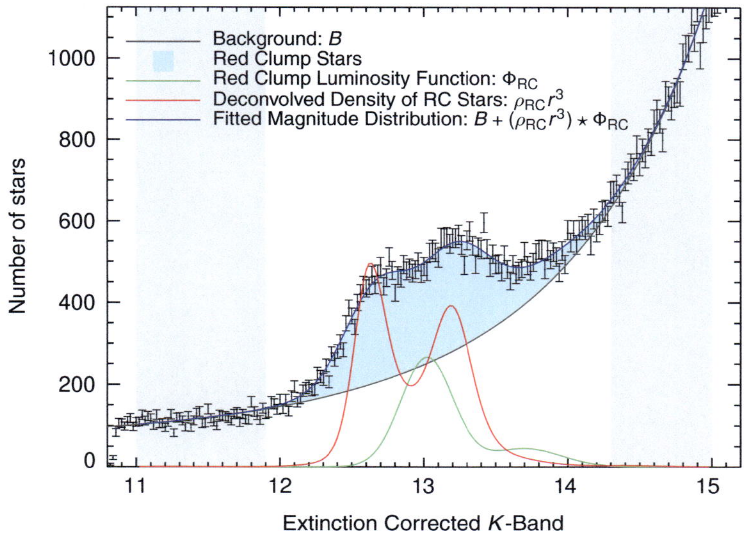

Fig. 5.2 The process of estimating the line-of-sight density of RCGs from the star counts in one bulge field. RCGs (*blue*) are statistically identified as stars above the smooth nearly exponential background of non-RCGs (*grey*). Then a variant of the Lucy-Richardson algorithm is then used to estimate the density (*red*) which when convolved with the luminosity function (*green*) is consistent with the data (Adapted from [14])

5.3 Made-to-Measure Dynamical Models of the Galactic Bulge

Because RCGs are good tracers of the stellar mass [11], their 3D density measurment can be used as a constraint on the stellar density in building dynamical models of the Galactic bulge. This has been done by [8] using the Made-to-Measure (M2M) modelling technique, which consists of adapting the N-body weights of an initial particle model until it reproduces a given set of data. We started from a set of five N-body models of a barred disk evolved in different dark matter halos, and fitted the stellar density to the RCGs density map as well as kinematic data from the BRAVA survey [6]. The kinematic data allow to recover the a priori unknown normalization factor between RCGs number density and the stellar mass density. This is shown in Fig. 5.4 where the velocity dispersion of the model matches the BRAVA data only for a specific value of the stellar mass in the bulge, which depends mostly on the dark matter mass present in the bulge. After M2M fitting, our set of models provides five equally good dynamical models of the bulge, even though their dark matter fractions range from 10 % to 30 %. We find that the total (stellar + dark matter) mass in the bulge, defined as the VVV RCG measurement box, is very well constrained by the RCGs density and the BRAVA data to $(1.84 \pm 0.07) \times 10^{10} M_{\odot}$.

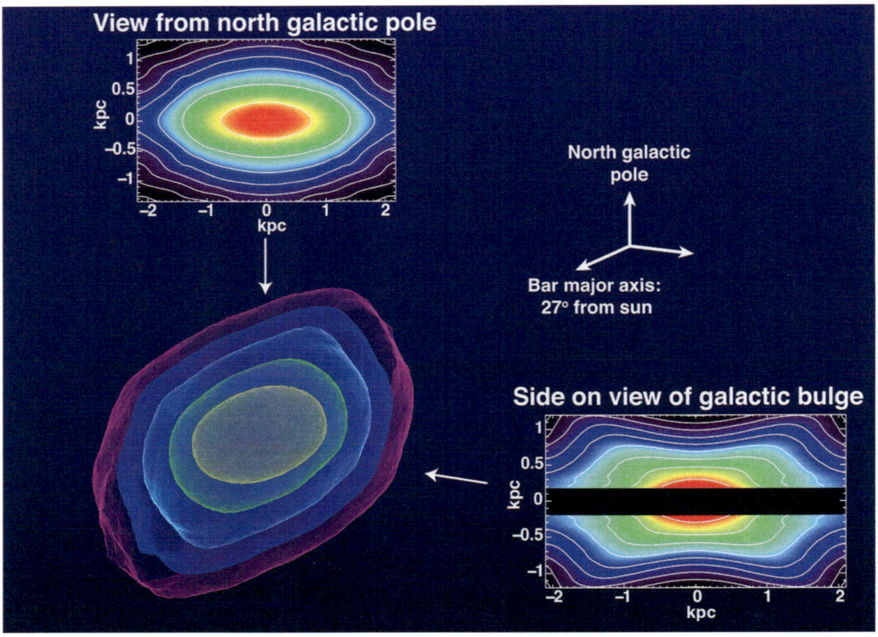

Fig. 5.3 The RCG density of the bulge. In the *center* we show the three-dimensional iso-density contours. The projections show the surface density from the north Galactic pole, where we see a bar with $\approx$2:1 axis ratio, and the view of the bulge from the side where we see a strongly peanut shaped bulge (Adapted from [15])

In addition, we used the COBE K-band measurements to compute the mass-to-light ratio of the models and compared them to predictions from population synthesis models. We measure stellar mass-to-light ratios ranging from 0.8 to 1.1, depending on the dark matter fraction. These are broadly consistent with predictions from the Initial Mass Functions (IMF) of [17] or [5], but are inconsistent with the [12] IMF that we therefore rule out for the Galactic bulge. The IMF from [17], directly measured from the bulge luminosity function, requires a large dark matter fraction in the bulge of about 40 %.

5.4 The Structure of the Bar Outside the Bulge: The Long Bar

While the structure of the barred bulge of the Milky Way has been extensively studied, the details of the bar outside this are very uncertain. The vertical scale height is smaller than in the bulge and it has therefore been termed the long bar. It can be seen in the surface density of stars in Fig. 5.1 as the enhancement close to the galactic mid-plane at positive longitudes. First investigated by [4] it has

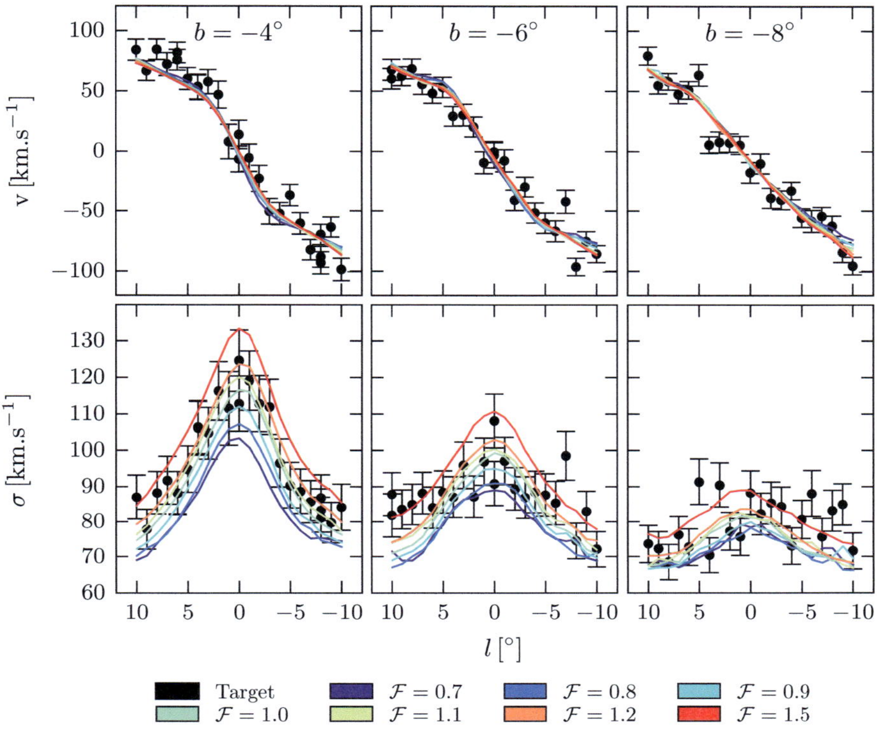

Fig. 5.4 Results of made-to-measure modeling of the Galactic bulge in [8] compared to the BRAVA data [6] in three latitude slices. The pattern speed of the bar and 3D density mostly determine the mean velocity in the *upper panels*. The stellar mass is varied by the $\mathcal{F}$ factor which primarily impacts the dispersion shown in the *bottom panels* (Adapted from [8])

subsequently been confirmed using RCGs with increasingly powerful NIR [e.g. 2] and longer wavelength GLIMPSE data [1]. Uncertainty remains however over the details of its structure and the relationship between the bar inside and outside the bulge.

We have matched and combined data from VVV, 2MASS and UKIDSS to get the widest possible view of the long bar covering $|l| < 40°, |b| < 9°$ [16]. By identifying RCGs in extinction free K-band magnitude distributions and using them as standard candles we have investigated the bar outside the bulge both directly in the data, and by constructing parametric density models which match the magnitude distributions. The combination of data allows us for the first time a large scale view of the long bar. A number of features are seen.

- The bar extends to $b \sim 5°$ at $l < 25°$, well outside the bulge region.
- Both in the data directly, and in our parametric density models, the bar outside the bulge has bar angle $\approx 29°$, consistent with alignment with the barred bulge.

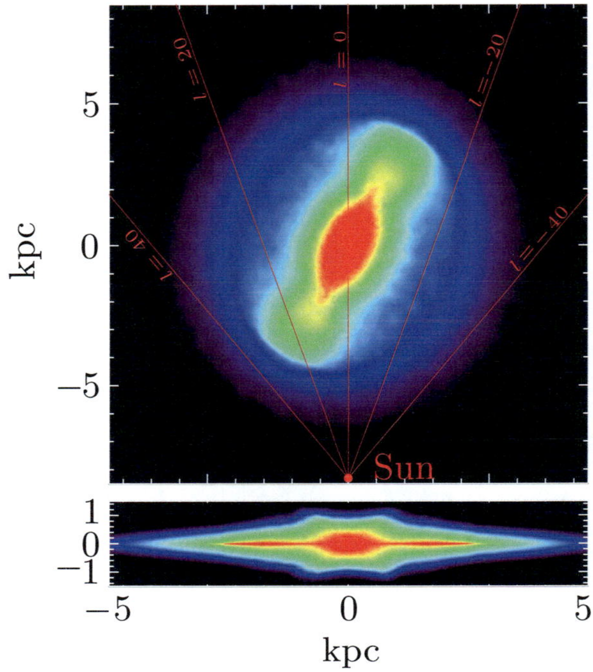

Fig. 5.5 The model density of the Milky Way viewed from above (*upper panel*) and from the side (*lower panel*). The density was constructed by adjusting the parameters of the density until the predicted magnitude distribution of RCGs matched the data across the region $-10° < l < 30°$ (Adapted from [16])

- The scale height of the RCGs smoothly transitions between a thicker scale height in the B/P bulge to the much thinner long bar. This scale height transition, and the alignment with the barred bulge, arises naturally in N-body models of barred galaxies and seems common in external galaxies [3].
- There is evidence for two scale heights in the long bar. We find a ∼180 pc thin bar component reminiscent of the old thin disk near the Sun, and a ∼45 pc super-thin bar component which exists predominantly towards the bar end.
- Constructing parametric models for the RCG magnitude distributions, we find a bar half length of 5.0 ± 0.2 kpc for the 2-component bar, and 4.6 ± 0.3 kpc for the 180 pc thickness bar component alone.

We show in Fig. 5.5 the face-on and side-on projections of one of our best fitting density models of the bar.

References

1. Benjamin, R.A., et al.: First GLIMPSE results on the stellar structure of the galaxy. ApJ **630**, L149 (2005)
2. Cabrera-Lavers, A., Gonzalez-Fernandez, C., Garzon, F., Hammersley, P.L., Lopez-Corredoira, M.: The long Galactic bar as seen by UKIDSS Galactic plane survey. A&A **491**, 781 (2008)

3. Erwin, P., Debattista, V.P.: Peanuts at an angle: detecting and measuring the three-dimensional structure of bars in moderately inclined galaxies. MNRAS **431**, 3060 (2013)
4. Hammersley, P.L., Garzon, F., Mahoney, T., Calbet, X.: Detection of the old stellar component of the major Galactic bar. MNRAS **269**, 753 (1994)
5. Kroupa, P.: On the variation of the initial mass function. MNRAS **322**, 231 (2001)
6. Kunder, A., et al.: The Bulge radial velocity assay (BRAVA). II. Complete sample and data release. AJ **143**, 57 (2012)
7. Lucas, P.W., et al.: The UKIDSS Galactic plane survey. MNRAS **391**, 136 (2008)
8. Portail, M., Wegg, C., Gerhard, O., Martinez-Valpuesta, I.: Made-to-measure models of the Galactic box/peanut bulge: stellar and total mass in the bulge region. MNRAS **448**, 713 (2015)
9. Rich, R.M.: The Galactic bulge. In: Planets, Stars and Stellar Systems. Volume 5: Galactic Structure and Stellar Populations, pp. 271–346. Springer Netherlands, Dordrecht (2013)
10. Saito, R.K., et al.: VVV DR1: the first data release of the Milky Way bulge and southern plane from the near-infrared ESO public survey VISTA variables in the Vía Láctea. A&A **537**, 107 (2012)
11. Salaris, M., Girardi, L.: Population effects on the red giant clump absolute magnitude: the K band. MNRAS **337**, 332 (2002)
12. Salpeter, E.E.: The luminosity function and stellar evolution. ApJ **121**, 161 (1955)
13. Stanek, K.Z., Mateo, M., Udalski, A., Szymanski, M., Kaluzny, J., Kubiak, M.: Color-magnitude diagram distribution of the bulge red clump stars: evidence for the galactic bar. ApJ **429**, L73 (1994)
14. Wegg, C., Gerhard, O.: Mapping the three-dimensional density of the Galactic bulge with VVV red clump stars. MNRAS **435**, 1874 (2013)
15. Wegg, C., Gerhard, O.: The Milky Way's box/peanut bulge: measuring its three-dimensional structure using the VVV survey. The Messenger **154**, 54 (2013)
16. Wegg, C., Gerhard, O., Portail, M.: The structure of the Milky Way's bar outside the bulge. MNRAS **450**, 4050 (2015)
17. Zoccali, M., Cassisi, S., Frogel, J.A., Gould, A., Ortolani, S., Renzini, A., Rich, R.M., Stephens, A.W.: The initial mass function of the Galactic bulge down to $\sim$0.15 M°. ApJ **530**, 418 (2000)

Chapter 6
The VMC Survey

M.-R.L. Cioni and The VMC team

Abstract The VISTA survey of the Magellanic Clouds system (VMC) is a public survey project of ESO. It is collecting near-infrared data across large areas of the Large and Small Magellanic Clouds, the Bridge and a few fields in the Stream. The combination of great sensitivity to stars below the old main sequence turn off, and the multiplicity at K_s band (at least 12 epochs) make VMC highly suitable for the determination of the spatially resolved star formation history and three-dimensional geometry, using variable stars such as Cepheids and RR Lyrae stars. The VMC observations, to be completed in 2018, have a high legacy value and many science results have already been published while others are in preparation.

6.1 Survey Status

The near-infrared YJK_s survey of the Magellanic Clouds system (VMC; [1]; www.star.herts.ac.uk/~mcioni/vmc) began in 2009 and will be completed in 2018. It covers about 170 deg^2 in total distributed as follows: 105 deg^2 on the Large Magellanic Cloud (LMC), 42 deg^2 on the Small Magellanic Cloud (SMC), 21 deg^2 on the Magellanic Bridge connecting the two Clouds, and 3 deg^2 on the Magellanic Stream. The sensitivity of the observations allows us to detect sources as faint as the main sequence turn off ($\sim$22 mag in the Vega scale), sampling the entire age range of stellar populations and allowing us to derive accurate star formation histories. Multi-epochs are collected at each filter: 3 at Y and J, and at least 12 at K_s. The latter allows us to obtain accurate mean magnitudes for variable stars (e.g. Cepheids and RR

M.-R.L. Cioni (✉)
Institut für Physik und Astronomie, Universität Potsdam, Karl-Liebknecht-Str. 24/25, 14476 Potsdam, Germany

Leibniz-Institut für Astrophysik Potsdam, An der Sternwarte 16, 14482 Potsdam, Germany

Physics Astronomy and Mathematics, University of Hertfordshire, College daue, Hatfield AL10 9AB, UK
e-mail: mcioni@aip.de

The VMC team
http://www.star.herts.ac.uk/~mcioni/vmc/

N.R. Napolitano et al. (eds.), *The Universe of Digital Sky Surveys*, Astrophysics and Space Science Proceedings 42, DOI 10.1007/978-3-319-19330-4_6

Lyrae stars) that are used to measure the three-dimensional geometry of the system. The VMC data have a high legacy value not only for studies of the Magellanic Clouds but also for studies of the Milky Way foreground and of background galaxies (including quasars, [2]).

VMC data are obtained from the VISTA telescope in Chile and are processed at both the Cambridge Astronomy Survey Unit (CASU, [10]) in Cambridge and at the Wide-Field Astronomy Unit (WFAU, [8]) in Edinburgh using the VISTA Data Flow System pipeline (VDFS; [6]). CASU performs the standard processing of the individual images, and produces source catalogues at each epoch of observation for each detector (the VISTA camera has 16 detectors in the field-of-view = pawprint) and for each tile (the mosaic of six pawprints to fill a contiguous area of sky = tile). WFAU links the individual epochs and combines the observations of the same tiles re-extracting sources and producing deep catalogues. We also derive point-spread function photometry to detect fainter sources and improve detections in crowded regions. The VMC data are periodically released to the community via the VISTA Science Archive [5] and via the ESO archive.

VMC observations are performed in service mode which guarantees a high level of homogeneity of the data. The resulting FWHM of sources in the densest regions is $\leq 1''$ at each filter while for sources in the outermost regions of the system it is $0.1'' - 0.2''$ higher. Individual tiles correspond to 1^{h}-long observations and reach a sensitivity of $Y = 21.2$, $J = 21.4$, and $K_s = 19.6$ mag. The overall completion of the survey is $>60\,\%$: the two Stream tiles are fully observed, the Bridge and SMC are $>80\,\%$ observed, and the LMC is $\sim 50\,\%$ observed. Five tiles in the LMC, including tiles overlapping the South Ecliptic Pole (SEP) and the 30 Doradus regions, and two tiles in the SMC are already publicly available.

The VMC survey is the only near-infrared survey of the Magellanic system that is complementary to other ongoing large-scale optical surveys: the OGLE-IV survey that will provide the variability information across an unprecedented area of sky, the SMASH survey that searches for stellar debris in the outermost regions of the system, the STEP survey at the VST that focuses on the star formation history of the SMC and Bridge, Skymapper observations of the southern sky with a few repeats, and Gaia observations that will provide accurate positions and proper motions for all stars down to a magnitude below that of the horizontal branch, including RR Lyrae stars. The combination between VMC and these surveys will be highly valuable to investigate the origin and evolution of the studied systems.

6.2 Science Highlights

Within the past 5 years, 15 articles have demonstrated the high quality and versatility of the VMC data. We studied planetary nebulae [12] and asymptotic giant branch stars [7], we measured the star formation history [20] and the parameters of star clusters [15, 16], we analysed the periodicity of Cepheids [17–19] and set the stage for further variability studies [13], we identified eclipsing binary stars [14] and quasar candidates [2], we built a map of the dust distribution [22], and studied population gradients [11] of Milky Way star clusters. We report below a brief summary of some recent and ongoing investigations.

The most accurate star formation history map of the SMC has been obtained using the VMC data [21]. It was derived from the analysis of 10 tiles covering the main body. Each tile was subdivided into 16 equal regions, and for each region colour-magnitude diagrams were created. These showed a range of stellar populations that sample the entire history of the galaxy. We found, for each diagram, the best combination of 'partial' models [20] that describes the stellar populations using the STARFISH [9] code. The reddening and the distance corresponding to each region were derived from a range of input values and then the associated star formation rate and iron abundance were analysed as a function of time. Overall, the star formation rate across the SMC was modest at ages <5 Gyr ($0.15 M_{\odot} yr^{-1}$) until a major burst that was responsible for the creation of most of the galaxy mass. Another star formation peak, at 1.5 Gyr, agrees with peaks in the formation of star clusters and in the LMC. The most recent star formation took place in the centre and in the eastern part of the galaxy.

The VMC survey has already covered a large fraction of the LMC providing counterparts for about 40 % of the known classical Cepheids (Ripepi et al. in prep.). These stars follow a tight period-luminosity-colour relation that is a powerful distance indicator. Previous works on smaller samples of classical Cepheids in the SEP and 30 Doradus regions [17], and on both anomalous and type II Cepheids, indicate dispersions $\leq$0.1 mag [18, 19]. The 13 VMC data points sample well the K_s-band light-curves of Cepheids and for most of the stars the 4 J-band data points are also well representing the expected sinusoidal variation.

RR Lyrae stars, that are about 3.5 mag fainter than classical Cepheids, show a lower K_s-band amplitude, but their light-curve is well sampled by the VMC measurements. Their dispersion around the period-luminosity relation is larger than that obtained for Cepheids and this is probably due to the stars belonging to a spheroid rather than to a disk, like in the case of Cepheids. In the SEP region there are 109 confirmed RR Lyrae stars with a VMC counterpart (79 RRab, 23 RRc, and 7 RRd), Moretti et al. (in prep.).

Projected onto the SMC background and to the North-West of the galaxy centre is the Milky Way star cluster 47 Tuc (Fig. 6.1). In this tile (1.5 deg^2 in size) we trace the stellar cluster density out to large radii, detect the increase in the star counts to the direction of the SMC centre and isolate the stars of the SMC star cluster NGC 121. In the colour-magnitude diagram, $(J - K_s)$ vs. K_s, these different stellar populations are easily distinguished from each other and from the large number of background galaxies, as well as from foreground stars. The VMC observations of this tile (SMC 5_2) have been used to measure the proper motion of the different stellar populations (Cioni et al. 2015). We used 10 separate K_s epochs obtained over a time-baseline of about 1 year. The proper motion of individual sources results from a linear fit to the pixel displacements among the different epochs calculated from individual VISTA detectors (16) and paw prints (6) placed on the same reference system defined by

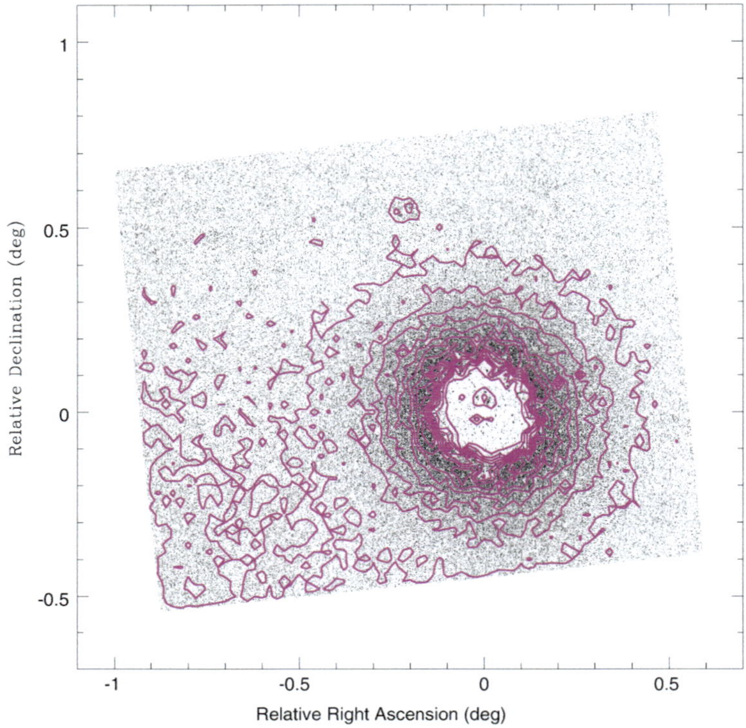

Fig. 6.1 Distribution of VISTA sources over tile SMC 5_2. The Milky Way star cluster 47 Tuc occupies about 1/4 of the area while the SMC star cluster NGC 121 is located to the North of it. The centre of the figure corresponds to the centre of the 47 Tuc cluster at $(\alpha_0, \delta_0) = (6^\circ.023625, -72^\circ.081283)$. The SMC stellar population density increases to the South-East of 47 Tuc

background galaxies (there are a few hundreds galaxies per detector). Then, the proper motion for different stellar populations, selected from the colour magnitude diagram, e.g. SMC red clump stars, SMC upper and lower red giant branch stars, Milky Way stars, 47 Tuc upper and lower main sequence stars, and 47 Tuc red giant branch stars are derived by averaging the values for all stars within the given regions applying a sigma clipping method (see Cioni et al. 2014 for details). Proper motion results indicate a clear separation between the SMC, the Milky Way, and the 47 Tuc populations. Furthermore, uncertainties are sufficiently small to distinguish between stellar populations of a different age. Combining all 5000 SMC stars and all 35 000 cluster stars we derive proper motion accuracies of 0.05 mas year^{-1} (25 km/s). This method is superior to the pilot study described in [3] because it is almost free from systematic uncertainties since it is based on the analysis of separate detectors, instead of using coordinates obtained after reconstructing a tile (each detector has its own distortion pattern that is difficult to evaluate after it is combined with the others in the tile mosaic).

6.3 Conclusions

The VMC survey is continuing to observe the Magellanic system acquiring data of unprecedented quality at near-infrared wavelengths. The survey will be completed in 2018 and some data area already publicly available. The accuracy of the VMC data permits the study of different stellar populations, derive their age, metallicity, reddening, distance, and absolute proper motions with respect to a reference system made of background galaxies. The VMC survey has a high legacy value and provides a wealth of targets for future spectroscopic investigations, e.g. with the 4MOST (at VISTA) and MOONs (at the VLT) spectrographs in the 2020s. The VMC survey is a public survey project of ESO and the VMC team welcomes enthusiastic astronomers who would like to exploit this rich data set.

Acknowledgements MRLC acknowledges support by the German Academic Exchange Service (DAAD).

References

1. Cioni, M.-R.L., Clementini, G., Girardi, L., et al.: The VMC survey I. Strategy and first data. A&A **527**, A116 (2011)
2. Cioni, M.-R.L., Kamath, D., Rubele, S., et al.: The VMC survey VI. Quasars behind the Magellanic system. A&A **549**, A29 (2013)
3. Cioni, M.-R.L., Girardi, L., Moretti, M.I., et al.: The VMC survey IX. Pilot study of the proper motion of stellar populations in the LMC from 2MASS and VISTA data. A&A **562**, A32 (2014)
4. Cioni, M.-R.L., Bekki, K., Girardi, L., et al.: The VMC Survey. XVII. The proper motion of the small magellanic cloud and of the milkey way globular cluster 47 Tucanae. A&A, accepted (2015)
5. Cross, N.J.C., Collins, R.S., Mann, R.G., et al.: The VISTA science archive. A&A **548**, A119 (2012)
6. Emerson, J.P., Irwin, M.J., Lewis, J., et al.: VISTA data flow system: overview. SPIE Conf. Ser. **5493**, 401 (2004)
7. Gullieuszik, M., Groenewegen, M.A.T, Cioni, M.-R.L., et al.: The VMC survey III. Mass-loss rates and luminosities of LMC AGB stars. A&A **537**, A105 (2012)
8. Hambly, N.C., Mann, R.G., Bond, I., et al.: VISTA data flow system survey access and curation: the WFCAM science archive. SPIE Conf. Ser. **5493**, 423 (2004)
9. Harris, J., Zaritsky, D.: A method for determining the star formation history of a mixed stellar population. ApJS **136**, 25 (2001)
10. Irwin, M.J., Hodgkin, S., et al.: VISTA data flow system: pipeline processing for WFCAM and VISTA. SPIE Conf. Serr. **5493**, 411 (2004)
11. Li, C., de Grijs, R., Deng, L., et al.: The VMC survey XI. Radial stellar population gradients in the galactic globular cluster 47 Tucanae. ApJ **790**, 35 (2014)
12. Miszalski, B., Napiwotzki, R., Cioni, M.-R.L., Groenewegen, M.A.T., Oliveira, J.M., Udalski, A.: The VMC survey II. A multi-wavelength study of LMC planetary nebulae and their mimics. A&A **531**, A157 (2011)
13. Moretti, M.I., Clementini, G., Muraveva, T., et al.: The VMC survey X. Cepheids, RR Lyrae stars and binaries as probes of the Magellanic System's structure. MNRAS **437**, 2702 (2014)

14. Muraveva, T., Clementini, G., Maceroni, C., et al.: Eclipsing binary stars in the Large Magellanic Cloud: results from the EROS-2, OGLE and VMC surveys. MNRAS **443**, 432 (2014)
15. Piatti, A., Guandalini, R., Ivanov, V., et al.: The VMC survey XII. Star cluster candidates in the Large Magellanic Cloud. A&A **570**, A74 (2014)
16. Piatti, A., de Grijs, R., Rubele, S., Cioni, M.-R.L., Ripepi, V., Kerber, L.: The VMC survey XV. The Small Magellanic Cloud-Bridge connection history as traced by their star cluster populations. MNRAS **450**, 552 (2015)
17. Ripepi, V., Moretti, M.I., Marconi, M., et al.: The VMC survey V. First results for Classical Cepheids. MNRAS **424**, 1807 (2012)
18. Ripepi, V., Marconi, M., Moretti, M.I., et al.: The VMC survey VIII. First results for Anomalous Cepheids. MNRAS **437**, 2307 (2014)
19. Ripepi, V., Moretti, M.I., Marconi, M., et al.: The VMC survey XIII. Type II Cepheids in the Large Magellanic Cloud. MNRAS **446**, 3034 (2015)
20. Rubele, S., Kerber, L., Girardi, L., et al.: The VMC survey IV. The LMC star formation history and disk geometry from four VMC tiles. A&A **537**, 106 (2012)
21. Rubele, S., Girardi, L., Kerber, L., et al.: The VMC survey XIV. First results on the look-back time star formation rate tomography of the Small Magellanic Cloud. MNRAS **449**, 639 (2015)
22. Tatton, B., van Loon, J.Th., Cioni, M.-R.L., et al.: The VMC survey VII. Reddening map of the 30 Doradus field and the structure of the cold interstellar medium. A&A **554**, 33 (2013)

Chapter 7
Galactic Plane Hα Surveys: IPHAS and VPHAS+

N.J. Wright

Abstract The optical Galactic Plane Hα surveys IPHAS and VPHAS+ are dramatically improving our understanding of Galactic stellar populations and stellar evolution by providing large samples of stars in short lived, but important, evolutionary phases, and high quality homogeneous photometry and images over the entire Galactic Plane. Here I summarise some of the contributions these surveys have already made to our understanding of a number of key areas of stellar and Galactic astronomy.

7.1 Introduction

Hα emission is an important tracer of many critical phases of stellar evolution, including pre- and post-main-sequence (MS) stars, interacting binaries and massive stars, as well as tracing diffuse ionised nebulae. Despite their importance these evolutionary phases are generally short-lived, and therefore stars in these phases are rare and difficult to find. The Galactic Plane surveys IPHAS (INT Photometric Hα Survey, [3]) and VPHAS+ (VST Photometric Hα Survey [4]) represent a concerted effort to overcome this deficiency, to improve our understanding of these short-lived evolutionary phases and to facilitate large-scale stellar population and Galactic structure studies.

The two surveys are imaging the entire Northern (IPHAS) and Southern (VPHAS+) Galactic Plane and bulge using the Isaac Newton Telescope (INT) and VLT Survey Telescope (VST). Together with their blue-filter partner survey UVEX (UV Excess Survey [5]) on the INT they are covering an area of $\sim$3,800 deg^2, across the entire Galactic Plane ($0° \leq l < 360°$, $|b| < 5°$) and bulge ($10° > l > 350°$, $|b| < 10°$), using the filters u, g, r, i, and Hα. The Hα filters used by IPHAS

N.J. Wright, on behalf of the IPHAS and VPHAS+ consortia, see the survey websites www.iphas.org and www.vphasplus.org for a full list of collaborators.

N.J. Wright (✉)
Centre for Astrophysics Research, University of Hertfordshire, Hertfordshire, UK
e-mail: nick.nwright@gmail.com

N.R. Napolitano et al. (eds.), *The Universe of Digital Sky Surveys*, Astrophysics and Space Science Proceedings 42, DOI 10.1007/978-3-319-19330-4_7

($\lambda_c = 6{,}568$ Å) and VPHAS+ ($\lambda_c = 6{,}589$ Å) are broad enough (FWHM of 95 Å and 100 Å respectively) to capture most Doppler shifts due to Galactic motion.

The observations are taken in blue ($u/g/r$) and red ($r/i/\mathrm{H}\alpha$) blocks, providing co-eval multi-band photometry necessary for compiling the many colour-colour diagrams facilitated by the surveys that can efficiently be used to identify and characterise rare types of object. The observations have been performed under good conditions (VPHAS+ median seeing $= 0.8''$) and reach a 5σ depth of $g, r \sim 22$ mag, and saturation at ~11–13 mag. The IPHAS second data release [2] can be obtained from the survey website (www.iphas.org/dr2), while VPHAS+ data is available from the ESO science archive (http://www.eso.org/sci/observing/phase3/data_releases.html).

7.2 Star Formation and Young Stars

Hα emission is a common feature of pre-MS stars due to accretion from a circumstellar disc onto the star itself, and therefore can be used as a diagnostic of youth [13, 16]. Furthermore the r–Hα colour also provides a measure of the equivalent width of the Hα emission line (see Fig. 7.1), which can then provide an estimate of the accretion rate. Barentsen et al. [1] used IPHAS photometry for stars towards the H II region IC 1396, identifying 158 pre-MS candidates with masses between 0.2 and 2.0 $M_\odot$. They measured accretion rates for these stars and found a power-law dependency between the stellar mass and the accretion rate with a slope of $\alpha = 1.1 \pm 0.2$. They also found evidence for an age gradient, manifested through accretion rates and infrared excesses, with younger stars lying further from the central ionising O-type star of the region, which provides evidence that the formation of these stars was sequentially triggered by the O star.

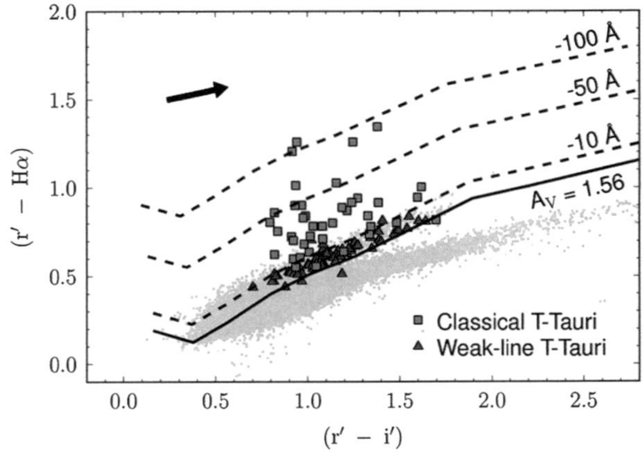

Fig. 7.1 The IPHAS ($r' - \mathrm{H}\alpha$) vs ($r' - i'$) colour-colour diagram showing the identification of Hα-emitting classical T-Tauri stars in IC 1396 relative to known weak-line T-Tauri stars. Also shown are a reddened main-sequence (*solid line*) and the positions of stars at increasing levels of Hα emission. *Grey dots* shown field stars in the region (Figure from [1])

The Hα imagery from these surveys also facilitates large-scale studies of the diffuse emission across H II regions, and has already led to the discovery of a unique class of young stellar object experiencing feedback from nearby OB stars [18].

7.3 Massive Stars

Massive O and early B stars can be efficiently identified and characterised using the blue filters employed by the UVEX and VPHAS+ surveys (see Fig. 7.2). Mohr-Smith et al. [6] have used this method to identify 356 new O and early B star candidates across a 2 deg^2 region of the Carina spiral arm of the Milky Way at distances of 3–6 kpc. The photometry can be used to fit spectral types and reddening parameters, which were verified using known OB stars in the field. This allowed the authors to map out the distribution of massive stars across the region and study the variation of the reddening law, which requires non-standard values of $3.5 < R_V < 4$.

The Hα imagery from these surveys can also be used to study the environments of massive stars to probe their mass-loss history. Wright et al. [19] presented Hα images of an ionised nebula surrounding the red supergiant (RSG) W26 in the massive star clusters Westerlund 1 consisting of a circumstellar shell or ring ∼0.1 pc in diameter and a triangular nebula ∼0.2 pc from the star. The authors hypothesise that the circumstellar nebula likely formed from a previous episode of mass-loss from the RSG, while the triangular nebula may be a flow of material channelled off the circumstellar nebula by the cumulative cluster wind and radiation field.

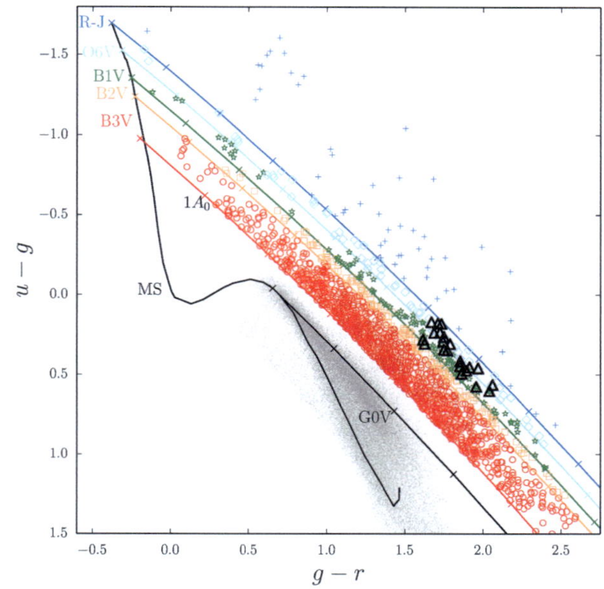

Fig. 7.2 The VPHAS+ $(u - g)$ vs $(g - r)$ colour-colour diagram showing the identification of O and B stars (*coloured circles*, as per the spectral types listed) and known OB stars in Westerlund 2 (*black triangles*). Also shown are an unreddened main sequence (MS, *black line*) and $R_V = 3.8$ reddening lines (*coloured lines*) extending from various labelled spectral types and from the position of a Rayleigh-Jeans spectrum (Figure adapted from [6])

7.4 Evolved Stars and Stellar Remnants

The filters employed by these surveys are very effective for identifying evolved stars and stellar remnants such as asymptotic giant branch stars [17], white dwarfs [11], supernova remnants [8], and planetary nebulae (PNe) [7]. In the latter category the Hα filter employed by these surveys has been valuable for discovering many new compact [12] and extended [9] PNe, as well as improving our understanding of those already known [14]. The blue filters employed by these surveys also allows the central stars of many existing PNe to be identified for the first time [4]. Finally the legacy value of these deep and detailed surveys should not be underestimated, particularly as we enter an era giving more emphasis to time-domain studies, such as for identifying and studying the progenitors of novae [15].

7.5 Galactic Structure

The combination of precise and high spatial resolution photometry, the unique ability of r–Hα to discriminate intrinsic stellar colour, and the wide area coverage facilitates detailed studies of the spatial distribution of stars, gas and dust across our Galaxy. Farnhill et al. (*in prep*) produced calibrated stellar density maps from IPHAS data to test Galactic population synthesis models and 3D extinction maps. The authors found a number of discrepancies between the models and observations, with the main problem being an under-prediction of extinction at low Galactic longitudes ($l \simeq 30°$). Sale et al. [10] used IPHAS photometry and a hierarchical Bayesian model to compile a 3D map of extinction across the northern Galactic plane at an unprecedented spatial ($\sim$10 arcmin) and distance (100 pc) resolution that will be vital for studies of Galactic stellar populations.

7.6 Summary

The Hα Galactic Plane surveys IPHAS and VPHAS+ have brought many areas of stellar and Galactic astronomy into the modern era by providing high quality and homogeneous photometry for hundreds of millions of stars across our Galaxy. The science discussed here highlights the contribution these surveys have already made to many key areas of astrophysics. Once completed these surveys will provide a valuable resource for many future studies, particularly when combined with astrometry from ESA's Gaia satellite.

Acknowledgements NJW acknowledges a Royal Astronomical Society Research Fellowship and is grateful to all those within IPHAS and VPHAS+ who have made these surveys possible and contributed to them. Thanks to M. Mohr-Smith for Fig. 7.2.

References

1. Barentsen, G., Vink, J.S., Drew, J.E., et al.: T Tauri candidates and accretion rates using IPHAS: method and application to IC 1396. MNRAS **415**, 103 (2011)
2. Barentsen, G., Farnhill, H.J., Drew, J.E., et al.: The second data release of the INT photometric Hα survey of the northern galactic plane (IPHAS DR2). MNRAS **444**, 3230 (2014)
3. Drew, J.E., Greimel, R., Irwin, M.J., et al.: The INT photometric Hα survey of the northern galactic plane (IPHAS). MNRAS **362**, 753 (2005)
4. Drew, J.E., Gonzalez-Solares, E., Greimel, R., et al.: The VST photometric Hα survey of the southern galactic plane and bulge (VPHAS+). MNRAS **440**, 2036 (2014)
5. Groot, P.J., Verbeek, K., Greimel, R., et al.: The UV-excess survey of the northern galactic plane (UVEX). MNRAS **399**, 323 (2009)
6. Mohr-Smith, M., Drew, J.E., Barentsen, G., et al.: New OB star candidates in the carina arm around Westerlund 2 from VPHAS+. MNRAS (2015, in press). arXiv 1504.04342
7. Sabin, L., Zijlstra, A.A., Wareing, C., et al.: New candidate planetary nebulae in the IPHAS survey: the case of planetary nebulae with ISM interaction. PASA **27**, 166 (2010).
8. Sabin, L., Parker, Q.A., Contreras, M.E., et al.: New galactic supernova remnants discovered with IPHAS. MNRAS **431**, 279 (2013)
9. Sabin, L., Parker, Q.A., Corradi, R.L.M., et al.: First release of the IPHAS catalogue of new extended planetary nebulae. MNRAS **443**, 3388 (2014)
10. Sale, S.E., Drew, J.E., Barentsen, G., et al.: A 3D extinction map of the northern galactic plane based on IPHAS photometry. MNRAS **443**, 2907 (2014)
11. Verbeek, K., Groot, P.J., Nelemans, G., et al.: A determination of the space density and birth rate of hydrogen-line (DA) white dwarfs in the Galactic plane, based on the UVEX survey. MNRAS **434**, 2727 (2013)
12. Viironen, K., Greimel, R., Corradi, R.L.M., et al.: Candidate planetary nebulae in the IPHAS photometric catalogue. A&A **504**, 291 (2009)
13. Vink, J.S., Drew, J.E., Steeghs, D., et al.: IPHAS discoveries of young stars towards Cyg OB2 and its southern periphery. MNRAS **387**, 308 (2008)
14. Wareing, C.J., O'Brien, T.J., Zijlstra, A.A., et al.: The shaping of planetary nebula Sh2-188 through interaction with the interstellar medium. MNRAS **366**, 387 (2006)
15. Wesson, R., Barlow, M.J., Corradi, R.L.M., et al.: A planetary nebula around Nova V458 Vulpeculae undergoing flash ionization. ApJL **688**, 21 (2008)
16. Witham, A.R., Knigge, C., Drew, J.E., et al.: The IPHAS catalogue of Hα emission-line sources in the northern Galactic plane. MNRAS **384**, 1277 (2008)
17. Wright, N.J., Greimel, R., Barlow, M.J., et al.: Extremely red stellar objects revealed by IPHAS. MNRAS **390**, 929 (2008)
18. Wright, N.J., Drake, J.J., Drew, J.E., et al.: Photoevaporating proplyd-like objects in Cygnus OB2. ApJL **746**, 21 (2012)
19. Wright, N.J., Wesson, R., Drew, J.E., et al.: The ionised nebula surrounding the red supergiant W26 in Westerlund 1. MNRAS **437**, 1 (2014)

Chapter 8
Science Archives: Facilitating Survey Science

M. Read, B. Mann, R. Blake, R. Collins, N. Cross, C. Davenhall, M. Holliman, and E. Sutorius

Abstract In this paper we discuss the role of science archives and data centres in supporting survey astronomy. We start by describing the work of the Wide Field Astronomy Unit (WFAU) at the Institute for Astronomy of the University of Edinburgh, which in recent years has built science archives for the current large area and deep infrared surveys (UKIDSS and VISTA). We then go on to briefly discuss, in more general terms, how archives should operate and service the current and ever increasing volume of digital survey data.

8.1 The Wide Field Astronomy Unit

The Wide Field Astronomy Unit (WFAU) is part of the Institute for Astronomy at the University of Edinburgh. Prior to its formation, WFAU had operated under the guise of the UK Schmidt Telescope Unit (UKSTU) and operated the SuperCOSMOS measuring machine. WFAU is currently staffed by a mixture of astronomers, software developers and systems administrators and has been serving data and supporting survey science for over 25 years.

8.1.1 WFAU Science Archives

The first digital science archive created and hosted by WFAU was based on SuperCOSMOS scans of Schmidt photographic survey plates [1]. Using the Sloan Digital Sky Survey (SDSS) [2] as a model, the SuperCOSMOS Science Archive

M. Read (✉) • B. Mann • R. Blake • R. Collins • N. Cross • C. Davenhall • M. Holliman • E. Sutorius
Institute for Astronomy, University of Edinburgh, Royal Observatory, Blackford Hill, Edinburgh EH9 3HJ, UK
e-mail: mar@roe.ac.uk

N.R. Napolitano et al. (eds.), *The Universe of Digital Sky Surveys*, Astrophysics and Space Science Proceedings 42, DOI 10.1007/978-3-319-19330-4_8

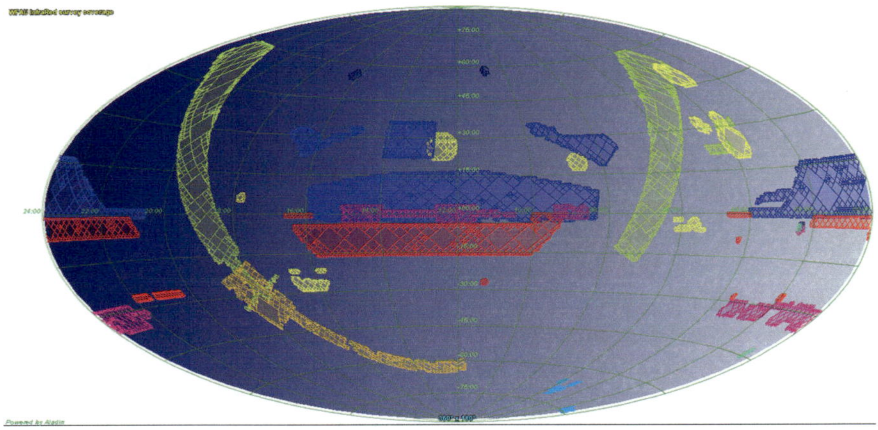

Fig. 8.1 Coverage of the UKIDSS and VISTA surveys archived by WFAU

(SSA[1]), is housed in a relational database management system (Microsoft SQL Server). Related to its SSA work, WFAU also designed and hosts the 6dF Galaxy Redshift Survey (6dFGS) archive.[2]

As part of the VISTA Data Flow System, WFAU then went on to develop archives for the recent and ongoing major infrared surveys carried out by WFCAM on the United Kingdom Infrared Telescope (UKIRT) and by the Visible and Infrared Survey Telescope for Astronomy (VISTA). With the WFCAM Science Archive (WSA[3]) [3] primarily providing users with access to the UKIRT Infrared Deep Sky Surveys (UKIDSS [4]): Large Area Survey (LAS), Galactic Plane Survey (GPS), Galactic Clusters Survey (GCS), Deep Extragalactic Survey (DXS) and Ultra Deep Survey (UDS) and the VISTA Science Archive (VSA[4]) hosting data from the VISTA public surveys [5]: VISTA Hemisphere Survey (VHS), VISTA Variables in the Via Lactea (VVV), VISTA Kilo-Degree Infrared Galaxy Survey (VIKING), VISTA Kilo-Degree Infrared Galaxy Survey (VMC) and VISTA Deep Extragalactic Observations Survey (VIDEO). In addition to archiving the surveys carried out by UKIRT and VISTA, WFAU supports astronomers by providing data releases for their own programmes and also assists VISTA survey heads in providing catalogue and images for ingest into the ESO Science Archive (Fig. 8.1).

WFAU also operates the OmegaCAM Science Archive (OSA) for the ATLAS survey carried out by the VLT Survey Telescope (VST) and the Gaia-ESO Survey (GES[5]) Science Archive which serves spectroscopic data and analysis. In the near future WFAU will publish data from the UKIRT Hemisphere Survey (UHS).

[1]http://surveys.roe.ac.uk/ssa

[2]http://www-wfau.roe.ac.uk/6dFGS

[3]http://surveys.roe.ac.uk/wsa

[4]http://surveys.roe.ac.uk/vsa

[5]http://ges.roe.ac.uk

8.1.2 Data Flow System

For the archives currently been curated by WFAU (WSA, VSA, OSA and GES) the initial, nightly, pipeline processing is carried out by the Cambridge Astronomy Survey Unit (CASU). The photometrically calibrated images and catalogues are then transferred up to Edinburgh where the observational metadata and object catalogues are ingested into the relevant load database. Transfers and ingests are typically run on a monthly basis as this allows for the accumulation of sufficient calibration observations to accurately reduce the data. Following quality control and flagging, additional data products are created, for example band merged catalogues, deep stacks and variability analysis tables and neighbour tables with other surveys.

Periodic data releases are made from the load database which are then copied to a static release database accessed by the user.

8.1.3 Data Volumes and Access

WFAU archives currently total over 1.0 PB in size, with 0.35 Pb of that being held in databases. The tables house over 10^{12} rows and provide calibrated photometry and astrometry for single-band and band-merged sources. Neighbour tables with external surveys are provided to allow fast cross-referencing and variability measurements for multi-epoch programmes are also generated.

The VISTA VVV survey, with its observations of the Galactic Plane, accounts for around 30 % of the data and presents the most challenges for curation. The largest table in the archives, with over 19 billion rows, is the catalogue of VVV sources.

Data releases are often initially proprietary and users are required to login to access these databases (there are over 1,000 registered UKIDSS users). Databases are made publicly accessible once any proprietary period has passed.

Users submit over 50,000 queries each month to the archives, with the queries returning over 2 billion rows of results. Users are able to directly submit SQL queries through the browser user-interface thus providing a powerful and flexible way to mine the data. Thousands of images, mainly cut-outs, are also served out daily.

8.1.4 Current Development Work

WFAU is currently focussing its development work on providing users with an even richer environment for exploring and querying its archives. This work is driven by user requirements for developing complex queries iteratively, the need to query across more than one archive and increasing data volumes.

Using Virtual Observatory infrastructure[6] at its core the system being developed will support distributed queries and allow users to stage and share results in their own user database. An updated, browser based, user-interface will be the access point for the system. A prototype of which is available for the OSA.[7]

8.2 Supporting Survey Science

To allow astronomers to get the best science from current and future surveys, science archives and data centres should exhibit several attributes:

1. High quality science ready data releases e.g. calibrated and quality controlled.
2. Intuitive/powerful user interfaces and access through the Virtual Observatory.
3. Good user support e.g. documentation and helpdesk
4. Add value to the data e.g. neighbour table with other surveys, coverage maps

These requirements are borne out by WFAU's interaction with the astronomical community over two decades.

Larger data volumes and rates produced by current and planned surveys will increase the demands on archives. Flexible queries, the staging of results and querying across archives will go some way in supporting users access these datasets. However more of the analysis will need to be carried out at the data centres and archives will need to develop systems for users to upload and run their own algorithms.

8.3 Conclusion

We have described the archives operated by WFAU and the work in development to further facilitate the science carried out on the survey data held in these archives. Data centres have an increasingly important role to play in getting the best out of the current and future sky surveys. It is important to have involvement throughout the data life-cycle and understand and support users in the way they work.

[6] http://www.ivoa.net

[7] http://osa.roe.ac.uk

References

1. Hambly, N., MacGillivray, H.T., Read, M.A., et al.: The SuperCOSMOS sky survey – I. Introduction and description. MNRAS **326**, 1279–1294 (2001)
2. Szalay, A.S., Kunszt, P.Z., Thakar, A.R., et al.: The Sloan Digital Sky Survey and its archive. ASPC **216**, 405 (2000)
3. Hambly, N., Collins, R.S., Cross, N.J., et al.: The WFCAM science archive. MNRAS **384**, 637–662 (2008)
4. Dye, S., et al.: The UKIRT infrared deep sky survey early data release. MNRAS **372**, 1227 (2006)
5. Emerson, J.P., et al.: The visible & infrared survey telescope for astronomy. ESO Messenger **117**, 27–32 (2004)

Chapter 9
Science Archives at the Wide Field Astronomy Unit

R. Blake, M. Read, E. Sutorius, N. Hambly, N. Cross, R. Collins, M. Holliman, and B. Mann

Abstract The Wide Field Astronomy Unit (WFAU) at the Royal Observatory, Edinburgh (ROE) has been producing archives of astronomy data for more than a decade. It houses a collection of over 80 billion individual detections spread across five major astronomical surveys dating back over 60 years. As well as these surveys, we also host copies of external surveys to allow the cross-referencing of sources in our surveys with those detected with other instruments. This article details the data held by WFAU and the services we provide to our users.

9.1 The Wide Field Astronomy Unit

Beginning life as the UK Schmidt Telescope Unit, which developed SuperCOSMOS [7], the Wide Field Astronomy Unit (WFAU), based at the Royal Observatory, Edinburgh (ROE), has helped develop the VISTA Data Flow System (VDFS) [4] which handles data from the Visible and Infrared Survey Telescope for Astronomy (VISTA) [5] and the UKIRT Infrared Deep Sky Survey (UKIDSS) [8]. More recently, WFAU have become involved with other data archive projects such as the Gaia-ESO Science Archive (GES) and the OmegaCAM Science Archive (OSA).

9.2 The Science Archives at WFAU

As a result of the last decade or more of research, WFAU now houses a collection of different online science archives. A summary of the available data can be seen in Table 9.1, while a more detailed description of each archive follows below.

R. Blake (✉) • M. Read • E. Sutorius • N. Hambly • N. Cross • R. Collins • M. Holliman • B. Mann
Institute for Astronomy, University of Edinburgh, Royal Observatory, Blackford Hill, Edinburgh, EH9 3HJ, UK
e-mail: rpb@roe.ac.uk

N.R. Napolitano et al. (eds.), *The Universe of Digital Sky Surveys*, Astrophysics and Space Science Proceedings 42, DOI 10.1007/978-3-319-19330-4_9

Table 9.1 A summary of data held in WFAU science archives

Survey	Main filters	# of detections	# of sources	Time period
SSA	B, R, I	6,354,116,532	1,899,775,021	1949–2002
WSA	Z, Y, J, H, K	9,510,576,982	1,445,199,704	2005–
VSA	Z, Y, J, H, Ks	62,317,337,036	1,325,979,729	2009–
OSA	u, g, r, i, z	4,178,356,501	88,689,046	2011–
GES		(Spectra) 65,046	33,887	2012–

The SuperCOSMOS Science Archive (SSA) WFAU's first major data archive resulted from the SuperCOSMOS project to digitise photographic plates. The ESO/SERC Southern Sky Survey was a survey of the southern sky performed between 1974 and 1994 taken by the UK Schmidt Telescope (UKST) in Australia [1] (blue band observations in range 3,950–5,400 Å) and the ESO-Schmidt telescope in Chile [12] (red band observations in range 6,300–6,900 Å). A second red band epoch was then taken by the UKST between 1984 and 2000.

To complement this, plates from the second Palomar Observatory Sky Survey (POSS II)[10] were also digitised, providing a complete northern hemisphere survey in blue (4,000–5,700 Å), red (6,100–7,000 Å) and near-IR (7,400–9,400 Å) bands taken between 1987 and 1999. A second epoch was provided by scanning the earlier POSS I [9] plates taken between 1949 and 1958.

A database was created to contain all of these data and it was made available online (http://surveys.roe.ac.uk/ssa).

The WFCAM Science Archive (WSA) After creating the SSA, WFAU became involved in the VDFS project [4]. The first stage of this was to create a processing pipeline and online archive of data from WFCAM [3] for UKIDSS [8], commencing in May 2005. Over the next 7 years, 10 incremental releases of data were created. Initially viewable only by the survey teams, all ten releases, as well as many non-survey data releases, are now publically viewable online (http://surveys.roe.ac.uk/wsa).

As UKIDSS was ending, UKIRT began the UKIRT Hemisphere Survey (UHS), a large area J-band survey following on from the UKIDSS observations which commenced in May 2012. At the date of writing, these data are still being collected and fed into the WSA with a view to a future data release.

The VISTA Science Archive (VSA) The VISTA telescope [5] came online in 2009 and data collection began in November of that year. All the data observed by VISTA are fed into the VSA, and we are currently releasing regular consortium releases, which later become publically available, for 5 of the 6 surveys. Non-survey data releases are also created on a slightly more ad hoc basis. A full list of the available data releases can be found on the VSA website (http://surveys.roe.ac.uk/vsa).

The OmegaCAM Science Archive (OSA) The OSA hosts data observed with the OmegaCAM mounted on the VLT Survey Telescope (VST) [2] in Chile. Primarily we are responsible for the data from the VST ATLAS survey [11]. As with the

WSA and VSA, the OSA is used to create consortium database releases, with the first public release now available (http://surveys.roe.ac.uk/osa).

The Gaia-ESO Science Archive (GES) In a slight departure from previous archives, GES consists of a collection of calibrated one- and two-dimensional spectra for stars observed in the Gaia-ESO Survey [6], and many astrophysical parameters derived from these spectra. Starting from 2012, GES now contains 65,046 spectra for 33,887 individual sources.

9.3 Value Added Services at WFAU

As well as creating the searchable databases of astronomical data and providing access to the original pixel files, as multi-extension FITS files, WFAU also provide extra data products for our users. One example is variability information which we include in our databases for sources which have multi-epoch data and which show obvious variability. We combine multiple observations to create merged filter catalogues of sources. Where multiple epochs of the same filter and pointing exist, we stack the images to create deep products in order to enable the study of fainter sources.

Flags are applied to the data to warn users of reasons why the data might be imperfect, such as deblended sources, saturated source images, or the source being close to a detector boundary. Users can then easily filter out affected sources if they feel this would be a problem.

We provide neighbour tables which makes it easy to compare sources between archives. The "neighbour table" is a table in the database that matches the identifier for a source to the identifier from another database for a source in the same position (within a matching radius). As well as neighbour tables for our own surveys (e.g. to match the VSA with the SSA), we provide neighbour tables for up to 35 external surveys. These include different SDSS releases and many popular surveys such as 2MASS, WISE, DENIS, EROS, AKARI, SPITZER, and XCS. To ensure our users can access these other surveys quickly and easily, we maintain copies of these survey databases on our servers in Edinburgh.

References

1. Cannon R.D.: Sky surveys with the UK 1.2m Schmidt telescope. In: Capaccioli, M. (ed.) Astronomy with Schmidt-Type Telescopes. Proceedings of IAU Colloquium, vol. 78, pp. 25–36. Kluwer, Dordrecht (1984)
2. Capaccioli, M., Mancini, D., Sedmak, G.: VST: The VLT survey telescope. Mem. S.A.It. **74**, 450–451 (2003)
3. Casali, M., et al.: The UKIRT wide-field camera. A&A **467**, 777–784 (2007)

4. Emerson, J.P., et. al.: VISTA data flow system: overview. In: Quinn, P.J., Bridger, A. (eds.) Optimizing Scientific Return for Astronomy through Information Technologies. Proceedings of the SPIE, vol. 5493, pp. 401–410. SPIE, Bellingham (2004)
5. Emerson, J.P., Sutherland, W.J., McPherson, A.M., Craig, S.C., Dalton, G.B., Ward, A.K.: The visible & infrared survey telescope for astronomy. ESO Messenger **117**, 27–32 (2004)
6. Gilmore, G., et al.: The Gaia-ESO public spectroscopic survey. ESO Messenger **147**, 25–31 (2012)
7. Hambly, N.C., Miller, L., MacGillivray, H.T., Herd, J.T., Cormack, W.A.: Precision astrometry with SuperCOSMOS. MNRAS **298**, 897–904 (1998)
8. Lawrence, A., et al.: The UKIRT infrared deep sky survey (UKIDSS). MNRAS **379**, 1599–1617 (2007)
9. Minkowski, R.L., Abell G.O.: The national geographic society – Palomar Observatory sky survey. In: Strand, K.A. (ed.) Basic Astronomical Data. Stars & Stellar Systems, vol. III, pp. 481–487. University of Chicago Press, Chicago (1963)
10. Reid, I.N., et al.: The second Palomar sky survey. PASP **103**, 661–674 (1991)
11. Shanks, T., et al.: VSA ATLAS first science results. ESO Messenger **154**, 38–40 (2013)
12. West R.M.: The ESO sky surveys. In: Capaccioli M. (ed.) Astronomy with Schmidt-type Telescopes. Proceedings of IAU Colloquium, vol. 78, pp. 13–24. Kluwer, Dordrecht (1984)

Chapter 10
HELP: The Herschel Extragalactic Legacy Project and The Coming of Age of Multi-wavelength Astrophysics

M. Vaccari

Abstract How did galaxies form and evolve? This is one of the most challenging questions in astronomy today. Answering it requires a careful combination of observational and theoretical work to reliably determine the observed properties of cosmic bodies over large portions of the distant Universe on the one hand, and accurately model the physical processes driving their evolution on the other. Most importantly, it requires bringing together disparate multi-wavelength and multi-resolution spectro-photometric datasets in an homogeneous and well-characterized manner so that they are suitable for a rigorous statistical analysis. The Herschel Extragalactic Legacy Project (HELP) funded by the EC FP7 SPACE program aims to achieve this goal by combining the expertise of optical, infrared and radio astronomers to provide a multi-wavelength database for the distant Universe as an accessible value-added resource for the astronomical community. It will do so by bringing together multi-wavelength datasets covering the 1,000 deg^2 mapped by Herschel extragalactic surveys in an homogeneous and well-characterized manner, creating a joint lasting legacy from several ambitious sky surveys.

10.1 The HELP Project and Its Science Objectives

How did galaxies form and evolve? This is one of the most challenging questions in astronomy today. Although astronomers now have a good understanding of the background cosmology and of the formation of the large-scale structure of the dark matter [14, 16], the complex astrophysics that leads to the variety and numbers of galaxies observed within dark matter halos is still very poorly understood. Anecdotal clues to these questions can be found through focussed studies of individual galaxies. However, the fundamental requirement for rigorous testing of

M. Vaccaric (✉)
Astrophysics Group, Physics Department, University of the Western Cape, Private Bag X17, 7535, Bellville, Cape Town, South Africa

INAF – Istituto di Radioastronomia, via Gobetti 101, 40129 Bologna, Italy
e-mail: mattia@mattiavaccari.net

N.R. Napolitano et al. (eds.), *The Universe of Digital Sky Surveys*, Astrophysics and Space Science Proceedings 42, DOI 10.1007/978-3-319-19330-4_10

any theories of galaxy formation and evolution is a complete statistical audit or census of the stellar content and star formation rates of galaxies in the Universe at different times and as a function of the mass of the dark matter halos that host them. This audit requires many elements. We need unbiased maps of large volumes of the Universe made with telescopes that probe the different wavelengths at which the different physical processes of interest manifest themselves. We need catalogues of the galaxies contained within these maps with photometry estimated uniformly from field-to-field, from telescope-to-telescope and from wavelength-to-wavelength. We need to understand the probability of a galaxy of given properties appearing in our data sets. We need the machinery to bring together these various data sets and calculate the value-added physical data of primary interest, e.g. the distances (or redshifts), stellar masses, star-formation rates and the actual number densities of the different galaxy populations.

The Herschel Extragalactic Legacy Project (HELP, PI : Seb Oliver, University of Sussex, http://herschel.sussex.ac.uk) brings together several of the teams that have been undertaking ambitious coordinated multi-wavelength digital sky surveys to study large volumes of the distant Universe over the past decade. These observational projects are mature enough that we are now able to undertake the necessary homogenization and thus provide the first representative and comprehensive census of the galaxy populations in the distant Universe. HELP was therefore funded by the European Commission FP7-SPACE-2013-1 Scheme (Grant Agreement 607254) to produce over the 2014–2017 4-year funding period a database for the distant Universe as an accessible value-added resource for the astronomical community. HELP will thus provide a lasting legacy of many thousands of hours on different space telescopes and thousands of nights of ground based telescope and a solid foundation for future space missions and ground-based observatory projects.

10.2 The Herschel Satellite and Its Mission

Galaxies emit electromagnetic radiation over a very wide wavelength range, but most of it is absorbed by the Earth's atmosphere and thus cannot be studied from the ground. With the development of satellite missions, however, astronomers have gradually been able to explore galaxy emission over the full electromagnetic spectrum, from γ-rays to radio waves, in an uninterrupted manner. Comparing the properties of galaxies observed at different wavelengths have thus enabled substantial progress in the understanding of the physical processes driving their formation and evolution.

The Herschel satellite [9] was developed by the European Space Agency and carried out its 3.5 year science mission between 2009 and 2013. Herschel was the first 4-m class space telescope and the first far-infrared and sub-millimeter telescope sensitive enough to detect galaxies in the distant Universe, vastly improving the state of observations in this poorly explored band. Herschel thus signaled the completion of an ambitious program jointly carried out by ESA and NASA to map galaxy evolution across most of the life of the Universe at all wavelengths. In particular, the Herschel imaging instruments SPIRE [4] and PACS [10] fully constrain the peak of the far-infrared and sub-millimeter background with six photometric channels covering the 70–500 μm wavelength range. Herschel thus allows us to thoroughly investigate the sources of the infrared background radiation and characterize their total obscured star formation as a function of cosmic time through systematic observing programs such as HerMES [8] and H-ATLAS [2].

However, the large size of the Herschel beam (6–37″ across the 70–500 μm wavelength range) means that the objects that can be clearly seen as individual sources only make up a small fraction of the cosmic infrared background. In order to unlock the full information available in Herschel maps one must therefore develop new methods to estimate the short-wavelength sources most likely to contribute to emission in Herschel bands in a statistically rigorous manner. HELP therefore brings together leading scientists working on Herschel extragalactic surveys with experts in multi-wavelength data reduction, analysis and homogenization to make sure that we can effectively build upon existing best practices to deliver the highest-quality multi-wavelength datasets and selection functions.

10.3 HELP Data Products and Herschel Scientific Legacy

Many of the outstanding problems in galaxy evolution require a multi-wavelength approach to properly account for different physical processes over the full life-cycle of galaxies and black holes. Modeling tools to account for the Spectral Energy Distribution of galaxies have become sufficiently advanced that mis-matched photometry (i.e. photometry at different wavelengths which samples different areas of a galaxy) may be the dominant source of error. Thus in all wavebands it is now common to produce "aperture-matched" catalog. In this approach the source detection is carried out on either a single image or, for the greatest sensitivity, on a combined image, but then the photometry is carried out on all images with a common aperture or a common model for the galaxy light profile. However, statistical studies of galaxy populations also require a detailed understanding and modeling of the selection processes in the basic data products and the derived properties. These "selection functions" are seldom available for individual datasets and rarely, if ever, published for derived physical properties, but are crucial to estimate source number densities in a reliable manner [1, 3, 15].

The starting point for our project will be the multi-wavelength images and catalogs that have been produced and publicly released by our team as well as by others based on more than a decade of concerted observational projects [5–7]. Our first task is to homogenize these data, so that photometry and calibration is consistent from catalog to catalog, field to field, and from wavelength to wavelength, building upon the framework developed as part of the Spitzer Data Fusion (http://www.mattiavaccari.net/df/). The benchmark for this will be that we can determine accurate photometric redshifts, or distances, adopt consistent galaxy

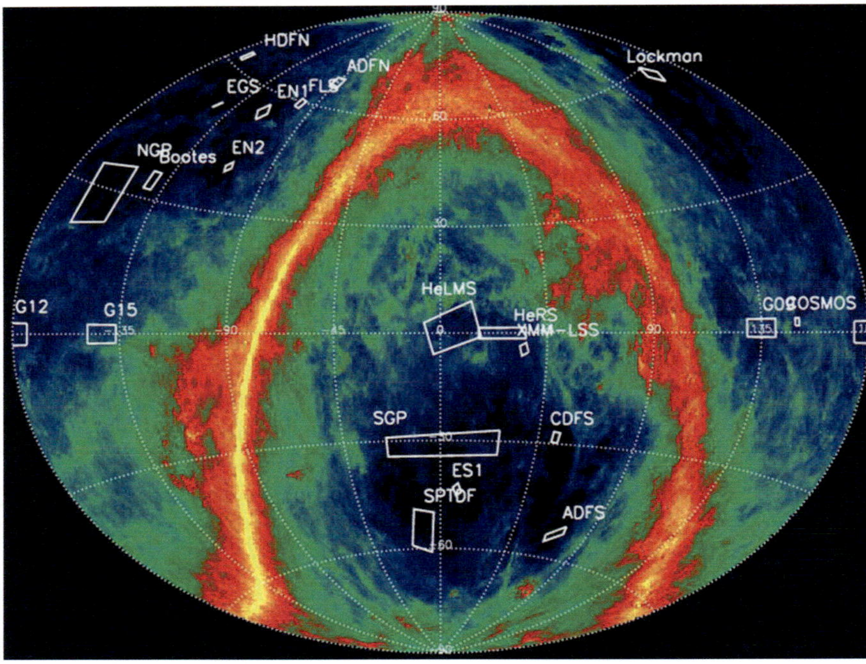

Fig. 10.1 The HELP fields overlaid on the IRAS/COBE dust maps by [13] in ecliptic coordinates

spectral templates and characterize the physical properties of the galaxies [11, 12]. For this purpose we will also assemble available optical spectroscopy from e.g. SDSS, 2dFGRS, GAMA, PRIMUS, VVDS, zCOSMOS and provide homogeneous reliability flags for both spectroscopic and photometric redshifts. We will thus produce a key input catalogs for future spectroscopic surveys with e.g. WEAVE and WAVES.

HELP will thus produce and publicly release multi-wavelength datasets (catalogs and images) and selection functions as well as physical parameters for individual galaxies over the 1,000 deg^2 covered by the Herschel extragalactic surveys (See Fig. 10.1). The data products as well as the techniques and tools that we will produce will enable astronomers to fully capitalize on datasets provided by Herschel as well as other surveys (See Table 10.1). This will extend the kinds of scientific investigations made possible a decade ago in the nearby Universe by the SDSS into the early Universe and provide a lasting legacy for surveys and facilities in the future. In so doing, HELP will provide a complementary view to the AstroDeep project focusing on five small sky areas totaling 2 deg^2 and thus sampling fainter galaxies but not the full range of environments, rare objects and short-lived phases in galaxy evolution (See Fig. 10.2).

Table 10.1 A selection of relevant multi-wavelength survey projects within HELP fields

Wavelength	Telescope/instrument	Observing band	Survey project
Ultraviolet	GALEX	FUV & NUV	DIS, MIS, AIS
Visible	PS1, SDSS, DECAM, VST, CFHT, INTWFC	*ugrizy*	PS1, SDSS, DES, ATLAS, KIDS, VOICE, INTWFS
Near-Infrared	2MASS, UKIRT, VISTA	*ZYJHK*	UKIDSS, VIDEO, VIKING, VHS
Mid-Infrared	IRAC, WISE	3.6/4.5/5.8/8.0/12.0 μm	WISE, SWIRE, SERVS, S-COSMOS, SpUDS, SPLASH
Far-Infrared	WISE, MIPS, PACS	22/24/70/100/160 μm	WISE, SWIRE, PEP, HerMES
(Sub-)Millimeter	SPIRE, SCUBA, SCUBA2, LABOCA, AzTEC, ACT, SPT	250/350/500/850 μm	HerMES, SHADES, S2LS, ACT, SPT
Radio	ATCA, GMRT, LOFAR, MeerKAT, JVLA	0.6, 1.4, 3, 5 GHz	ATLAS, LOFAR, WODAN, MIGHTEE, VLASS

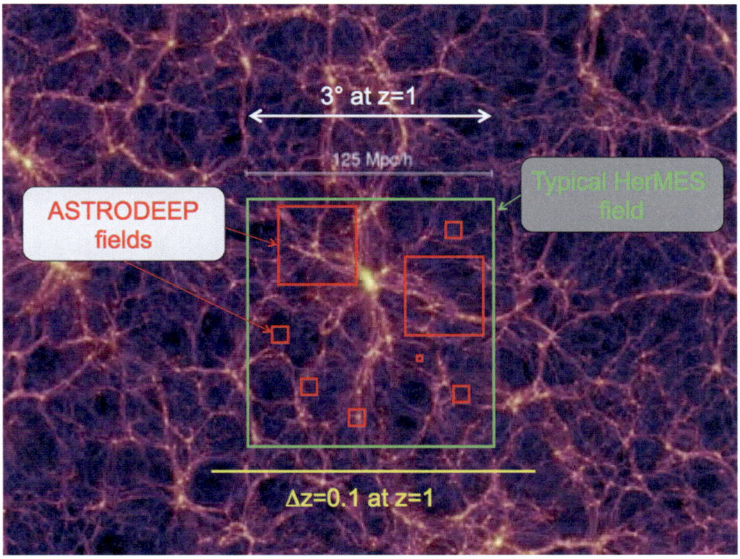

Fig. 10.2 A slice of the dark matter in the universe today from the millennium simulation of the universe by [14]. Overlaid are various scale markers, including the footprints of the AstroDeep fields and of one of $3 \times 3\,\text{deg}^2$ HerMES fields, illustrating how much of the progenitors of these structures they would sample at $z = 1$. The smallest H-ATLAS field is about twice the size of this entire image. HELP covers an area 15 times as large as the whole image

10.4 Conclusion

HELP (Herschel Extragalactic Legacy Project, http://herschel.sussex.ac.uk) will produce an unrivaled public database of multi-wavelength sky images, catalogs of individually detected sources and their physical properties as well as related selection functions over the 1,000 $\deg^2$ of the extragalactic sky covered by the Herschel satellite. It will bring together in a concerted manner space-based and ground-based wide-area galaxy surveys for the first time, enabling highly-accurate statistical studies of galaxy properties and their evolution with cosmic time and providing a lasting legacy for the astronomical community to mine for decades to come and a template for future space science data curation projects in the petascale and exascale era.

Acknowledgements HELP is funded by the EC REA (FP7-SPACE-2013-1 GA 607254 – Herschel Extragalactic Legacy Project). Mattia Vaccari is also supported by the Square Kilometre Array South Africa Project, the South African NRF and DST (DST/CON 0134/2014) and Italy's MAECI (PGR GA ZA14GR02 – Mapping the Universe on the Pathway to SKA)

References

1. Burgarella, D., Buat, V., Gruppioni, C., et al.: Herschel PEP/HerMES: the redshift evolution ($0 \leq z \leq 4$) of dust attenuation and of the total ($UV + IR$) star formation rate density. A&A **554**, 70 (2013)
2. Eales, S.A., Dunne, L., Clements, D., et al.: The Herschel ATLAS. PASP **122**, 499 (2010)
3. Eales, S.A., Raymond, G., Roseboom, I.G., et al.: First results from HerMES on the evolution of the submillimetre luminosity function. A&A **518**, L23 (2010)
4. Griffin, M.J., Abergel, A., Abreu, A., et al.: The Herschel-SPIRE instrument and its in-flight performance. A&A **518**, L3 (2010)
5. Jarvis, M.J., Bonfield, D.G., Bruce, V.A., et al.: The VISTA Deep Extragalactic Observations (VIDEO) survey. MNRAS **428**, 1281 (2013)
6. Lonsdale, C.J., Smith, H.E., Rowan-Robinson, M., et al.: SWIRE: The SIRTF Wide-Area Infrared Extragalactic Survey. PASP **115**, 897 (2003)
7. Mauduit, J.-C., Lacy, M., Farrah, D., et al.: The Spitzer Extragalactic Representative Volume Survey (SERVS): survey definition and goals. PASP **124**, 714 (2012)
8. Oliver, S.J., Bock, J., Altieri, B., et al.: The Herschel Multi-tiered Extragalactic Survey: HerMES. MNRAS **424**, 1614 (2012)
9. Pilbratt, G., Riedinger, J.R., Passvogel, T., et al.: Herschel Space Observatory. An ESA facility for far-infrared and submillimetre astronomy. A&A **518**, L1 (2010)
10. Poglitsch, A., Waelkens, C., Geis, N., et al.: The Photodetector Array Camera and Spectrometer (PACS) on the Herschel Space Observatory. A&A **518**, L2 (2010)
11. Rowan-Robinson, M., Babbedge, T., Oliver, S., et al.: Photometric redshifts in the SWIRE Survey. MNRAS **386**, 697 (2008)
12. Rowan-Robinson, M., Gonzalez-Solares, E., Vaccari, M., Marchetti, L.: Revised SWIRE photometric redshifts. MNRAS **428**, 1958 (2013)

13. Schlegel, D.J., Finkbeiner, D.P., Davis, M.: Maps of dust infrared emission for use in estimation of reddening and cosmic microwave background radiation foregrounds. ApJ **500**, 525 (1998)
14. Springel, V., White, S.D.M., Jenkins, A., et al.: Simulations of the formation, evolution and clustering of galaxies and quasars. Nature **435**, 629 (2005)
15. Vaccari, M., Marchetti, L., Franceschini, A., et al.: The HerMES SPIRE submillimeter local luminosity function. A&A **518**, L20 (2010)
16. Vogelsberger, M., Genel, S., Springel, V., et al.: Properties of galaxies reproduced by a hydrodynamic simulation. Nature **509**, 177 (2014)

Chapter 11
Cepheids and the Distance Ladder

I. Musella

Abstract Classical Cepheids plays a key role in the calibration of the extragalactic distance scale. In spite of their importance, some uncertainties related to their properties remain. In particular, a general consensus on the possible dependence on the metallicity of the host galaxy of the Cepheid properties has not been reached yet. These uncertainties could produce significant systematic errors in the calibration of the secondary distance indicators we need to reach cosmologically significant distances and in turn in the evaluation of the Hubble constant H_0. Possible solutions are discussed.

11.1 Introduction

Classical Cepheids (CCs) are population I pulsating stars located in the Instability Strip (IS), a defined region of the color-magnitude diagram. They are very bright ($-2 < M_V < -7$ mag) and are characterized by light-curves with large amplitudes so that they have been observed by the *Hubble Space Telescope (HST)* out to about 25 Mpc [7]. In particular, they obey to a characteristic period-luminosity relation (PLR) that is used as distance indicator and to calibrate secondary distance indicators, observed with *HST* in external galaxies, and in turn to obtain an estimate of H_0 (see e.g. [6, 19]). The PLR was traditionally considered linear and "universal" [5, 9], assuming the slope of the Large Magellanic Cloud (LMC) CC sample and setting the zero-point to the LMC distance modulus or on the basis of Galactic CCs with independent distance estimates. Actually, the advent of HST with the observation of many spiral galaxies in the Virgo Cluster with a large spread in metallicity, as well as new studies of the difference between Galactic and Magellanic CC properties (see e.g. [18, 21] and references therein), showed much evidence that the CC properties depend on the metallicity of the host galaxy. Despite the many efforts, from the theoretical and observational point of view, performed to understand the dependence of the CC properties on the chemical

I. Musella (✉)
INAF – Osservatorio Astronomico di Capodimonte, Salita Moiariello, 16, I-80131 Napoli, Italy
e-mail: ilaria.musella@oacn.inaf.it

N.R. Napolitano et al. (eds.), *The Universe of Digital Sky Surveys*, Astrophysics and Space Science Proceedings 42, DOI 10.1007/978-3-319-19330-4_11

composition, there are still many doubts both on the sign and the size of this effect ([1, 2, 10, 11, 13, 18, 20] and references therein). Many studies were aimed at understanding the systematic errors in the extragalactic distance scale based on the CC PLR. In particular, the main error sources are: (i) the dispersion of the PLR due to intrinsic factors such as the finite width of the IS, but also to the possible differential reddening in the studied galaxy and to a poor sampling of the CC light curves; (ii) the errors due to the metallicity and reddening effects that can reach 15 % and (iii) the use of LMC as anchor galaxy, in spite of its large spread in reddening and metallicity and the remaining uncertainties on its distance. Possible solutions to the previous problems are: (i) to have a solid theoretical scenario; (ii) to use the near -infrared bands where the effects due to metallicity and reddening are low or negligible and (iii) to adopt a different anchor galaxy with a metallicity more similar to the majority of the galaxies observed by HST. In particular an interesting alternative to the LMC is the galaxy NGC 4258 that has an independent and very accurate (3 %) geometrical measurement of the distance based on the kinematics of the hosted water masers ([8, 11, 15–17] and references therein).

11.2 Pulsational Models

From the theoretical point of view, extensive sets of nonlinear convective pulsational models were computed ([3, 13] and references therein) for different chemical compositions — $0.0004 < Z < 0.04$, $0.25 < Y < 0.33$, $0.5 < \Delta Y/\Delta Z < 4$ — and for typical CC masses ranging from 3 to 13 $M_{\odot}$. These models allow us to obtain all the CC observables (light curves, periods, mean magnitudes), to predict the position of the CC IS in a color-magnitude diagram and to build theoretical PLRs and period-luminosity-color relations for different filters, that can be compared with the observed properties to check the models themselves, but also to obtain a theoretical calibration of the extragalactic distance scale. These models, in agreement with observational results (see e.g. [12]), predict that the CC ISs and the PLRs in different photometric filters are dependent on the adopted chemical composition, with the PLR getting flatter as Z increases. Moreover, the optical PLRs are not linear. Both these effects, as well as the intrinsic dispersion, decrease as the wavelength increases from the optical to the NIR bands. At very low metallicity, the dependence on the chemical composition becomes negligible with the predicted PLR at $Z = 0.0004$ very similar to the one for $Z = 0.004$ [14]. These models combined with observational data can represent a powerful tool. An example of this application is presented in the next section.

11.3 Comparison with the Riess Sample

In order to obtain accurate (at 3 %) and solid constraints to the Hubble constant H_0, [17] have observed with ACS@HST and WFC3@HST large CC samples in NGC 4258 and in eight SNIa host galaxies, both in the optical and near-infrared bands. In this work, they adopt NGC 4258 as anchor galaxy for the calibration of the distance scale. Pulsational models were transformed in the same photometric filters adopted by these authors. Figure 11.1 shows a very good agreement between models and observations, in particular with the models for Z = 0.02, adopting the modulus obtained by [17] using a metal-dependent Wesenheit relation. For the other galaxies of the sample, in the left panel of Fig. 11.2 the theoretical ISs, in the plane F160W-(V-I), for different chemical compositions, are compared with the Riess sample, finding that for the more distant galaxies CCs appear bluer and brighter. This could be due to an increasing crowding at larger distances. Applying the theoretical F160W-band PLR to the Riess CC samples, the moduli obtained are slightly different from those found by [17]. The right panel of Fig. 11.2 shows these differences versus the [17] moduli, pointing out that the differences increase at larger distances. Adopting the results obtained from the application of the theoretical PLRs to the Riess CC samples (without using any secondary indicator), it is possible to obtain for H_0 a value of 76.0 ± 1.9 km s^{-1} Mpc^{-1} [4], in good agreement, within the errors, with the value found by [17], using the SNIa distance scale calibrated on the CCs, but with a smaller error.

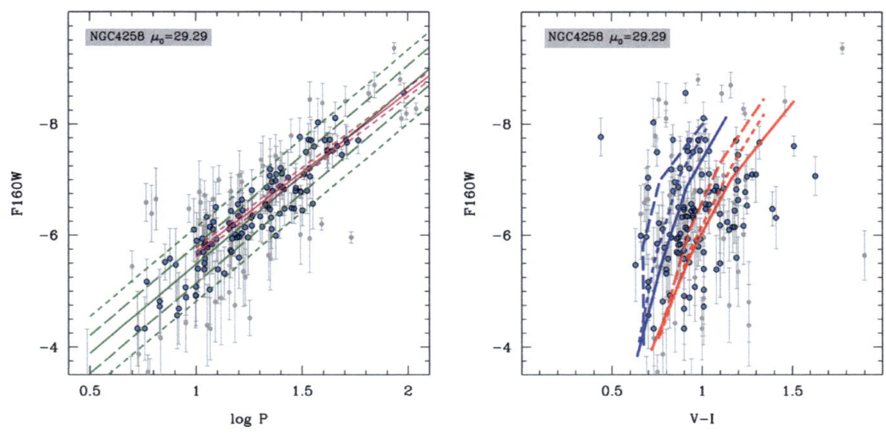

Fig. 11.1 *Left*: The PLR for the NGC4258 CCs in the F160W band. The *dark blue* and *grey symbols* have been used for good and rejected observed CCs (see [17] for details). In this panel, we also show the theoretical (Z = 0.02, Y = 0.28 and *logP* > 1; *dark magenta solid line*) and the observed (*green solid line*) F160W PLRs together with their confidence lines (*dashed* for the whole CC sample and *long-dashed* for the good ones). *Right*: comparison between observational and theoretical (*blue* and *red* boundaries) ISs in the plane (F160W, V–I) for NGC 4258 CCs. For the theoretical ISs, we adopt three different chemical compositions: Z = 0.02; Y = 0.28 (*solid*), Z = 0.02; Y = 0.31 (*dashed*) and Z = 0.008; Y = 0.25 (*long dashed*)

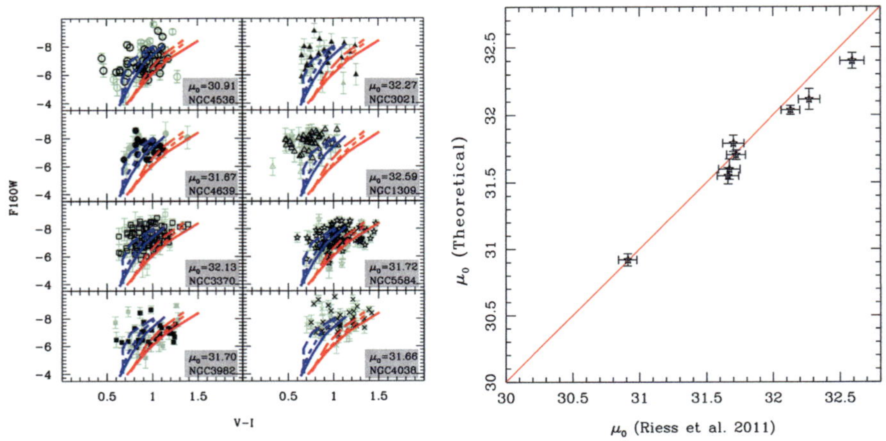

Fig. 11.2 *Left*: As in Fig. 11.1 (*right panel*) for all the other CC samples. In this case, *dark* and *light green symbols* represent good and rejected CCs (see [17] for details), respectively. *Right*: the comparison between the distance moduli obtained using our theoretical F160W PL for Z = 0.02 Y = 0.28 and for periods larger than 10 d and those obtained by [17]

Acknowledgements I'm grateful to Massimo Capaccioli who inspired my activity in the extragalactic distance scale field and I thank the SOC for inviting me to this conference in honor of Massimo's seventieth birthday.

References

1. Bono, G., Caputo, F., Fiorentino, G., Marconi, M., Musella, I.: Cepheids in external galaxies. I. The maser-host galaxy NGC 4258 and the metallicity dependence of period-luminosity and period-Wesenheit relations. ApJ **684**, 102 (2008)
2. Bono, G., Caputo, F., Marconi, M., Musella, I.: Insights into the Cepheid distance scale. ApJ **715**, 277 (2010)
3. Fiorentino, G., Caputo, F., Marconi, M., Musella, I.: Theoretical models for classical Cepheids. VIII. Effects of helium and heavy-element abundance on the Cepheid distance scale. ApJ **576**, 402 (2002)
4. Fiorentino, G., Musella, I., Marconi, M.: Cepheid theoretical models and observations in HST/WFC3 filters: The effect on the Hubble constant H0. MNRAS **434**, 2866 (2013)
5. Freedman, W.L., Madore, B.: An empirical test for the metallicity sensitivity of the Cepheid period-luminosity relation. ApJ **365**, 186 (1990)
6. Freedman, W.L., et al.: Final results from the Hubble space telescope key project to measure the Hubble constant. ApJ **553**, 47 (2001)
7. Freedman, W.L., Madore, B.: The Hubble constant. ARA&A **48**, 673 (2010)
8. Humphreys, E.M.L., Reid, M.J., Moran, J.M., Greenhill, L.J., Argon, A.L.: Toward a new geometric distance to the active galaxy NGC 4258. III. Final results and the Hubble constant. ApJ **775**, 13 (2013)
9. Iben, I., Renzini, A.: Single star evolution I. Massive stars and early evolution of low and intermediate mass stars. PhR **105**, 329 (1984)

10. Kennicutt, R.C., et al.: The Hubble space telescope key project on the extragalactic distance scale. XIII. The metallicity dependence of the Cepheid distance scale. ApJ **498**, 181 (1998)
11. Macri, L.M., Stanek, K.Z., Bersier, D., Greenhill, L.J., Reid, M.J.: A new Cepheid distance to the maser-host galaxy NGC 4258 and its implications for the Hubble constant. ApJ **652**, 1133 (2006)
12. Madore, B., Freedman, W.L.: The Cepheid distance scale. PASP **103**, 933 (1991)
13. Marconi, M., Musella, I., Fiorentino, G.: Cepheid pulsation models at varying metallicity and $\Delta Y/\Delta Z$. ApJ **632**, 590 (2005)
14. Marconi, et al.: Pulsation models for ultra-low ($Z = 0.0004$) metallicity classical Cepheids. ApJ **713**, 615 (2010)
15. Riess, A.G., et al.: A redetermination of the Hubble constant with the Hubble space telescope from a differential distance ladder. ApJ **699**, 539 (2009)
16. Riess, A.G., et al.: Cepheid calibrations of modern type Ia supernovae: Implications for the Hubble constant. ApJS **183**, 109 (2009)
17. Riess, A.G., et al.: A 3% solution: determination of the Hubble constant with the Hubble space telescope and wide field camera 3. ApJ **730**, 119 (2011)
18. Romaniello, M., et al.: The influence of chemical composition on the properties of Cepheid stars. II. The iron content. A&A **488**, 731 (2008)
19. Saha, A., et al.: Cepheid calibration of the peak brightness of type Ia supernovae. XI. SN 1998aq in NGC 3982. ApJ **562**, 314 (2001)
20. Sakai, S., Ferrarese, L., Kennicutt, R.C., Jr., Saha, A.: The effect of metallicity on Cepheid-based distances. ApJ **608**, 42 (2004)
21. Sandage, A., Tammann, G.A., Reindl, B.: New period-luminosity and period-color relations of classical Cepheids. III. Cepheids in SMC. A&A **493**, 471 (2009)

Chapter 12
The Environment of Barred Galaxies Revisited

B. Cervantes Sodi, C. Li, C. Park, and L. Wang

Abstract We present a study of the environment of barred galaxies using galaxies drawn from the SDSS. We use several different statistics to quantify the environment: the projected two-point cross-correlation function, the background-subtracted number counts of neighbor galaxies, the overdensity of the local environment, the membership of our galaxies to galaxy groups to segregate central and satellite systems, and, for central galaxies, the stellar to halo mass ratio (M_*/M_h). When we split our sample into early- and late-type galaxies, we see a weak but significant trend for early-type galaxies with a bar to be more strongly clustered on scales from a few 100 kpc to 1 Mpc when compared to unbarred early-type galaxies. This indicates that the presence of a bar in early-type galaxies depends on the location within their host dark matter halos. This is confirmed by the group catalog in the sense that for early-types, the fraction of central galaxies is smaller if they have a bar. For late-type galaxies, we find fewer neighbors within ~50 kpc around the barred galaxies when compared to unbarred galaxies from the control sample, suggesting that tidal forces from close companions suppress the formation/growth of bars. For central late-type galaxies, bars are more common on galaxies with high M_*/M_h values, as expected from early theoretical works which showed that systems with massive dark matter halos are more stable against bar instabilities. Finally, we find no obvious correlation between overdensity and the bars in our sample, showing that galactic bars are not obviously linked to the large-scale structure of the universe.

B. Cervantes Sodi (✉)
Centro de Radioastronomía y Astrofísica – UNAM, Apdo. Postal 3-72 (Xangari), 58089 Morelia, Mich., Mexico

Korea Institute for Advance Study, Dongdaemun-gu, Seoul 130-722, Korea
e-mail: b.cervantes@crya.unam.mx

C. Li • L. Wang
Partner Group of the Max Planck Institute for Astrophysics and Key Laboratory for Research in Galaxies and Cosmology of Chinese Academy of Sciences, Shanghai Astronomical Observatory, Nandan Road 80, Shanghai 200030, China

C. Park
Korea Institute for Advanced Study, Dongdaemun-gu, Seoul 130-722, Korea

N.R. Napolitano et al. (eds.), *The Universe of Digital Sky Surveys*, Astrophysics and Space Science Proceedings 42, DOI 10.1007/978-3-319-19330-4_12

12.1 Background

With the outcome of large galaxy surveys, the role of environment in triggering or suppressing the formation of stellar bars in galaxies has been a popular topic of research, with a large variety of results, depending on the methods employed [3, 4, 6, 7].

In this work we employ a large volume-limited sample of ∼30,000 galaxies to study the likelihood for galaxies to host bars as a function of environment, using several statistics to study different scales, with the aim of giving a complete study for galaxies in the nearby universe. For a more detailed description of the sample and a complete discussion of the results, we refer the reader to [1, 5].

12.2 Main Results

Our first approach is to compute the two-point cross correlation function (2PCCF, $w_p(r_p)$) of our galaxy sample with respect to a reference sample of the general galaxy population, from scales of a few tens of kiloparsecs up to a few tens of megaparsecs. We then construct a control sample of unbarred galaxies with matched stellar mass, color and surface mass density ($\triangle \log M_* \leq 0.08$, $\triangle(g-r) \leq 0.025$, $\triangle \log \mu_* \leq 0.08$). In Fig. 12.1, first panel, we present the 2PCCF for the barred sample (AB), the control sample (AU) as well as the ratio between them. In this

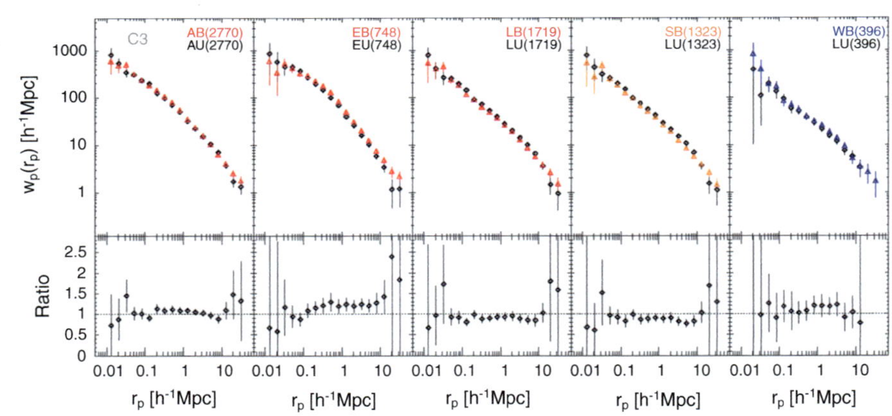

Fig. 12.1 Projected 2PCCF for barred $w_p^{bar}(r_p)$ and unbarred $w_p^{unbar}(r_p)$ galaxies (*top panels*) and $w_p^{bar}(r_p)$ to $w_p^{unbar}(r_p)$ ratio (*bottom panels*) for the different subsamples of our control sample C3 where the galaxies share a common distribution of stellar mass M_*, color g−r, and stellar surface mass density μ_*. Panels from left to right correspond to the whole galaxy sample of barred (AB *orange triangles*) plus unbarred (AU *black diamonds*) galaxies, early-types barred (EB *orange triangles*) and unbarred (EU *black diamonds*), late-types barred (LB *orange triangles*) and unbarred (LU *black diamonds*), strongly barred late-types (SB *yellow triangles*) and LU, and weakly barred late-types (WB *blue triangles*) and LU

case, no significant overclustering is detected for the full sample of barred galaxies over unbarred ones. If we segregate the samples by morphology, the 2PCCF for early-type barred galaxies (EB) shows a weak but significant trend of overclustering on scales from a few 100 kpc to 1 Mpc, when compared with unbarred early-type galaxies (EU), a trend not found for the late-type galaxies, regardless if their bars are classified as strong or weak. This result implies that the presence of bars in early-type galaxies depends on the location of the galaxies within their host dark matter halos.

With the fiber collision problem suffered by the SDSS at small separation angles, we explore the clustering of our samples at small separations by computing the background-subtracted neighbor counts, $N_C(r_p < R_p)$, which is the number of galaxies in the photometric reference sample within the projected radius R_p of the barred or unbarred galaxies, with the effect of chance projection being statistically corrected [5]. In Fig. 12.2 we plot N_C: panels top to bottom show the results for the whole sample, the early-type and late-type subsamples respectively and from left to right, the results for different apparent r-band limiting magnitude applied to the photometric reference sample, ranging from $r_{lim} = 18$ for the left-most panels to $r_{lim} = 21$ for the right-most panels. For early-type galaxies, the overclustering of barred galaxies becomes again evident for scales larges than 100 kpc, but this time the subsample including only late-type galaxies shows a lack of neighbors in the vicinity of barred galaxies on scales $\lesssim$50 kpc when compared to unbarred galaxies,

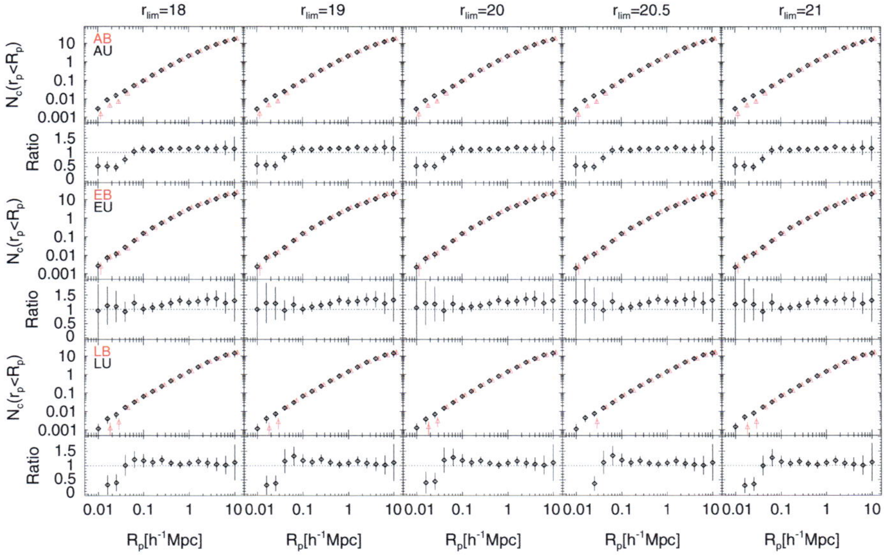

Fig. 12.2 Average counts of galaxies in the photometric sample within a given projected radius R_P from the galaxies in our samples of barred and unbarred control samples . Each line corresponds to different apparent magnitude limits in r-band ($r \leq 18,19,20,20.5,21$) for the galaxies in the photometric sample

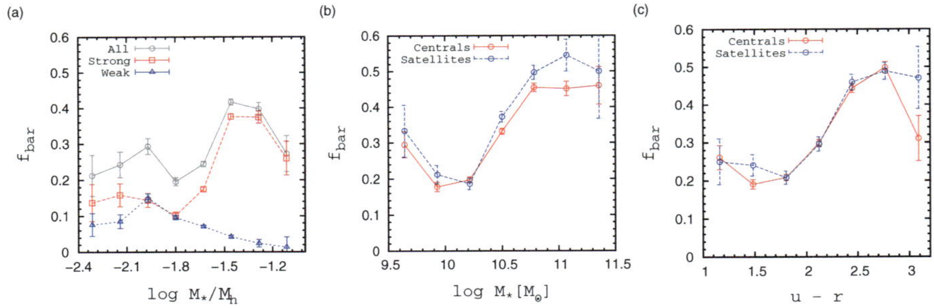

Fig. 12.3 The fraction of barred galaxies $f_{\rm bar}$ as a function of *(a)* stellar-to-halo mass ratio, for strong, weak and strong plus weak bars, *(b)* stellar mass for central and satellite galaxies and, *(c)* color for centrals and satellites

suggesting that tidal forces from close companions suppress the formation/growth of bars.

Using the galaxy catalog by [8], we classified our galaxies as central or satellite galaxies. For the late-type centrals, we estimated the stellar-to-halo mass ratio (M_*/M_h), using the stellar mass data from the MPA/JHU SDSS database [2] and the halo masses from the same galaxy catalog. Figure 12.3a shows that the bar fraction is a strong function of this ratio, with strong bars more commonly found in galaxies with high M_*/M_h, as expected form theoretical works which show that galaxies with massive dark matter halos are more stable against bar formation, while weak bars follow the opposite trend.

With our galaxies segregated into centrals and satellites, we compare the bar fraction of both subsamples as a function of stellar mass and color in Fig. 12.3b, c. At fixed stellar mass we find that the bar fraction of satellite galaxies is slightly higher than that of central galaxies, but at fixed color this difference vanishes. We interpret this as follows; the color of satellite galaxies, on average, is redder than the color of centrals at the stellar mass ranges involved in our study. Given that the bar fraction is higher for redder galaxies, the difference of the bar fraction between centrals and satellites is not directly due to the group environment, but indirectly through its dependence on internal morphology.

References

1. Cervantes Sodi, B., Li, C., Park, C.: Dark matter halos of barred disk galaxies. ApJ **807**, 111 (2015)
2. Kauffmann, G., Heckman, T.M., White, S.D.M., et al.: Stellar masses and star formation histories for 10^5 galaxies from the Sloan Digital Sky Survey. MNRAS **341**, 33 (2003)
3. Lee, G.-H., Park, C., Lee, M.G., Choi, Y.-Y.: Dependence of barred galaxy fraction on galaxy properties and enviornment. ApJ **745**, 125 (2012)
4. Li, C., Gadotti, D.A., Mao, S., Kauffmann, G.: The clustering of barred galaxies in the local universe. MNRAS **397**, 726 (2009)

5. Lin, Y., Cervantes Sodi, B., Li, C., Wang, L., Wang, E.: The environment of barred galaxies in the low-redshift universe. ApJ **796**, 98 (2014)
6. Skibba, R.A., Masters, K.L., Nichol, R.C., et al.: Galaxy Zoo: the environmental dependence of bars and bulges in disc galaxies. MNRAS **423**, 1485 (2012)
7. van den Bergh, S.: Bar galaxies and their environments. ApJ **124**, 782 (2002)
8. Yang, X., Mo, H.J., van den Bosch, F.C., et al.: Galaxy groups in the SDSS DR4. I. The catalog and basic properties. ApJ **671**, 153 (2007)

Chapter 13
Using Gamma Regression for Photometric Redshifts of Survey Galaxies

J. Elliott, R.S. de Souza, A. Krone-Martins, E. Cameron, E.E.O. Ishida, and J. Hilbe

Abstract Machine learning techniques offer a plethora of opportunities in tackling *big data* within the astronomical community. We present the set of Generalized Linear Models as a fast alternative for determining photometric redshifts of galaxies, a set of tools not commonly applied within astronomy, despite being widely used in other professions. With this technique, we achieve catastrophic outlier rates of the order of $\sim$1%, that can be achieved in a matter of seconds on large datasets of

J. Elliott (✉)
Harvard-Smithsonian Center for Astrophysics, 60 Garden Street, Cambridge, MA 02138, USA

Max-Planck-Institut für extraterrestrische Physik, Giessenbachstraße 1, 85748, Garching, Germany
e-mail: jonathan.elliott@cfa.harvard.edu

R.S. de Souza
MTA Eötvös University, EIRSA "Lendulet" Astrophysics Research Group, Budapest 1117, Hungary
e-mail: rafael.2706@gmail.com

A. Krone-Martins
SIM, Faculdade de Ciências, Universidade de Lisboa, Ed. C8, Campo Grande, 1749-016, Lisboa, Portugal
e-mail: algol@sim.ul.pt

E. Cameron
Department of Zoology, University of Oxford, Tinbergen Building, South Parks Road, Oxford, OX1 3PS, UK
e-mail: dr.ewan.cameron@gmail.com

E.E.O. Ishida
Max-Planck-Institut für Astrophysik, Karl-Schwarzschild-Str. 1, 85748 Garching, Germany
e-mail: emille@mpa-garching.mpg.de

J. Hilbe
Arizona State University, 873701,Tempe, AZ 85287-3701, USA

Jet Propulsion Laboratory, 4800 Oak Grove Dr., Pasadena, CA 91109, USA
e-mail: j.m.hilbe@gmail.com

for the COIN collaboration

N.R. Napolitano et al. (eds.), *The Universe of Digital Sky Surveys*, Astrophysics and Space Science Proceedings 42, DOI 10.1007/978-3-319-19330-4_13

size ~1,000,000. To make these techniques easily accessible to the astronomical community, we developed a set of libraries and tools that are publicly available.

13.1 Introduction

Generalized Linear Models [GLMs; 10] are widely used throughout other scientific disciplines such as: biology [1], medicine [8], and economics [12], and is available within the overwhelming majority of contemporary statistical software packages. However, they have been very little used within the astronomical community

There are plenty of opportunities to apply GLMs within astronomy, and one particularly important problem is the estimation of photometric redshifts (photo-z).

Galaxy spectra are made up of many of its physical properties, including morphology, age, metallicity, star formation history, merging history, and a host of other confounding factors in addition to its redshift. This makes robust estimation of photo-zs a difficult task. Estimation is usually done in two ways, by template fitting, or by using machine learning techniques.

There exist several studies that have investigated the advantages of the publicly available codes that estimate the photo-z of galaxies [for a glimpse on the diversity of existent methods, see [7], and references therein]. The overall performance of most codes is good, demonstrating catastrophic errors from 5 to 9 %, which is considered reliable within the field. There are also a number of growing techniques that implement a hybrid approach of template and machine learning techniques [3].

Despite the current advancement within this field, there still exist large practical difficulties. In the next years there are a large number of surveys that will start having *big data* catalogues, e.g., the *Large Synoptic Survey Telescope*[1] [9], *EUCLID*[2] [13] or the *Wide-Field Survey Infrared Telescope*[3] [6]. Current techniques will become difficult to employ if they require large training sets, and as such, this warrants the need for fast and reliable photo-z methods that are capable of robustly estimating redshifts quickly, and on large training datasets.

We introduce GLMs as a new technique to quickly and robustly estimate galaxy photo-zs. We show that it can run in a matter of seconds on a single core computer, even for millions of objects. As part of the COsmostatistics INitiative (COIN[4]) collaboration, we created and distributed easy to use software, and web-applications for use of the wider community in estimating photo-zs.[5]

---

[1] http://www.lsst.org/lsst

[2] http://sci.esa.int/euclid

[3] http://wfirst.gsfc.nasa.gov

[4] https://asaip.psu.edu/organizations/iaa/iaa-working-group-of-cosmostatistics

[5] https://github.com/COINtoolbox

13.2 Methodology

We will not go into the details of deriving GLMs or the formula that is used in our technique, we instead encourage the reader to see the details in [5] and references therein. However, we note the importance of GLMs, such that they allow you to choose the type of distribution you want to model. GLMs are applicable to the entire set of *exponential families* of distributions: Gaussian/normal, gamma, inverse Gaussian, Bernoulli, binomial, Poisson, and negative binomial. For example, in this study, we want to predict the photo-zs of galaxies from multi-wavelength photometry. For such a study, the gamma distribution is favourable, as a redshift is positive and continuous, as is the gamma distribution. To then use the gamma distribution to predict redshifts, we utilised the following machine learning methodology:

1. The data was randomly split into training and test sets.
2. Robust principal component analysis (e.g., [2, 4]) was carried out on the complete data set.
3. We utilised a gamma family distribution to reflect the fact that measured redshifts are positive and continuous.
4. The predicted photo-z for the test data was calculated using the principal component projections of the test data set and the best-fit GLM using the training sample.
5. To measure how well the photo-zs were estimated, we employed a metric commonly used in the literature, specifically, the catastrophic error.

13.3 Data Samples

We used two publicly available galaxy datasets to test the technique outlined in the previous section. The first was the *PHoto-z Accuracy Testing* (PHAT), an international initiative to identify the most promising photo-z methods. We used their publicly available simulated datasets that contains 169,520 simulated galaxies with redshifts ranging from $z = 0.02 - 2.24$, and magnitudes in 11 filters (u, g, r, i, z, Y, J, H, K, $IRAC1$, and $IRAC2$).

Given that this dataset was purely synthetic, we also used a real dataset acquired from the *Sloan Digital Sky Survey* [SDSS; 14]. For details on the query see [5]. The sample used contained 1,347,640 galaxies with a redshift range of $z = 0 - 1.0$, with magnitudes in 5 filters (u', g', r', i', and z'). Comparisons with dereddened values showed no inconsistencies

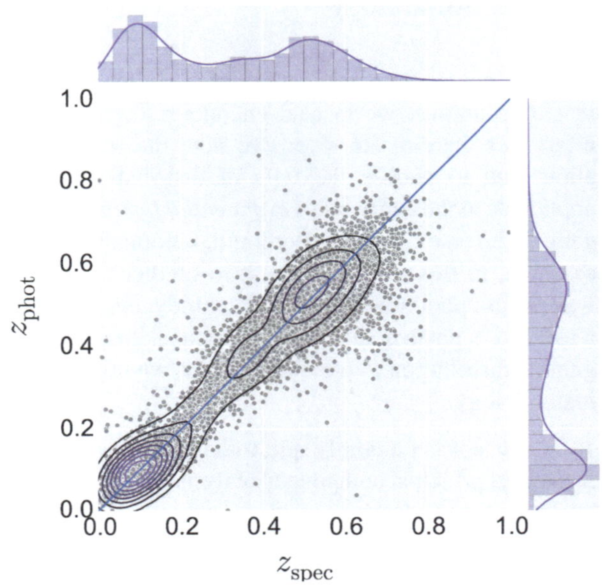

Fig. 13.1 The 2D probability density of the predicted redshift from the GLM fit vs. the spectroscopic redshift (*central plots*). The *upper* and *right subplots* in each panel depict the redshift distribution along photo-z and z_{spec}, respectively

13.4 Results

Both data sets were fit using an AMD Athlon X2 Dual-Core QL-64 processor with 1.7 GB RAM on the Ubuntu 10.04 operating system, which represents an old laptop at today's standards. We achieved catastrophic errors of 1.4 % for the PHAT0 data set and 8 % for the SDSS data set, within ∼1,200, and 10 s. Lower catastrophic errors of ∼1 % could be achieved when using more principal components, but would take longer computational time, ∼5,000 s. We plot the best-fit GLM models for SDSS datasets in Fig. 13.1.

13.5 Conclusions

The astronomical community has left Generalized Linear Models relatively untouched, despite its use throughout the academic world. We have demonstrated their ease of use and quick applicability to estimate the photo-zs of galaxies. This technique has been shown to be competitive with current techniques implemented, that can require larger training sets and longer time for their algorithms to learn. Such properties of this technique will become important in the close future when upcoming wide field sky surveys, such as the LSST, will start collecting data at enormous rates per day.

To make GLMs more accessible to the community, we developed a set of software libraries written in Python and R, that can be easily implemented into people's own work. A web application is also available to be used instantly without the need for installation.

Acknowledgements We thank V. Busti, E. D. Feigelson, M. Killedar, J. Buchner, and A. Trindade for interesting suggestions and comments. JE, RSS and EEOI thank the SIM Laboratory of the *Universidade de Lisboa* for hospitality during the development of this work. Cosmostatistics Initiative (COIN)[6] is a non-profit organisation whose aim is to nourish the synergy between astrophysics, cosmology, statistics and machine learning communities. This work was partially supported by the ESA VA4D project (AO 1-6740/11/F/MOS). AKM thanks the Portuguese agency *Fundação para Ciência e Tecnologia – FCT*, for financial support (SFRH/BPD/74697/2010). EEOI is partially supported by the Brazilian agency CAPES (grant number 9229-13-2). Work on this paper has substantially benefited from using the collaborative website AWOB developed and maintained by the Max-Planck Institute for Astrophysics and the Max-Planck Digital Library. This work was written on the collaborative `WriteLatex` platform, and made use of the GitHub a web-based hosting service and `git` version control software. This work made use of the cloud based hosting platform `ShinyApps.io`. This work used the following public scientific Python packages `scikit-learn v0.15` [11], `seaborn v0.3.1`, and `statsmodels v0.6.0`. Funding for SDSS-III has been provided by the Alfred P. Sloan Foundation, the Participating Institutions, the National Science Foundation, and the U.S. Department of Energy Office of Science.

References

1. Brown, D., Rothery, P., et al.: Models in Biology: Mathematics, Statistics and Computing. John Wiley, New York (1993)
2. Candès, E.J., Li, X., Yi Ma, Wright J.: Robust principal component analysis? J. ACM **58**(3), 11:1–11:37 (2011). ISSN 0004-5411. doi: 10.1145/1970392.1970395. http://doi.acm.org/10.1145/1970392.1970395
3. Carrasco Kind, M., Brunner, R.J.: Exhausting the information: novel Bayesian combination of photometric redshift PDFs. MNRAS **442**, 3380–3399 (2014). doi: 10.1093/mnras/stu1098
4. de Souza, R.S., Maio, U., Biffi, V., Ciardi, B.: Robust PCA and MIC statistics of baryons in early minihaloes. MNRAS **440**, 240–248 (2014). doi: 10.1093/mnras/stu274
5. Elliott, J., de Souza, R.S., Krone-Martins, A., Cameron, E., Ishida, E.E.O., Hilbe, J.: The overlooked potential of generalized linear models in astronomy-II: gamma regression and photometric redshifts. Astron. Comput. **10**, 61–72 (2015). doi: 10.1016/j.ascom.2015.01.002
6. Green, J., Schechter, P., Baltay, C., Bean, R., Bennett, D., Brown, R., Conselice, C., Donahue, M., et al.: Wide-Field InfraRed Survey Telescope (WFIRST) Final Report (2012). arxiv:1208.4012
7. Hildebrandt, H., Arnouts, S., Capak, P., Moustakas, L.A., Wolf, C., Abdalla, F.B, Assef, R.J., Banerji, M., et. al.: PHAT: PHoto-z accuracy testing. A&A **523**, A31 (2010). doi: 10.1051/0004-6361/201014885
8. Lindsey, J.K.: A review of some extensions to generalized linear models. Stat. Med. **18**(17–18), 2223–2236 (1999). ISSN 0277-6715. http://view.ncbi.nlm.nih.gov/pubmed/10474135
9. LSST Science Collaboration, Abell, P.A., Allison, J., Anderson, S.F., Andrew, J.R., Angel, J.R.P., Armus, L., Arnett, D., Asztalos, S.J., Axelrod, T.S., et al.: LSST Science Book, Version 2.0 (2009). arxiv:0912.0201

[6] https://asaip.psu.edu/organizations/iaa/iaa-working-group-of-cosmostatistics

10. Nelder, J.A., Wedderburn, R.W.M.: Generalized linear models. J. R. Stat. Soc., Ser. A, Gen. **135**, 370–384 (1972)
11. Pedregosa, F., Varoquaux, G., Gramfort, A., Michel, V., Thirion, B., Grisel, O., Blondel, M., Prettenhofer, P., Weiss, R., Dubourg, V., Vanderplas, J., Passos, A., Cournapeau, D., Brucher, M., Perrot, M., Duchesnay, E.: Scikit-learn: machine learning in Python. J. Mach. Learn. Res. **12**, 2825–2830 (2011)
12. Pindyck, R.S., Rubinfeld, D.L.: Econometric Models and Economic Forecasts, vol. 4. Irwin/McGraw-Hill, Boston (1998)
13. Refregier, A., Amara, A., Kitching, T.D., Rassat, A., Scaramella, R., Weller, J., Euclid Imaging Consortium, f. t.: Euclid Imaging Consortium Science Book (2010). arxiv:1001.0061
14. York, D.G., Adelman, J., Anderson, J.E., Jr., Anderson, S.F., Annis, J., Bahcall, N.A., Bakken, J.A., Barkhouser, R., et al., SDSS Collaboration: The sloan digital sky survey: technical summary. AJ **120**, 1579–1587 (2000). doi: 10.1086/301513

Chapter 14
The Nature of Faint Blue Stars in the PHL and Ton Catalogues Based on Digital Sky Surveys

H. Andernach, F. Romero Sauri, W. Copó Córdova, and I. del C. Santiago-Bautista

Abstract We determined accurate positions for 3,000 of the "faint blue stars" in the PHL (Palomar-Haro-Luyten) and Ton/Ton S catalogues. These were published from 1957 to 1962, and, aimed at finding new white dwarfs, provide approximate positions for ∼10,750 blue stellar objects. Some of these "stars" had become known as quasars, a type of objects unheard-of before 1963. We derived subarcsec positions from a comparison of published finding charts with images from the first-epoch Digitized Sky Survey. Numerous objects are now well known, but as of February 2015 neither their PHL or Ton numbers, nor their discoverers, are recognized in current databases. A comparison with modern radio, IR, UV and X-ray surveys leads us to suggest that the fraction of extragalactic objects in the PHL and Ton catalogues is at least 15 %. However, as we failed to locate the original PHL plates or finding charts, it may be impossible to correctly identify the remaining 7,726 PHL objects.

14.1 Introduction and Motivation for This Work

In 1947 Humason and Zwicky [10] searched for faint blue stars using 4-filter photography, with the aim to find new white dwarfs (WDs). Motivated by this, Luyten [14] surveyed the north Galactic cap for faint blue stars with red and blue plates. This, in turn, prompted G. Haro to search systematically for blue stars using

H. Andernach (✉)
Deptartamento de Astronomía, DCNE, Universidad de Guanajuato, Guanajuato, Mexico
e-mail: heinz@astro.ugto.mx

F. Romero Sauri
Universidad Autónoma de Yucatán, Mérida, Mexico

W. Copó Córdova
Instituto Tecnológico Superior de Centla, Tabasco, Mexico

I. del C. Santiago-Bautista
Departamento de Astronomía, DCNE, Universidad de Guanajuato, Guanajuato, Mexico

N.R. Napolitano et al. (eds.), *The Universe of Digital Sky Surveys*, Astrophysics and Space Science Proceedings 42, DOI 10.1007/978-3-319-19330-4_14

three exposures on the same plate. This resulted in (a) the "Ton" and "Ton S" lists of 2008 faint blue objects in the north [4, 12] and south [3] Galactic caps from Tonantzintla Schmidt telescope plates, and (b) a list of 8,746 "PHL" objects from 49 Palomar 48-in Schmidt plates [9]. One year later, the first radio-loud quasar was reported [17], and within another 2 years, radio-quiet quasars were found to be $\sim$10 times more common than radio-loud ones [16]. Some of the PHL or Ton objects became very famous, like, e.g., the high-redshift QSOs PHL 2871 (3C 9) or PHL 957. In fact, 25 PHL objects have over 200 references in Simbad [18] ($\sim$50 % are Seyfert galaxies or QSOs, and the remainder are Galactic stars).

While Simbad contains all 8,746 PHL objects, 94.6 % of these are listed with their poor original position ($\pm \sim 1.5'$), and only 490 (5.6 %) appear with more precise positions (150 QSOs or AGN, 160 WDs, and 180 stars of other types). On the other hand, NED [15] recognizes only 302 (3.5 %) of all PHL objects (113 QSOs, 21 galaxies, 144 stars, 8 WDs and 16 of unknown type). As of February 2015, neither NED nor Simbad quote the detection paper [9] for *any* of the PHL/Ton objects they contain. This motivated us to try to determine precise positions for those PHL and Ton objects with published finding charts (FCs) in order to: (a) assess how many are already known (with other names) in astronomical databases, (b) estimate the fraction of extragalactic objects among them, and (c) give credit to the original discoverers.

14.2 Procedure and Results

While the Ton and Ton S catalogues [3, 4, 12] included FCs for all 2008 objects, only for 1,020 (12 %) of all PHL objects FCs were published [5–8]. To identify the correct PHL or Ton object, we displayed the published FCs side-by-side to a similar-sized ($16' \times 16'$) image of the Digitized Sky Survey (DSS) centered on the published position, and then retrieved the position of the marked object to $< 1''$ precision. We chose the blue, first-epoch "DSS1", for being most similar to the published FCs in both color and epoch, to avoid displacements for high proper motion objects.

We were able to identify all 1,020 PHL objects, but, owing to some very poor FCs, only $\sim$97 % of the Ton/TonS objects; a few objects were found with the help of astrometry.net [13]. The mean positional offsets for the PHL and Ton/Ton S objects ("published minus DSS") are $21''$ and $7''$ in R.A., and $4''$ and $-11''$ in Decl., with a dispersion in position of $\sim 1'$ for PHL and $\sim 2'$ for Ton/Ton S.

During our work we found several curiosities, like *very large positional offsets*, often due to typos which can be resolved using [13]; *sign errors*: e.g., the Decl. sign for PHL 1 should be negative, since only with this choice there is a blue object near to its published position; *large proper motions* of $\geq 3''$ between the epochs of DSS1 ($\sim$1950) and SDSS ($\sim$2000; [1]) implying proper motions in excess of 60 mas yr^{-1} for several objects; *variability*: a few objects appear on some plates, but not on others, e.g., PHL 6287 is present on DSS1 images, but absent in DSS2 and

SDSS. We found some of the cross-identifications with PHL objects proposed by [2] to be false, as can be seen on FCs in [5–8], which were published *after* [2].

14.3 Conclusions

For the first time since their publication, we derived accurate positions for all 3,000 PHL/Ton objects with available finding charts (FCs). We found erroneous identifications for PHL objects in some articles that were published before a FC had become available. Thus, the latter are crucial for an unambiguous identification. Many PHL/Ton objects have become well-known objects, but, as of February 2015 the discovery papers are cited in NED or Simbad. Curiously, the best-known PHL objects are *not* those with published FCs which suggests that their positions had either been found from the approximate original ones, or that FCs had been circulated privately by Haro or Luyten [9]. No traces could be found of the 49 PHL plates taken at Palomar, neither at INAOE (Tonantzintla, Mexico, where Haro and Chavira worked), nor at Univ. of Minnesota (where Luyten worked), nor of the corresponding FCs for the remaining 7,726 (88 % of all) PHL objects. This makes the correct identification of these objects virtually impossible, except perhaps for the ~1,400 "very blue" ones in table 2 of [9].

Extragalactic objects may be distinguished from Galactic stars using color indices from IR to UV wavelengths (e.g. [11]), or from proper motions (comparing DSS with SDSS positions), or via a detection in radio surveys. Our preliminary cross-correlation with object catalogues in these wavelength ranges leads us to estimate that at least 15 % of the PHL and Ton objects are likely to be extragalactic.

Acknowledgements HA & ISB were supported by DAIP-UG grant #318/13, and ISB by grants from CONACyT and Universidad de Guanajuato (UG). FRS & WCC are grateful to Academia Mexicana de Ciencias (AMC) and UG for summer research fellowships. Irina Andernach helped identifying objects on the DSS, and Roger Coziol made useful comments on the manuscript.

References

1. Ahn, C.P., Alexandroff, R., Allende Prieto, C., et al.: The ninth data release of the Sloan Digital Sky Survey: first spectroscopic data from the SDSS-III Baryon Oscillation Spectroscopic Survey. ApJS **203**, 21 (2012)
2. Berger, J., Fringant, A.-M.: A search for faint blue stars in high galactic latitudes. III twelve PSS fields at declinations -6 and -12 deg near the South Galactic Pole. A&AS **58**, 565 (1984)
3. Chavira, E.: Bol. Estrellas Azules en el Casquete Galactico Sur. Obs. Tonantzintla y Tacubaya **2q**, 15 (Ton S) (1958)
4. Chavira, E.: Estrellas azules en el casquete galactico norte II. Bol. Obs. Tonantzintla y Tacubaya **2r**, 3 (Ton II) (1959)
5. Chavira, E.: Identification charts of PHL objects II. A field of approximately 40 square degrees centered at 0^h04^m and +6.5deg (1950). Rev. Mex. Astron. Astrof. **16**, 123 (1988)

6. Chavira, E.: Identification charts of PHL objects. III. Plate centered at 2^h28^m and -5.5deg. Rev. Mex. Astron. Astrof. **20**, 47 (1990)
7. Chavira, E.: Identification charts of PHL objects. IV. Plate centered at 22^h30^m and -11.5deg. Rev. Mex. Astron. Astrof. **24**, 139 (1992)
8. Haro, G., Chavira, E.: On the nature of the PHL objects – identification charts of PHL objects in a field of approximately 40 square degrees centered at 0 H 53 M and -11.5 deg (1950). Rev. Mex. Astron. Astrof. **15**, 107 (1987)
9. Haro, G., Luyten, W.J.: Faint blue stars in the region near the South Galactic Pole. Bol. Obs. Tonantzintla y Tacubaya **3**, 37 (PHL) (1962)
10. Humason, M., Zwicky, F.: A search for faint blue stars. ApJ **105**, 85 (1947)
11. Hutchings, J.B., Bianchi, L.: A catalog of 19,100 quasi-stellar object candidates with redshift 0.5-1.5. AJ **140**, 1987 (2010)
12. Iriarte, B., Chavira E.: Estrellas Azules en el Casquete Galactico Norte. Bol. Obs. Tonantzintla y Tacubaya **2p**, 3 (Ton I) (1957)
13. Lang, D.: Astrometry.net: blind astrometric calibration of arbitrary astronomical images. AJ **139**, 1782 (2010). http://astrometry.net
14. Luyten, W.J.: A further search for faint blue stars and University of Minnesota Monographs. AJ **58**, 75 (1953)
15. NASA/IPAC Extragalactic Database (NED; http://ned.ipac.caltech.edu)
16. Sandage, A.: The existence of a major new constituent of the Universe: the quasistellar galaxies. ApJ, **141**, 1560 (1965)
17. Schmidt, M.: 3C 273 : a star-like object with large red-shift. Nature **197**, 1040 (1963)
18. SIMBAD Astronomical Database. http://simbad.u-strasbg.fr/simbad/

Chapter 15
The Construction of a Reference Star Catalog for the Euclid Mission

B. Bucciarelli, R. Drimmel, M.G. Lattanzi, R.L. Smart, A. Spagna, A. Bacchetta, and A. Bosco

Abstract Optimization of scientific throughput from the ESA Euclid mission imposes stringent requirements on the performance of satellite absolute pointing. This will be achieved by the on-board Fine Guidance Sensor (FGS) with the aid of a specific stellar catalog, which must comply with well-defined, FGS-driven astrometric and photometric properties. By means of the OATo Star Catalog Database, used for the production of the Initial Gaia Source List, we present a preliminary assessment of the methods of construction of such a reference catalog.

15.1 Satellite Attitude Requirements and the FGS Reference Star Catalog

The ESA Euclid mission will investigate the nature of dark energy, dark matter and gravity by detecting weak gravitational lensing and barionic acoustic oscillations. Euclid will cover the extragalactic sky at $|b| > 30°$ (with foreseen extension to the entire sky) by observing strips of adjacent fields along great circles of roughly constant ecliptic longitude. To meet the demanding pointing performance, the satellite Attitude and Orbital Control System will be endowed with a Fine Guidance Sensor instrument (FGS). The FGS must achieve an Absolute Pointing Error of $< 2.5''$ ($1\,\sigma$), and a Relative Pointing Error over 700 s of $< 0.025''$ ($1\,\sigma$) in the X and Y telescope axes. In Absolute Tracking Mode, targets will be selected using an input

B. Bucciarelli (✉) • R. Drimmel • M.G. Lattanzi • R.L. Smart • A. Spagna
INAF, Osservatorio Astrofisico di Torino (OATo), Turin, Italy
e-mail: bucciarelli@oato.inaf.it

A. Bacchetta • A. Bosco
Thales Alenia Space, Turin, Italy
e-mail: andrea.bacchetta@thalesaleniaspace.com

N.R. Napolitano et al. (eds.), *The Universe of Digital Sky Surveys*, Astrophysics and Space Science Proceedings 42, DOI 10.1007/978-3-319-19330-4_15

star catalog, which will be derived from the FGS *Reference Star Catalog*, complying with three main requirements: (a) 5 (with a minimum of 3) or more stars brighter than $R_{FGS} = 19$ mag per Field-of-View (FGS),[1] the limit operating magnitude; (b) astrometric accuracy of $0.4''$ absolute position (per coordinate) at epoch 2025; and (c) more than $2''$ distance from nearest neighbor, or with a difference in brightness of more than 2 mag.

15.2 Selection and Analysis of Test Fields from the OATo Database

The catalog database hosted at OATo, a compilation of the major astronomical catalogs presently available, represents a resource for various scientific and technical activities. The cross-matched object list is mapped onto the celestial sphere using a level-6 HEALPIX[2] scheme; in practice, the sphere is pixelized in 49,152 equal areas of 0.84 deg^2 each. This database provided the basis for the production of the Initial Gaia Source List[3] (IGSL) that is being used as starting point for the Initial Data Treatment of Gaia's observations, and includes the following parameters, when available: positions, object classification, proper motions, magnitude and color information of all objects brighter than Gaia magnitude 21 over the entire sky.

We have extracted from the IGSL basic data some representative sky regions 11×11 deg^2 in size, each sampling 10,000 independent FGS FOV areas of $7' \times 7'$. Then, we have selected all the potential FGS reference stars complying with the requirements specified above. A color-coded stellar density map of each 100×100 FOV areas is a useful tool for identifying lack of stars, anomalous gradients, or peculiar features not obviously attributable to the real sky (see figures). We analyzed a few of these cases by overlaying the catalog objects onto the digitized astronomical images available from the ALADIN Sky Atlas (aladin.u-strasb.fr). Figures 15.1 and 15.2 show some results of the analysis of the Southern Ecliptic Pole (SEP) region centered at $\alpha = 90°$, $\delta = -67°$, and galactic latitude $b = -30°$.

[1]The FGS photometric bandpass can be assimilated to the photographic magnitude R_F, with variations below 0.1 mag, for all spectral types with the exclusion of very low-mass stars, for which R_{FGS} can be sensibly brighter (>0.5 mag) than R_F.

[2]http://healpix.jpl.nasa.gov/

[3]http://www.rssd.esa.int/SYS/docs/ll_transfers/project=PUBDB&id=3223578.pdf

SEP $\alpha = 90°$, $\delta = -67°$ (b=-30°)	$10<R_F<19$ stars	$10<R_F<19$ "single stars" $\sigma_{POS} < 0.4''$
Total number of stars (Ns)	4.25e+006	677393
Mean density per square degree	35131	5598
FOV mean density (Ns / FOV)	425	68
FOV density dispersion (Ns/FOV)	946	59
Min / Max counts per FOV	11 / 9176	9 / 387
Probability Ns / FOV ≥ 3	1.0	1.0

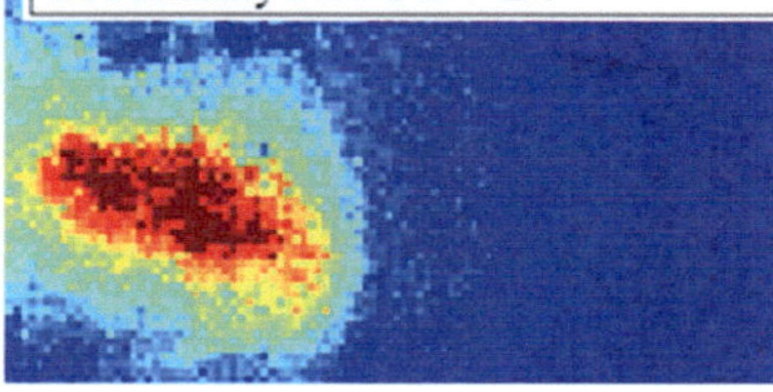

Fig. 15.1 Analysis of 100×100 FOVs in the SEP region. The color-coded map show the density of a sub-sample of "single" stars with astrometric accuracy better than 0.4″ (the *red*, i.e., denser area is in the direction of the large magellanic cloud). The table reports the complete objects statistics before and after FGS selection criteria are applied

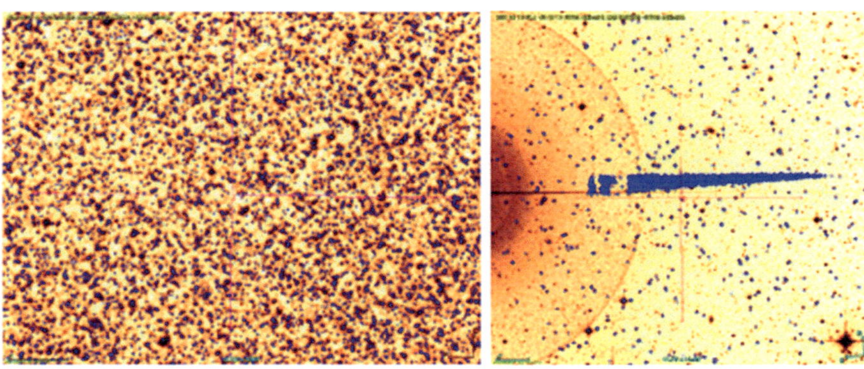

Fig. 15.2 ≈ 14 × 10 arcmin fields from the SEP region; catalog objects (*blue crosses*) are overplotted to the Anglo-Australian Schmidt survey plate image. *Left Panel:* In spite of the high crowding, some candidate reference stars are available but we must consider the limited reliability of the sources provided by the current astronomical catalogues *Right Panel:* Typical example of false object detection caused by the diffraction spike of a very bright star from a Schmidt image

15.3 Conclusions and Future Work

This preliminary assessment of the catalogue extracted from the OATo Database has highlighted some deficiencies, which must be cured in order to build an *operational Reference Star Catalog* for the Euclid FGS. We foresee better performances by including the new proper motions from GSC2.3 material (Qi Z. et al. 2015,

"Absolute Proper Motions outside the Plane (APOP), A step towards GSC2.4", Astron. Journal, 150:137) and, also, by paying special attention to critical sky areas. Moreover, the integration of novel data coming from ongoing ground-based CCD surveys such as PAN STARSS, Sky Mapper, VST, and ultimately Gaia, will help improve count completeness as well as astrometric and photometric accuracy.

Acknowledgements This work has been partially supported by contract TASI – OATo, purchase order no.1520037410

Chapter 16
Singular Spectrum Analysis for Astronomical Time Series: Constructing a Parsimonious Hypothesis Test

G. Greco, D. Kondrashov, S. Kobayashi, M. Ghil, M. Branchesi, C. Guidorzi, G. Stratta, M. Ciszak, F. Marino, and A. Ortolan

Abstract We present a data-adaptive spectral method – Monte Carlo Singular Spectrum Analysis (MC-SSA) – and its modification to tackle astrophysical problems. Through numerical simulations we show the ability of the MC-SSA in dealing with $1/f^{\beta}$ power-law noise affected by photon counting statistics. Such noise process is simulated by a first-order autoregressive, AR(1) process corrupted by intrinsic Poisson noise. In doing so, we statistically estimate a basic stochastic variation of the source and the corresponding fluctuations due to the quantum nature of light. In addition, MC-SSA test retains its effectiveness even when a significant percentage of the signal falls below a certain level of detection, e.g., caused by the instrument sensitivity. The parsimonious approach presented here may be broadly applied, from the search for extrasolar planets to the extraction of low-intensity coherent phenomena probably hidden in high energy transients.

G. Greco (✉) • M. Branchesi • G. Stratta
Università degli Studi di Urbino "Carlo Bo" – DiSBeF, I-61029 Urbino (PU), Italy

INFN, Sezione di Firenze, Via Sansone 1, I-50019 Sesto Fiorentino (FI), Italy
e-mail: giuseppe.greco@uniurb.it

D. Kondrashov
University of California – AOS and IGPP, Los Angeles, CA 90095-1565, USA

M. Ghil
University of California – AOS and IGPP, Los Angeles, CA 90095-1565, USA

Ecole Normale Supérieure – CNRS and IPSL, F-75231 Paris Cedex 05, France

S. Kobayashi
ARI – Liverpool John Moores University, 146 Brownlow Hill, Liverpool, L3 5RF, UK

C. Guidorzi
Department of Physics and Earth Sciences, University of Ferrara, I-44122 Ferrara, Italy

M. Ciszak • F. Marino
CNR-Istituto Nazionale di Ottica, L.go E. Fermi 6, I-50125 Firenze, Italy

A. Ortolan
INFN, Laboratori Nazionali di Legnaro, I-35020 Legnaro (PD), Italy

N.R. Napolitano et al. (eds.), *The Universe of Digital Sky Surveys*, Astrophysics and Space Science Proceedings 42, DOI 10.1007/978-3-319-19330-4_16

16.1 Colored Noise and MC-SSA

$1/f^{\beta}$ power-law noise is known to be highly relevant in several astrophysical systems [3–5]. It includes the well-known white noise ($\beta = 0$), pink noise ($\beta = 1$) and red or Brownian noise ($\beta = 2$). Such kind of noises are very often not tied to instrumental disturbance but rather they are property of the observed sources that emit radiation varying in a stochastic manner. In the context of our work, it is sufficient to qualitatively characterize $1/f^{\beta}$ to generate a parsimonious hypothesis test in MC-SSA (see below). Subsequently, the dynamical origin of noise fluctuation or the extra-noise variance – according to acceptance or rejection of our null hypothesis – has to be determined by theoretical analysis.

Singular Spectrum Analysis (SSA) is an effective, data-adaptive and non-parametric method for the decomposition of a time series into a well-defined set of independent and interpretable components that include a non-linear trend, anharmonic, amplitude-modulated oscillations, and noise [6]. In its later developments, the Monte Carlo approach to signal-to-noise separation introduced by [1] has become known as Monte Carlo SSA (MC-SSA) [2]. The distinction between what is *noise* and what is *signal* is made through the measure of the resemblance of a given noise surrogate to the original data via eigendecomposition of the time-lagged covariance matrix. For this study the noise surrogates set is generated by using an AR(1) process. In practice, the AR(1) coefficients are estimated from the time-series under consideration by using a maximum-likelihood criterion. Subsequently, these AR(1) noise surrogates are corrupted by Poisson noise to mimic the effect of a photon counting detector.

16.2 Simulated $1/f^{\beta}$ Series Test

The procedure has been tested on a large sample of artificial time series of arbitrary colored noises. As a test of validity, we show several discrete colored noise vectors of length $N = 5{,}000$, with a power-law distribution of slope β ranging from 1 to 2 in steps of two. To obtain sequences of colored noises we use the Matlab library CNOISE.[1] A threshold of signal identification is incorporated by subtracting a mean value $\sim$35 % of the maximum peak-flux, C_{max}. In this way, we take into account episodes of emission during which the emission count rate drops to the background level. Finally, the time series are corrupted by Poisson noise to simulate the effect of the shot noise based on Poisson photon statistics. The resulting test series, are shown in the supplementary video.[2] The signal to noise ratio (SNR) of the artificial series test increases as the simulations run.

[1] http://people.sc.fsu.edu/~jburkardt/m$_$src/cnoise/cnoise.html

[2] https://vimeo.com/120699083

16.3 Conclusion and Future Work

We present a modified data-adaptive spectral method – Monte Carlo Singular Spectrum Analysis (MC-SSA) – in which a parsimonious (dynamic and instrumental) noise-model H_0 is adopted for astrophysical applications, i.e. $H_0 = \{AR(1) + \text{Poisson noise}\}$. AR(1) noise takes into account the long-term variability in the power-law slope $-\beta$ of a light-curve's spectral density, and the Poisson noise considers the short-term variability imposed by the quantum nature of light. Our simulations show a remarkable effectiveness of the model for colored noises with β values between 1.5 and 2 and with $C_{max} > 1{,}000$ counts; otherwise one should re-bin the light-curve to improve the signal-to-noise ratio of the source.

All analyses were performed by using SSA-MTM Toolkit freeware that has been developed by the Theoretical Climate Dynamics Group at UCLA.[3] To obtain presented results we relied on an advanced option in SSA-MTM Toolkit for MC-SSA that allows to read in ensemble of surrogates created off-line and corresponding to an arbitrary noise null-hypothesis H_0. In this regard, MC-SSA allows to iteratively refine the working hypothesis and adapt the null-hypothesis H_0 for various theoretical scenarios. We plan to extend further SSA-MTM Toolkit for astrophysical applications by directly including the algorithm for parsimonious H_0 in the freeware. Current and future work will focus on SSA as a promising technique for analysing complex spatio-temporal structures to detect possible periodicities and other statistical regularities due to various astrophysical processes.

Acknowledgements MB, GG and GS acknowledge the financial support of the Italian Ministry of Education, University and Research (MIUR) through grant FIRB 2012 RBFR12PM1F. CG acknowledges the PRIN MIUR project on "Gamma Ray Bursts: from progenitors to physics of the prompt emission process" (Prot. 2009 ERC3HT). MG and DK received support from the U.S. National Science Foundation (grant DMS-1049253) and from the U.S. Office of Naval Research (MURI grant N00014-12-1-0911).

References

1. Allen, M. R., & Smith, L. A. Monte Carlo SSA: detecting irregular oscillations in the presence of coloured noise. J. Climate. **9**, 3373–3404 (1996)
2. Ghil, M., Allen M. R., Dettinger M. D., et al. Advanced spectral methods for climatic time series. Rev. Geophys. **40**, 1–41 (2002)
3. Greco, G., Rosa, R., Beskin, G., et al. Evidence of deterministic components in the apparent randomness of GRBs: clues of a chaotic dynamic. Sci. Rep. **1**, 91 (2011)
4. Pont, F., Zucker, S., & Queloz, D. The effect of red noise on planetary transit detection. MNRAS **373**, 231–242 (2006)
5. Vaughan, S. A Bayesian test for periodic signals in red noise. MNRAS **402**, 307–320. (2010)
6. Vautard, R., & Ghil, M. Singular spectrum analysis in nonlinear dynamics, with applications to paleoclimatic time series. Physica D **35**, 395–424 (1989)

[3] http://www.atmos.ucla.edu/tcd/ssa/

Chapter 17
The Environment of Radio Sources in the VLA-COSMOS Survey Field

N. Malavasi, S. Bardelli, P. Ciliegi, O. Ilbert, L. Pozzetti, and E. Zucca

Abstract This work studies the correlation among environmental density and radio AGN presence up to $z = 2$. Using data from the photometric COSMOS survey and its radio 1.4 GHz follow-up (VLA-COSMOS), a sample of radio AGNs has been defined. The environment was studied using the richness distributions inside a parallelepiped with base side of 1 Mpc and height proportional to the photometric redshift precision. Radio AGNs are found to be always located in environments significantly richer than those around galaxies with no radio emission. Moreover, a distinction based on radio AGN power shows that the significance of the environmental effect is only maintained for low-power radio sources. The results of this work show that denser environments play a significant role in enhancing the probability that a galaxy hosts a radio AGN and, in particular, low-power ones.

17.1 Introduction

The problem of the transformation of the galaxy population from star-forming to quiescent is still an open one in modern astrophysics. General agreement has been reached on the fact that galaxy mass, galaxy environment and AGN feedback play a major role in star formation quenching. It has been suggested (see [4]) that the central AGN co-evolves with the host-galaxy: while the host-galaxy transforms from a star-forming to a quiescent one, the AGN passes from a quasar, X-ray emitter phase to a radio-galaxy one. These transformations happen at earlier epochs for haloes of higher mass, that were found to reside primarily in high-density

N. Malavasi (✉)
Department of Physics and Astronomy (DIFA), University of Bologna, v.le Berti Pichat 6/2, 40127 Bologna, Italy
e-mail: nicola.malavasi@unibo.it

S. Bardelli • P. Ciliegi • L. Pozzetti • E. Zucca
INAF–Osservatorio Astronomico di Bologna, via Ranzani 1, 40127 Bologna, Italy

O. Ilbert
LAM (Laboratoire d'Astrophysique de Marseille), UMR 7326, CNRS, Aix Marseille Université, 13388 Marseille, France

N.R. Napolitano et al. (eds.), *The Universe of Digital Sky Surveys*, Astrophysics and Space Science Proceedings 42, DOI 10.1007/978-3-319-19330-4_17

environments, where early-type galaxies dominate at low redshifts [3, 8]. Moreover, it was already known that many radio AGNs reside in early-type galaxies [6], that the probability that a galaxy hosts a radio AGN is increasing with stellar mass [1], and that the fraction of radio active early-type galaxies is an increasing function of local density [2]. In this work (see [7]) the environment of radio sources of the VLA-COSMOS survey [9], cross-identified with the COSMOS photometric redshift sample [5], is explored.

17.2 Data and Method

The analysis was performed on the environment of a selection of 272 radio AGNs from the VLA-COSMOS survey [9, 10]. This survey (which is composed of 1.4 GHz data, with a sensitivity of about 11 μJy r.m.s.) was cross-correlated with the COSMOS photometric survey [11] with photometric redshifts measured by [5]. The COSMOS survey catalogue is composed of optical galaxies down to $i^+ < 26.5$, which were used both as tracers for the environment around AGN sources and as extraction pool for the control samples. The accuracy of the photometric redshifts (z_p) is estimated to be $\sigma_{\Delta z/(1+z)} = 0.06$.

AGNs were extracted among radio sources as those hosted by massive and quiescent galaxies through a cut to $\log(M^*/M_{\odot}) \geq 10$ and $\log(SSFR/yr^{-1}) \leq -11$, as shown in the upper panel of Fig. 17.1. The control sample of normal galaxies QO has been extracted from the same lower-right region of the SSFR – M^* plane. In order to have a fully representative control sample, galaxies were randomly extracted with the same mass distribution of the radio AGNs.

The environment has been estimated around every AGN source and control galaxy by counting optical galaxies in a parallelepiped with a base side of 1 Mpc (comoving) and height $2 \cdot \Delta z = 2 \times 3 \times \sigma_{\Delta z/(1+z)} \times (1 + z_p)$, in three different redshift bins: $z \in [0.0-0.7[$, $[0.7-1.0[$, and $[1.0-2.0]$. The number of radio AGNs is respectively 119 sources, 100 sources, and 53 sources.

17.3 Results and Conclusions

It was found that the environment around radio AGNs is significantly denser than the environment around sources from the control sample (i.e. that show no sign of radio emission). This is visible in the lower panel of Fig. 17.1, which shows the overdensity richness distribution for the AGN sample and for the 100 independent extractions of the control sample in the farthest redshift bin. A Kolmogorov-Smirnov test between the distributions results in median values of the KS test probability value distribution of 8.6×10^{-5}, 6.0×10^{-6}, and 0.006 in each redshift bin, respectively.

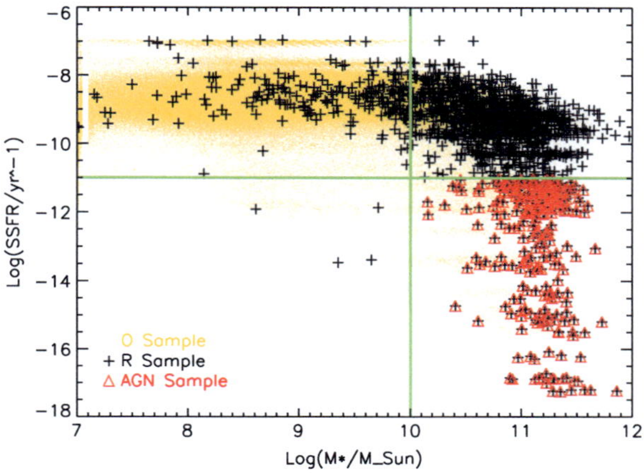

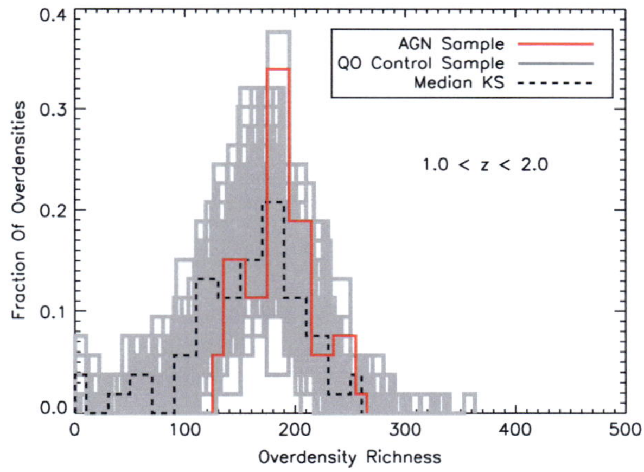

Fig. 17.1 *Top*: *SSFR - M* plane*. The *black crosses* refer to the full sample of VLA-COSMOS radio sources, the *red triangles* to the radio AGN sample, and the *green lines* to the cuts in stellar mass and SSFR described in text. In *yellow*, the full COSMOS optical sample is reported for comparison. *Bottom*: *Galaxy overdensity richness distribution, samples AGN and QO*. This plot refers to $1 \leq z \leq 2$. The *solid red line* represents the AGN sample, the *grey lines* the 100 extractions of the QO control sample. The *dashed black line* is the richness distribution of the control sample extraction corresponding to the median value of the KS probability value distribution

The AGN sample has been further divided according to its radio power: a high-power sub-sample ($\log(L_{1.4\,\mathrm{GHz}}) \geq 24.5$) and a low-power one ($24 \leq \log(L_{1.4\mathrm{GHz}}) < 24.5$) were created, with the distinction between the two that roughly corresponds to the canonical division between FRI and FRII objects.

It was found that the significance in the environmental segregation signal is maintained only for low-power radio AGNs in the lowest and intermediate redshift bins, while for the high-power radio AGNs no significant signal is present. Therefore, higher overdensity richness enhance the probability that a galaxy hosts a low-power radio AGN. In conclusion, we found a clear correlation between radio AGN presence and environment up to $z \sim 2$, consistent with the scenario sketched in [4].

References

1. Bardelli, S., Zucca, E., Bolzonella, M., et al.: The VVDS-VLA deep field. IV. Radio-optical properties. A&A **495**, 431 (2009)
2. Bardelli, S., Schinnerer, E., Smolčić, V., et al.: Properties and environment of radio-emitting galaxies in the VLA-zCOSMOS survey. A&A **511**, A1 (2010)
3. Chuter, R.W., Almaini, O., Hartley, W.G., et al.: Galaxy environments in the UKIDSS Ultra Deep Survey. MNRAS **413**, 1678 (2011)
4. Hickox, R.C., Jones, C., Forman, W.R., et al.: Host galaxies, clustering, Eddington ratios, and evolution of radio, X-ray, and infrared-selected AGNs. ApJ **696**, 891 (2009)
5. Ilbert, O., Capak, P., Salvato, M., et al.: Cosmos photometric redshifts with 30-bands for 2-deg^2. ApJ **690**, 1236 (2009)
6. Ledlow, M.J., Owen, F.N.: 20 CM VLA survey of Abell clusters of galaxies. VI. Radio/optical luminosity functions. AJ **112**, 9 (1996)
7. Malavasi, N., Bardelli, S., Ciliegi, P., Ilbert, O., Pozzetti, L., Zucca, E.: The environment of radio sources in the VLA-COSMOS survey field. A&A **576**, A101 (2015)
8. Quadri, R.F., Williams, R.J., Franx, M., Hildebrandt, H.: Tracing the star-formation-density relation to $z \sim 2$. ApJ **744**, 88 (2012)
9. Schinnerer, E., Smolčić, V., Carilli, C.L., et al.: The VLA-COSMOS survey. II. Source catalog of the large project. ApJS **172**, 46 (2007)
10. Schinnerer, E., Sargent, M.T., Bondi, M., et al.: The VLA-COSMOS survey. IV. Deep data and joint catalog. ApJS **188**, 384 (2010)
11. Scoville, N., Aussel, H., Brusa, M., et al.: The cosmic evolution survey (COSMOS): overview. ApJS **172**, 1 (2007)

Part II
News from VST

Chapter 18
The VLT Survey Telescope: What Stands Behind the Surveys

P. Schipani and the VST Team*

Abstract The VLT Survey Telescope (VST) is the 2.6-m survey telescope of the European Southern Observatory (ESO) for the visible band, the largest in the world designed to exclusively survey the skies in visible light. It is equipped with the OmegaCAM 268 Mpix 16×16 K CCD camera, covering a wide-field of 1 deg^2 with 0.21 arcsec/pixel resolution. We provide a short history of the project and an overview of the instrument, summarizing the delivered performance.

18.1 Introduction

The VST project originated from an idea of its PI M. Capaccioli, who promoted a joint venture between ESO and the Capodimonte Astronomical Observatory (OAC) in Naples, formerly an independent institution and since 2002 a part of the Italian National Institute for Astrophysics (INAF). The telescope was proposed to ESO in 1998 with the joint signature of a Memorandum of Understanding, where INAF-OAC committed itself to procure the telescope in return of a proportional

*VST Team
PI: M. Capaccioli (Università Federico II di Napoli & INAF – Capodimonte)
Commissioning:
M. Dall'Ora, S. D'Orsi and L. Marty (INAF – Capodimonte),
J. Farinato, D. Magrin and R. Ragazzoni (INAF – Padova),
C. Arcidiacono (INAF – Bologna), J. Argomedo (ESO), G. Umbriaco (Università di Padova)
GTO Project Office:
A. Grado, N. Napolitano (INAF – Capodimonte)
GTO Data Division:
F. Getman, L. Limatola (INAF – Capodimonte)
GTO PIs:
M. Marconi, P. Merluzzi and V. Ripepi (INAF – Capodimonte),
E. Cappellaro and E. Held (INAF – Padova),
G. Covone (Università Federico II di Napoli), M. D'Onofrio (Università di Padova)

P. Schipani (✉)
INAF – Osservatorio Astronomico di Capodimonte, Salita Moiariello 16, I-80131 Napoli, Italy
e-mail: pietro.schipani@oacn.inaf.it

N.R. Napolitano et al. (eds.), *The Universe of Digital Sky Surveys*, Astrophysics and Space Science Proceedings 42, DOI 10.1007/978-3-319-19330-4_18

share of the observing time, plus a number of nights at one VLT unit telescope. ESO committed itself to provide the rotating enclosure, to collaborate in setting specifications and interfaces of the telescope, to manage the transportations from Italy to Chile, to secure the CCD camera. Also, ESO took responsibility for the operation and maintenance of VST for a period of 10 years, adding it to the Paranal telescope family and thus using the existing observatory infrastructures and personnel. With this kind of agreement, INAF-OAC could invest its one-time funds for building the telescope, but was relieved from running costs.

The telescope was not committed to an industrial prime contractor: INAF owned the technical direction and system engineering of the project, doing in addition most of the design and realization work in-house, with its engineers. This strategy allowed the overall cost to stay within a smaller budget, and favoured an enormous increase of INAF internal competences in the telescope engineering area. Industry has been involved in manufacturing and design of specific parts, almost exclusively in the mechanics area (e.g. primary mirror cell mechanics – Tomelleri srl, hexapod – ADS International srl, rotating enclosure – EIE srl). ESO has always greatly supported the whole process.

It is worth mentioning that during the history of the project two major setbacks occurred, both during shipments: first the primary mirror was pulverized due to a transportation accident; afterwards its cell including axial and lateral actuators was invaded by water, arriving at Paranal extensively covered by rust. Despite the serious delays caused by these two remarkable disasters, the project has been finally a success.

In 2007 the telescope mount and the axes control were ready and shipped to Chile, while the active optics and some other subsystems needed additional work. At the end of 2010 the telescope was almost completely reintegrated in Chile. In 2011 the remaining integration tasks and the commissioning were carried out together in a remarkably short time, thus in October 2011 ESO granted the Preliminary Acceptance in Chile and the regular observations officially started [1, 7, 12]. According to the plans, after 3 years in 2014 the telescope got the Final Acceptance in Chile.

The Italian VST (Fig. 18.1) and its infrared UK companion VISTA are the two largest contributions by external consortia (i.e. among instruments not built by ESO) to the ESO instrumentation, in the world's most productive observatory of La Silla-Paranal (Fig. 18.2). VST is the only Italian led instrument realized for ESO so far [2].

18.2 System Overview

The VST is an optical F/5.5 telescope with an alt-azimuth mount. Its optics includes a field corrector which is essential to deliver a good optical quality over its wide-field of 1×1 deg. The corrector is composed of three lenses, two are in the telescope and the third is the curved dewar window of the camera. Strictly speaking, the

Fig. 18.1 The VLT Survey Telescope inside the enclosure in the ESO Paranal observatory, slightly inclined towards the observation sliding doors. The black secondary mirror box on the top hosts the hexapod; the primary mirror is hidden by its black metallic cover. Visible in this picture, the white electronic cabinets are shared between the observing floor and the telescope center piece. The wide-field camera is placed on the bottom of the mirror cell (Photo: P. Schipani/INAF)

VST does not belong to any classical category of telescopes (e.g. Ritchey-Chretien): the combination of mirrors and lenses has been designed in order to minimize the wavefront error across the whole field of view.

The 268 Mpix OmegaCAM camera with a pixel scale of 0.21 arcsec/pixel and a pixel size of 15 μm is the only focal plane instrument, provided by a dedicated international consortium. It is composed of a mosaic of 32 2×4 K scientific CCDs arranged in a 8×4 matrix [4].

Fig. 18.2 Telescopes on the ESO Paranal platform: the VST is on the right side, close to the *red* and *white* tower (Photo: G. Hudepohl/ESO)

18.3 Active Optics

The telescope is based on active optics [13, 14]. The primary mirror is a concave 2.6-m hyperbolic meniscus with a 60 cm central hole and 140 mm thickness (aspect ratio: 18.6) made of Astro-Sitall, axially supported by active supports and laterally by passive astatic levers.

The axial system supports the component of mirror weight along the optical axis (passive component), depending on zenith angle, and corrects the aberrations by tuning the optical surface (active component) [8]. The weight of the mirror (slightly less than 1,900 kg) is distributed between 84 axial supports (81 actuators and 3 fixed points) distributed in 4 rings of 12, 18, 24, 30 supports [9]. The radius and fraction of supported weight for each ring are the results of an optimization problem, where the criterion was to minimize the r.m.s deformation of the mirror surface when the mirror rests on its supports at zenith.

The lateral system supports the weight component of the mirror perpendicular to the optical and altitude axes. It is based on 24 astatic lever supports and three lateral fixed points defining the position of the mirror in its plane. The lateral forces are distributed according to the Schwesinger theory [16]. The lateral system force ratio is $\beta = 0.75$, i.e. 75 % of the lateral weight is supported tangentially and 25 % radially. The primary mirror is surrounded by a passive safety system for earthquake protection [6].

The convex hyperbolic 0.9-m secondary mirror is supported by an astatic levers system, i.e. a passive system based on counterweights. It takes the weight of the mirror self-adjusting the force components acting on the mirror supports when the altitude angle changes, interfacing with the back of the secondary mirror through 10 insert sleeves. The rigid body position of the mirror is controlled in 5 degrees of freedom by a hexapod device [10]. A fixed platform screwed to the telescope is connected to a moving platform supporting the mirror through six linear actuators, each controlled in position with $\leq 1\,\mu m$ precision. The mechanics of the VST device is a replica of the TNG implementation, while the control system is new and improved.

VST is also an ideal testbench for new active optics techniques dedicated to wide-field telescopes. Specific efforts are ongoing with promising perspectives [3, 5].

18.4 Pointing and Tracking

As the VST is an alt-azimuthal telescope, it has three tracking axes (azimuth, altitude and instrument rotator), that share the same control engineering concepts [11]. The overall tracking specification is on the stability of the image as seen on the guide probe, with autoguiding system active: the RMS (Root Mean Square) error should not exceed 0.1 arcsec. This specification imposes tight requirements on the servo errors of the individual axes, which should consequently be of the order of few hundredths of arc seconds RMS for azimuth and altitude, and of some arc seconds for the rotator. All these requirements are fully met.

The servo system of each axis minimizes the difference between the desired reference trajectory and the encoder feedback. This task is accomplished by appropriate servo controllers, implemented within the control software. Nevertheless, as the position encoders do not measure the orientation of the beam and no telescope mount is ideal, even a perfect servo control cannot perfectly stabilize the image on the detector: the physical imperfections are taken into account in a pointing model, included in the generation of the reference positions.

The position feedback is provided by high quality glass disc encoders with a circular scale (72,000 lines), with radial grating and reference marks, produced by Heidenhain GmbH. The overall resolution of the encoder system, averaging the four scanning heads, is about 0.001 arcsec, i.e. one order of magnitude smaller than the precision required to the control system. The velocity feedback is obtained by differentiating the encoder readings.

Both position and speed control loops are implemented within the software running on Local Control Units (LCU), based on the VME bus: this choice gives total freedom in the controller's design. The axes are moved by commercial Moog Inc. brushless servomotors powered by appropriate drives. Pairs of motors transmit the motion to gear wheels; the motor drives are configured to behave only as power amplifiers. The gear transmission might cause a backlash clearance at pinions: this potential problem is prevented by the application of a constant torque bias to the motors. A hydrostatic bearing system supports the telescope allowing a frictionless rotation around the azimuth axis.

18.5 Control System

The telescope local area network (LAN) is composed of a coordinating workstation running the Scientific Linux operating system, and a number of Local Control Units with related control electronics and application software, working with the VxWorks real-time operating system. All the graphic capabilities and the user interfaces reside on the telescope workstation. The LCUs are hosted in electrical cabinets and interface with the electronics through several electronic boards. In order to use as much commercial electronics as possible and simplify maintenance activities, the same standard boards have been used wherever the same functionality was needed.

The programming environment is the same as adopted by ESO in all similar telescope and instrumentation projects, that is periodically updated through the annual releases of the VLT Common Software. The supported languages are ANSI-C for LCU modules, C++ for workstation modules and Tcl/Tk for user interfaces. The GNU basic development tools are widely adopted.

The VST telescope control software [15] is composed of 61 modules concurrently working on the telescope workstation and on the eight Local Control Units. The complexity of this software package can be represented by the number of lines of code, slightly more than one million with more than 200 K lines of comments.

18.6 Image Quality

After more than 3 years of regular operations, the histograms of FWHM distribution rather closely follow the ones of Paranal seeing, i.e. the system can regularly deliver seeing-limited images with a uniform FWHM distribution across the whole field of view and with small ellipticities. In nights of good seeing, a FWHM of 0.5 arcsec or even less across the whole field is not uncommon. This means the telescope contribution to image blurring is very small, i.e. all subsystems do their job respecting the tight requirements imposed by the superb quality of the Paranal sky. Such image quality performance is one of the best among the existing survey machines. Work is ongoing to improve it further.

Acknowledgements We are grateful to all the people who have worked at this project in the past with team spirit, too many to be listed: thank you, you know who you are. But above all, definitely this telescope would not exist without M. Capaccioli and his unbelievable determination against all troubles: thank you Massimo, perseverance wins in the end.

References

1. Capaccioli, M., Schipani, P.: The VST opens to the sky: history of a commissioning. The Messenger **146**, 2–6 (2011)
2. D'Odorico, S.: Italian contributions to the ESO instrumentation, from La Silla to the E-ELT. Mem. SAIt **83**, 1134 (2012)

3. Holzloehner, R., Rakich, A., Noethe, L., Kuijken, K., Schipani, P.: Fast active optics control of wide-field telescopes based on science image analysis. Proc. SPIE **9151**, 915112I (2014)
4. Kuijken, K.: OmegaCAM: ESO's newest imager. The Messenger **146**, 8–11 (2011)
5. Noethe, L., Schipani, P., Holzloehner, R., Rakich, A.: A method for the use of ellipticities and spot diameters for the measurement of aberrations in wide-field telescopes. Adv. Opt. Technol. **3**, 315–334 (2014)
6. Perrotta, F., Schipani, P.: A simulation model for a telescope earthquake analysis: the VST primary mirror safety system. Exp. Astron. **29**(3), 189–206 (2011)
7. Schipani, P.: VST system engineering and management of commissioning. Proc. SPIE **8449**, 84491H (2012)
8. Schipani, P., Perrotta, F., Marty, L.: Active optics correction forces for the VST 2.6m primary mirror. Proc. SPIE **6273**, 1014–1025 (2006)
9. Schipani, P., D'Orsi, S., Ferragina, L., et al.: Active optics primary mirror support system for the 2.6-m VST telescope. Appl. Opt. **49**(8), 1234–1241 (2010)
10. Schipani, P., D'Orsi, S., Fierro, D., Marty, L.: Active optics control of VST secondary mirror. Appl. Opt. **49**(16), 3199–3207 (2010)
11. Schipani, P., Arcidiacono, C., Argomedo, J., et al.: The tracking control system of the VLT survey telescope. Rev. Sci. Instrum. **83**, 094501 (2012)
12. Schipani, P., Capaccioli, M., Arcidiacono, C., et al.: VST: from commissioning to science. Proc. SPIE **8444**, 84441C (2012)
13. Schipani, P., Noethe, L., Arcidiacono, C., et al.: Removing static aberrations from the active optics system of a wide-field telescope. J. Opt. Soc. Am. A **29**, 1359–1366 (2012)
14. Schipani, P., Magrin, D., Noethe, L., et al.: The active optics system of the VST: concepts and results. Proc. SPIE **8444**, 84444Z (2012)
15. Schipani, P., Marty, L., Argomedo, J., et al.: Software and control system for the VLT survey telescope. J. Instrum. **8**, P05013 (2013)
16. Schwesinger, G.: Lateral support of very large telescope mirrors by edge forces only. J. Mod. Opt. **38**, 1507 (1991)

Chapter 19
Galaxy Evolution Within the Kilo-Degree Survey

C. Tortora, N.R. Napolitano, F. La Barbera, N. Roy, M. Radovich, F. Getman, M. Brescia, S. Cavuoti, M. Capaccioli, G. Longo, and the KiDS collaboration

Abstract The ESO Public Kilo-Degree Survey (KiDS) is an optical wide-field imaging survey carried out with the VLT Survey Telescope and the OmegaCAM camera. KiDS will scan 1,500 $\deg^2$ in four optical filters (u, g, r, i). Designed to be a weak lensing survey, it is ideal for galaxy evolution studies, thanks to the high spatial resolution of VST, the excellent seeing and the photometric depth. The surface photometry has provided with structural parameters (e.g. size and Sérsic index), aperture and total magnitudes have been used to obtain photometric redshifts from Machine Learning methods and stellar masses/luminositites from stellar population synthesis. Our project aimed at investigating the evolution of the colour and structural properties of galaxies with mass and environment up to redshift $z \sim 0.5$ and more, to put constraints on galaxy evolution processes, as galaxy mergers.

19.1 Introduction

We start from the KiDS-Data Release 2 (DR2) multi-band source catalogs, based on source detection in the r-band images. Star/galaxy separation is based on the `CLASS_STAR` (star classification) and S/N (signal-to- noise ratio) parameters provided by S-Extractor [2, 12]. S-Extractor will provide with aperture photometry and Kron-like magnitude `MAG_AUTO`. From the original catalog of ~22 millions of sources, the star/galaxy separation leaves with a sample of ~7 millions of

C. Tortora (✉) • N.R. Napolitano • F. La Barbera • F. Getman • M. Brescia • S. Cavuoti
INAF – Osservatorio Astronomico di Capodimonte, Salita Moiariello, 16, 80131, Napoli, Italy
e-mail: ctortora@na.astro.it

M. Radovich
NAF – Osservatorio Astronomico di Padova, vicolo dell'Osservatorio 5, I-35122, Padova, Italy

N. Roy • M. Capaccioli
Department of Physics "Ettore Pancini", University "Federico II" of Naples, Italy

G. Longo
Department of Physics "Ettore Pancini", University "Federico II" of Naples, Italy

KiDS collaboration: http://kids.strw.leidenuniv.nl/team.php

N.R. Napolitano et al. (eds.), *The Universe of Digital Sky Surveys*, Astrophysics and Space Science Proceedings 42, DOI 10.1007/978-3-319-19330-4_19

galaxies. To perform galaxy evolution studies and determine reliable structural parameters, the highest-quality sources have to be taken [12, 13]. Thus, we have finally selected those systems with the highest r-band signal-to-noise ratio (S/N), $S/N \equiv 1/$MAGERR_AUTO_r > 50, where MAGERR_AUTO_r is the uncertainty of MAG_AUTO . This sample will consist of ∼380,000 galaxies. In addition to aperture photometry and Kron-like magnitudes, the relevant data for each galaxy are listed in the following.

19.2 Structural Parameters

Surface photometry is performed using the 2DPHOT environment. For each galaxy, a local PSF model is constructed by fitting the four closest stars to that galaxy. Galaxy images were fitted with PSF-convolved Sérsic models having elliptical isophotes plus a local background value [12]. The fit provides the following parameters: surface brightness μ_e, effective radius, R_e, Sérsic index, n, total magnitude, m_S, axis ratio, q, etc. Fig. 19.1 shows the 2DPHOT output for an example galaxy. For further details, see the contribution from Roy et al.

19.3 Photometric Redshifts

Photo-z's are derived from KiDS ugri photometry using a supervised machine learning model, the Multi Layer Perceptron with Quasi Newton Algorithm (MLPQNA, [3]) within the DAMEWARE (DAta Mining and Exploration Web Application REsource, [4]). Supervised methods use a knowledge base (in our case with spectroscopic redshifts) of objects for which the output (in our case the redshift) is known a-priori to learn the mapping function that transforms the input data (our optical magnitudes) into the desired output (i.e. the photometric redshift). For this reason, our sample is cross-matched with spectroscopic samples from SDSS [1] and GAMA [9] surveys, which provide a redshift coverage up to $z \sim 0.8$. The knowledge base consists of ∼60,000 objects. Sixty percent of these objects are used as train sample, and the remaining ones for the blind test set. The redshifts in the blind test sample resemble the spectroscopic redshifts quite well, with a scatter in the quantity $|z_{\rm spec}-z_{\rm phot}|/(1+z_{\rm spec})$ of ∼0.03. This approach reaches high accuracies with optical band only, and is far better than traditional spectral energy distribution (SED)-fitting methods. After these experiments the final catalogue of redshifts for our high-S/N sample is produced. See Cavuoti et al. [6] for details.

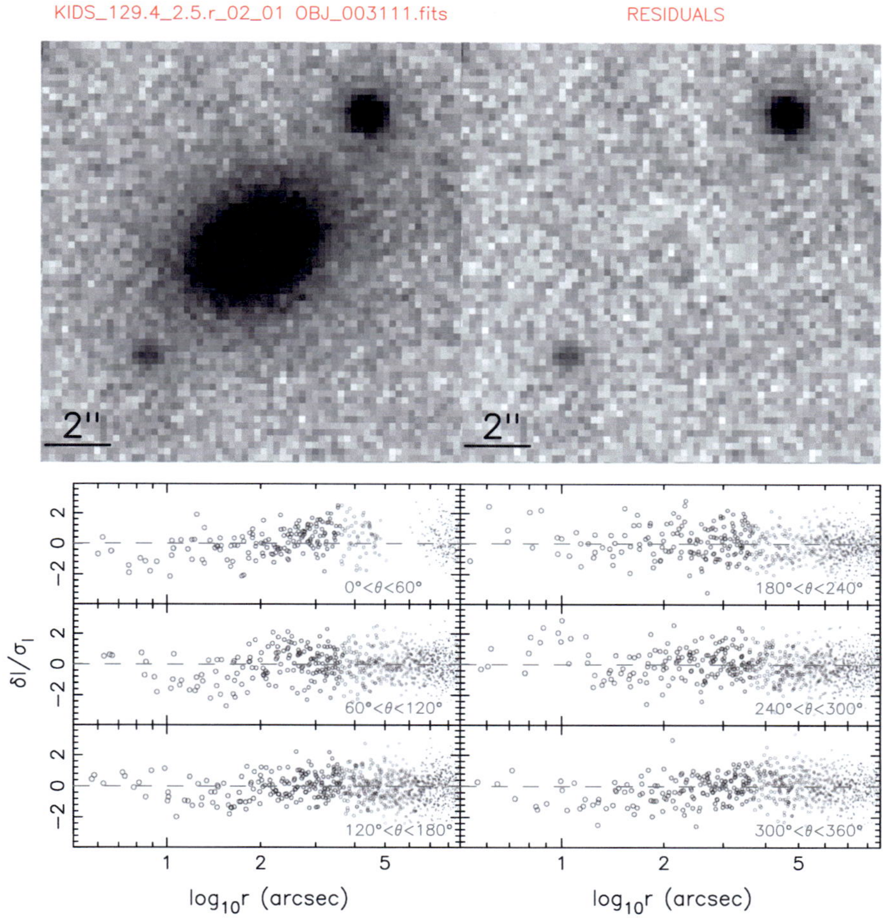

Fig. 19.1 2D fit output for an example galaxy from 2DPHOT. The *top panels* show the galaxy image (*left*) and the residual after the fit (*right*), while the six *bottom panels* provide residuals after subtraction, plotted as a function of the distance to the galaxy center, with each panel corresponding to a different bin of the polar angle. Residuals are normalized with the noise expected from the model in each pixel

19.4 Rest-frame Luminosities and Stellar Masses

Luminosities and stellar masses are made using the software `Le Phare` [10], which performs a simple χ^2 fitting method between the stellar population synthesis (SPS) theoretical models and data. Models from Bruzual and Charlot [5] and with a Chabrier IMF [7] are used. We adopt the observed *ugri* magnitudes (and related $1\,\sigma$ uncertainties) within a $3''$ aperture, which are corrected for Galactic extinction. Total Sérsic magnitudes, m_S, are used to correct the outcomes of `Le Phare` for missing flux.

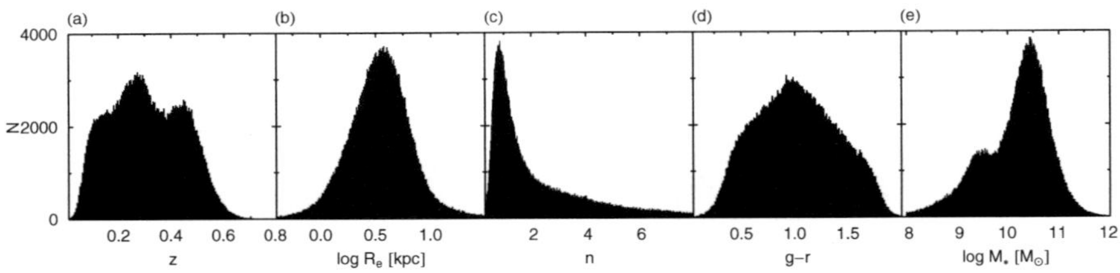

Fig. 19.2 Distribution of some galaxy parameters: photo-z, z (panel **a**), r-band effective radius $\log R_e/\text{kpc}$ (panel **b**), r-band Sérsic index, n (panel **c**), observer-frame g-r colour within an aperture of 3″ of radius (panel **d**) and Chabrier IMF-based stellar mass $\log M_\star/M_\odot$ (panel **e**)

This high-S/N sample is 90 % complete down to a magnitude of `MAG_AUTO_r` $\sim$ 21, which means that the sample is complete at masses $\gtrsim 5 \times 10^{10}$ up to redshift $z = 0.5$. The distribution of galaxy parameters for galaxies brighter than the completeness limit are shown in Fig. 19.2. The average redshift of the sample is $z = 0.3$, and the distribution reach the maximum redshifts of $z = 0.7$ (panel a). The distribution of effective radii is centered on ∼3.5 kpc average value (panel b), while the average Sérsic index is $\sim$ 1.5, but the distribution present a peak at $n \sim 0.8$, with a long tail to higher values (>10, panel c). Thus, our datasample is dominated by late-type systems, which are characterized by low Sérsic indices [11, 17]. In the panel (d) we show the distribution of the observer-frame g-r colour, which have a median value of $g-r = 1$ and a symmetric distribution. These colours, when converted to rest-frame values, confirm the considerations made looking at the distribution of Sérsic index, since the median rest-frame colour is ∼0.5, which is typical of a late-type blue galaxy. Finally, in panel (e) we plot the stellar mass distribution from SPS fitting, which presents a double-peaked distribution, with maxima at $\sim 2.5 \times 10^9$ and $\sim 2.5 \times 10^{10}\,\mathrm{M}_\odot$, the median mass of the sample is $1.9 \times 10^{10}\,\mathrm{M}_\odot$.

19.5 Aims of the Project

The present galaxy evolution project will aim to achieve different objectives: the analysis of the colour-magnitude evolution, colour gradients, as the size and mass accretion in late- and early-type systems. By the end of the survey, we plan to collect about 3.5 millions of galaxies with high-quality photometry, i.e. *the largest sample of galaxies with measured structural parameters in the u, g, r and i bands, up to redshift* $z = 0.7$. KiDS will provide the bridge to connect the local environment to the high-redshift Universe, by means of this strong characterization of the internal structure of the galaxies. The study of the mass, colour and structural evolution of galaxies, connected with the outcomes of theoretical models and cosmological simulations provide a variegate set of information about the main physical processes which shape the galaxy evolution. In particular, taking benefit from the good pixel scale and seeing, together with depth of VST telescope, we have scanned the first 156 $\deg^2$ of KiDS searching for massive ($M_\star > 8 \times 10^{10}\,M_\odot$) and compact ($R_e < 1.5\,$kpc) galaxies, relic of massive galaxies at redshift $z = 2$ (see Tortora et al. [21]). In some theoretical models the fraction of such massive and small objects, that could survive without having any significant transformation since $z \sim 2$, could reach a fraction about 1–10 % [14]. Previous observational works have not detected such old, compact and massive galaxies in the local universe [22]. In contrast, such peculiar objects are found at larger redshift [8]. A front-edge survey like KiDS will allow us to perform the census of these compact objects and trace their abundance and evolution in the last billions of years, contrasting the results with expectations from cosmological simulations. Finally, for a sample of KiDS galaxies with SDSS spectroscopy, we will perform the Jeans dynamical analysis of the measured velocity dispersions, to study dark matter fraction and Initial mass function in terms of redshift, mass, mass density and environment [18–20]. Testing the dependence of the Initial mass function by the mass density, using dynamics and spectral indices, can help to understand the results in Smith et al. [15], which find shallow IMF slopes in some high-velocity dispersion galaxies, in contrast with the current literature [16, 18].

Acknowledgements CT has received funding from the European Union Seventh Framework Programme (FP7/2007–2013) under grant agreement n. 267251 "Astronomy Fellowships in Italy" (AstroFIt).

References

1. Ahn, C.P., et al.: The tenth data release of the Sloan Digital Sky Survey: first spectroscopic data from the SDSS-III Apache Point observatory galactic evolution experiment. ApJs **211**, 17 (2014)
2. Bertin, E., Arnouts, S.: SExtractor: software for source extraction. Astron. Astrophys. **117**, 393 (1996)
3. Brescia, M., Cavuoti, S., Longo, G., De Stefano, V.: A catalogue of photometric redshifts for the SDSS-DR9 galaxies. Astron. Astrophys. **568**, A126 (2014)
4. Brescia, M., et al.: DAMEWARE: a web cyberinfrastructure for astrophysical data mining. PASP **126**, 783 (2014)
5. Bruzual, G., Charlot, S.: Stellar population synthesis at the resolution of 2003. MNRAS **344**, 1000 (2003)
6. Cavuoti et al.: Machine-learning-based photometric redshifts for galaxies of the ESO Kilo-Degree Survey data release 2. MNRAS, **452**, 3100 (2015)
7. Chabrier, G.: The galactic disk mass budget. I. Stellar mass function and density. ApJ **554**, 1274 (2001)
8. Damjanov, I., Hwang, H.S., Geller, M.J., Chilingarian, I.: The number density of quiescent compact galaxies at intermediate redshift. ApJ **793**, 39 (2014)
9. Driver, S.P., et al.: Galaxy and Mass Assembly (GAMA): survey diagnostics and core data release. MNRAS **413**, 971 (2011)
10. Ilbert, O., et al.: Accurate photometric redshifts for the CFHT legacy survey calibrated using the VIMOS VLT deep survey. Astron. Astrophys. **457**, 841 (2006)

11. Kauffmann, G., et al.: The dependence of star formation history and internal structure on stellar mass for 10^5 low-redshift galaxies. MNRAS **341**, 54 (2003)
12. La Barbera, F., et al.: SPIDER – I. Sample and galaxy parameters in the grizYJHK wavebands. PASP **120**, 681 (2008)
13. La Barbera, F., et al.: 2DPHOT: a multi-purpose environment for the two-dimensional analysis of wide-field images. MNRAS **408**, 1313 (2010)
14. Quilis, V., Trujillo, I.: Expected number of massive galaxy relics in the present day universe. ApJl **773**, L8 (2013)
15. Smith, R.J., Lucey, J.R., Conroy, C.: The SINFONI Nearby Elliptical Lens Locator Survey: discovery of two new low-redshift strong lenses and implications for the initial mass function in giant early-type galaxies. MNRAS **449**, 3441 (2015)
16. Spiniello, C., Barnabè, M., Koopmans, L.V.E., Trager, S.C.: (2015). Are the total mass density and the low-mass end slope of the IMF anticorrelated? MNRAS **452L**, 21 (2015)
17. Tortora, C., et al.: Colour and stellar population gradients in galaxies: correlation with mass. MNRAS **407**, 144 (2010)
18. Tortora, C., Romanowsky, A.J., Napolitano, N.R.: An inventory of the stellar initial mass function in early-type galaxies. ApJ **765**, 8 (2013)
19. Tortora, C., La Barbera, F., Napolitano, N.R., Romanowsky, A.J., Ferreras, I., de Carvalho, R.R.: Systematic variations of central mass density slopes in early-type galaxies. MNRAS **445**, 115 (2014)
20. Tortora, C., Napolitano, N.R., Saglia, R.P., Romanowsky, A.J., Covone, G., Capaccioli, M.: Evolution of central dark matter of early-type galaxies up to $z \sim 0.8$. MNRAS **445**, 162 (2014)
21. Tortora et al.: Towards a census of super-compact massive galaxies in the Kilo Degree Survey. arXiv:1507.00731 (2015)
22. Trujillo, I., et al.: Superdense massive galaxies in the nearby universe. ApJl **692**, L118 (2009)

Chapter 20
Strong Lens Search in the ESO Public Survey KiDS

N.R. Napolitano, G. Covone, N. Roy, C. Tortora, F. La Barbera, M. Radovich, F. Getman, M. Capaccioli, A. Colonna, M. Paolillo, G.A. Verdoes Kleijn, L.V.E. Koopmans, and the KiDS collaboration

Abstract We have started a systematic search for strong lens candidates in the ESO public survey KiDS based on the visual inspection of massive galaxies in the redshift range $0.1 < z < 0.5$. As a pilot program we have inspected $100\,\mathrm{deg}^2$, which overlap with SDSS and where there are known lenses to use as a control sample. Taking advantage of the superb image quality of VST/OmegaCAM, the colour information and accurate model subtracted images, we have found 18 new lens candidates, for which spectroscopic confirmation will be needed to confirm their lensing nature and study the mass profile of the lensing galaxies.

20.1 Dark Matter in Galaxy Cores

The dominant role of cold dark matter (CDM) in shaping structures in the Universe is now very well constrained and understood on the largest scales, but less on the scales of individual galaxy haloes. The assembly and evolution of CDM haloes can be studied in detail with N-body simulations, which predict that the CDM density profile, $\rho_{\mathrm{CDM}}(r)$, are well described by the so-called NFW profile with $\rho_{\mathrm{CDM}}(r) \propto r^{-3}$ in the outer regions, and $\rho_{\mathrm{CDM}}(r) \propto r^{-\alpha}$, in the centre (with $\alpha = -1$, [15];

N.R. Napolitano (✉) • C. Tortora • F. La Barbera • F. Getman
INAF – Osservatorio Astronomico di Capodimonte, Salita Moiariello, 16, 80131, Napoli, Italy
e-mail: napolita@na.astro.it

G. Covone • N. Roy • M. Capaccioli • A. Colonna • M. Paolillo
Department of Physics "Ettore Pancini", University "Federico II" of Naples, Italy

M. Radovich
INAF – Osservatorio Astronomico di Padova-vicolo Osservatorio 5, 35122 Padova, Italy

G.A. Verdoes Kleijn • L.V.E. Koopmans
Kapteyn Astronomical Institute, University of Groningen, P.O. Box 800, 9700 AV Groningen, The Netherlands

KiDS collaboration: http://kids.strw.leidenuniv.nl/team.php

N.R. Napolitano et al. (eds.), *The Universe of Digital Sky Surveys*, Astrophysics and Space Science Proceedings 42, DOI 10.1007/978-3-319-19330-4_20

$\alpha = -1.5$ [14]). However, N-body simulations only follow the evolution of CDM particles, not including the complex physics of gas and stars, which can alter the CDM profiles (e.g., via adiabatic contraction, [10]).

The study of the CDM profile in the inner regions of ETGs has received much interest, due to the large amount of available data samples (e.g., [11, 19]). However, it is made difficult by the degeneracy between the shape of the CDM profile and that of the stellar Initial Mass Function (IMF; [7, 20].)

20.2 Strong Lensing to Study the Mass Density of Galaxies

Strong lensing (SL hereafter) is an ideal tool to perform accurate studies of the CDM distribution of intermediate and massive galaxies (e.g. with velocity dispersions, $\sigma_v > 180$ km s^{-1}) independently of the dynamical status of the mass within the arc features. SL studies, often in combination with stellar dynamics in the galaxy central regions, have found that the light and halo profiles conspire to have a total mass density profile which is nearly isothermal [2, 3, 5, 12, 16], i.e. a total-mass density slope of $\alpha \sim -2$. This "conspiracy", once thought to be universal, has been recently found not to be so, as lower mass galaxies might have steeper α [8, 21]. The physical reasons for this systematic variation of the observed slope resides in the way baryons and CDM are mixed, after having been assembled over time [17]. Such detailed studies have been performed over samples of nearby early-type galaxies and a similar systematic study is ongoing at higher-z SL systems (e.g., by SLACS, [4]; SL2S, [18]), where SL is much more effective than standard dynamical studies. For this reason it is becoming of fundamental importance to enlarge the sample of SL systems both in the galaxy parameter space (e.g. toward less massive systems, like sub-L_*) and in redshift space.

20.3 Why Galaxy-Galaxy SL in the ESO Public Survey KiDS?

The ESO public survey KiDS (Kilo Degree Survey, [6]) is a multi-band (*ugri*) imaging survey, designed to take advantage of the superb seeing of Paranal and high image quality of the VST/OmegaCAM. Its primary usage is weak lensing measurements over 1,500 deg^2 of the sky.

As such, KiDS is particularly suitable for a systematic census of lensing galaxies based on the identification of arc-like structure from intermediate to massive galaxies (i.e., with velocity dispersion of the order of 200 km s^{-1} or larger), to galaxy groups and galaxy clusters, thus covering a wide range of lens masses. Indeed, the Einstein ring writes easily as a function of the velocity dispersion as

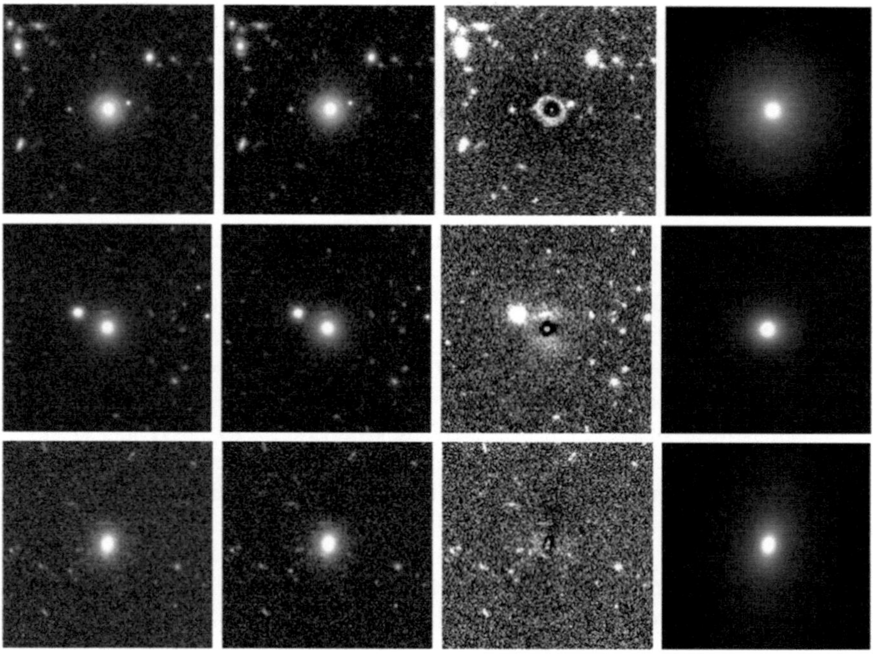

Fig. 20.1 Results from the visual inspection of SL candidates. Top rows show two high-confidence candidates, bottom is a a less secure one. At each row we show: (1) RGB colour image, (2) r—band image; (3) model subtracted image; (4) 2DPHOT galaxy model. In the top row, an Einstein ring is evident and shows up clearly in the residual image. In the middle row a blue arc is also evident in the colour image and nicely spotted into the residual image. In the bottom row a faint blue arc is seen in the colour image in the bottom edge of the redder central galaxy, which shows up slightly clearer in the model subtracted image. Estimated R_E are 6–8 arcsec

$R_{\rm E} \sim 1'' \times (\sigma_v/220\,{\rm km\,s^{-1}})^2$, and for a typical $FWHM \sim 0.7''$ in r-band, we can expect to observe lensing arcs of gravitational structures of $\sigma_v > 180\,{\rm km\,s^{-1}}$.

The search for SL in KiDS is expected to provide up to thousands galaxy rings (e.g. by extrapolating CFHT-LS numbers, [16]). This can be performed only with (semi-)automated techniques [1, 9], as potential candidates in KiDS can be of the order of thousands per square degree, making visual inspection inapplicable. However, lensing finding techniques still suffer from a number of issues yet to be solved, affecting the completeness and reliability of their catalogs and need to be trained over visually-inspected samples.

20.4 Selection of SL Candidates via Visual Inspection

Due to the yet limited extension of the area ($150\,{\rm deg}^2$) included in the KiDS ESO-Data Release 2 (ESO-DR2), we have decided to start a visual search based on simple criteria: (1) sources with SDSS spectroscopic redshifts between $z = 0.1$

and $z = 0.5$ and (2) sources brighter than $r = 20$, which together maximise the SL probability for the depth allowed by KiDS images. These criteria were restrictive and biased toward larger R_E easier to spot by visual inspection. In order to quantify the performance of our search, we decided to start on some control fields, overlapping with the SDSS, and selected ~600 lens candidates in ~100 deg^2, including a ten of SLACS lenses.

The visual inspection (performed by six observers) was based on (a) the galaxy r–band image (r–band is chosen as reference since it is the deepest and highest quality among KiDS wavebands), (b) a three-colour combined image to find a gradient in colour between the candidate lens galaxy and the arc, (c) the 2DPHOT model [13] subtracted image to detect residual arc features. Examples of lens candidates are in Fig. 20.1). Following this procedure, we ended up with a list of 18 new potential lens candidates, half of which have high significance. We also re-identified three SLACS lensing galaxies, while the other seven could not be found because arcs were too close to the galaxy centre to be uncovered. Due to the peculiarity of the SLACS selection procedure (i.e. very compact Einstein radii falling into the size of the SDSS spectra fiber, $R \sim 1.5''$) the fraction of the recovered SLACS systems (i.e. ~30 %) is a lower limit of the completeness of our selection.

The spectroscopic follow-up will provide us with the confirmation of the lensing nature of the lens candidates and the independent kinematic information to constrain the mass profile of the lensing galaxies.

References

1. Alard, C.: Gravitational arcs as a perturbation of the perfect ring. MNRAS **382**, L58 (2007)
2. Auger, M.W., Treu, T., Bolton, A.S., et al.: The Sloan Lens ACS Survey. X. Stellar, dynamical, and total mass correlations of massive early-type galaxies. ApJ **724**, 511 (2010)
3. Bolton, A.S., Burles, S., Koopmans, L.V.E., Treu, T., Moustakas, L.A.: Spectroscopically selected sample of massive early-type lens galaxies. ApJ **638**, 703 (2006)
4. Bolton, A.S., Burles, S., Koopmans, L.V.E., et al.: The Sloan Lens ACS Survey. V. The full ACS strong-lens sample. ApJ **682**, 964 (2008)
5. Chae, K.H., Bernardi, M., Kravtsov, A.V.: Modelling mass distribution in elliptical galaxies: mass profiles and their correlation with velocity dispersion profiles. MNRAS **437**, 3670 (2014)
6. de Jong, J.T.A., Kuijken, K., Applegate, D., et al.: The Kilo-Degree Survey. The Messenger **154**, 44 (2013)
7. Dutton, A.A., Mendel, J.T., Simard, L.: Evidence for a non-universal stellar initial mass function in low-redshift high-density early-type galaxies. MNRAS **422**, L33 (2012)
8. Dutton, A.A., Treu, T.: Elliptical galaxies: implications for the stellar initial mass function and halo response to Baryonic processes. MNRAS **438**, 3594 (2014)
9. Gavazzi, R., Marshall, P.J., Treu, T., Sonnenfeld, A.: RINGFINDER: automated detection of galaxy-scale gravitational lenses in ground-based multi-filter imaging data. ApJ **785**, 144 (2014)
10. Gnedin, O.Y., Kravtsov, A.V., Klypin, A.A., Nagai, D.: Response of dark matter halos to condensation of Baryons: cosmological simulations and improved adiabatic contraction model. ApJ **616**, 16 (2004)
11. Hyde, J.B., Bernardi, M.: The luminosity and stellar mass fundamental plane of early-type galaxies. MNRAS **396**, 1171 (2009)

12. Koopmans, L.V.E., Treu, T., Bolton, A.S., Burles, S., Moustakas, L.A.: The Sloan Lens ACS Survey. III. The structure and formation of early-type galaxies and their evolution since z $\sim$ 1. ApJ **649**, 599 (2006)
13. La Barbera, F., de Carvalho, R.R., Kohl-Moreira, J.L., et al.: 2DPHOT: a multi-purpose environment for the two-dimensional analysis of wide-field images. PASP **120**, 681 (2008)
14. Moore, B., Governato, F., Quinn, T., Stadel, J., Lake, G.: Resolving the structure of cold dark matter halos. ApJl **499**, L5 (1998)
15. Navarro, J.F., Frenk, C.S., White, S.D.M.: The structure of cold dark matter halos. ApJ **462**, 563 (1996)
16. Oguri, M., Rusu, C.E., Falco, E.E.: The stellar and dark matter distributions in elliptical galaxies from the ensemble of strong gravitational lenses. MNRAS **439**, 2494 (2014)
17. Remus, R.S., Burkert, A., Dolag, K., et al.: The dark halo—spheroid conspiracy and the origin of elliptical galaxies. ApJ **766**, 71 (2013)
18. Ruff, A.J., Gavazzi, R., Marshall, P.J., Treu, T., Auger, M.W., Brault, F.: The SL2S galaxy-scale lens sample. II. Cosmic evolution of dark and luminous mass in early-type galaxies. ApJ **727**, 96 (2011)
19. Tortora, C., La Barbera, F., Napolitano, N.R., de Carvalho, R.R., Romanowsky, A.J.: SPIDER – VI. The central dark matter content of luminous early-type galaxies: benchmark correlations with mass, structural parameters and environment. MNRAS **425**, 577 (2012)
20. Tortora, C., Romanowsky, A.J., Napolitano, N.R.: An inventory of the stellar initial mass function in early-type galaxies. ApJ **765**, 8 (2013)
21. Tortora, C., La Barbera, F., Napolitano, N.R., Romanowsky, A.J., Ferreras, I., de Carvalho, R.R.: Systematic variations of central mass density slopes in early-type galaxies. MNRAS **445**, 115 (2014)

Chapter 21
Early Type Galaxies and Structural Parameters from ESO Public Survey KiDS

N. Roy, N.R. Napolitano, F. La Barbera, C. Tortora, F. Getman, M. Radovich, M. Capaccioli, and the KiDS collaboration

Abstract The Kilo Degree survey (KiDS) is a large-scale optical imaging survey carried out with the VLT Survey Telescope (VST), which is the ideal tool for galaxy evolution studies. We expect to observe millions of galaxies for which we extract the structural parameters in four wavebands (u, g, r and i). This sample will represent the largest dataset with measured structural parameters up to a redshift $z = 0.5$. In this paper we will introduce the sample, and describe the 2D fitting procedure using the 2DPHOT environment and the validation of the parameters with an external catalog.

21.1 Introduction

The Kilo Degree Survey (KiDS) is a ESO public survey (PI. K. Kuijken) carried out with the VLT Survey Telescope (VST), at the ESO Paranal Observatory. It will cover 1,500 $\deg^2$ of the night sky in four optical bands (*ugri*). The large area covered, the depth, the good seeing and pixel scale make the VST the ideal tool to investigate the evolution of galaxies across the last billions of years. KiDS possess upper hand in different aspects compared to the previous surveys. E.g., with respect to the Sloan Digital Sky Survey (SDSS), which is the most successful survey in the galaxy evolution studies [4], KiDS has two major improvements that are crucial for its science goals: it is much deeper (by about 2 mag), and it has better image quality, particularly in the r-band with respect to SDSS.

N. Roy (✉) • M. Capaccioli
Department of Physics "Ettore Pancini", University "Federico II" of Naples, Italy
e-mail: roy@fisica.unina.it

C. Tortora • F. La Barbera • N.R. Napolitano • F. Getman
INAF – Osservatorio Astronomico di Capodimonte, Salita Moiariello, 16, 80131, Napoli, Italy

M. Radovich
INAF – Osservatorio Astronomico di Padova-vicolo Osservatorio 5, 35122 Padova, Italy

KiDS collaboration: http://kids.strw.leidenuniv.nl/team.php

N.R. Napolitano et al. (eds.), *The Universe of Digital Sky Surveys*, Astrophysics and Space Science Proceedings 42, DOI 10.1007/978-3-319-19330-4_21

We will study the structural properties of early-type galaxies, which, being the oldest and most massive systems in the universe, provide a unique way to trace the galaxy evolution and formation, and important physical processes like galaxy mergings.

21.2 The Sample

For the current study we are using a galaxy catalog extracted from the first 158 deg^2 of the KiDS Survey. Single band sourcelist have been extracted with a stand-alone procedure for the catalog extraction optimized for KiDS whose backbone is constituted by S-Extractor [1] software for the source detection and extraction which is fed with KiDS image, weighting maps (from Astro-Wise pipeline, [10]) and external masks for bad pixels, saturated stars and stars haloes obtained with an automated procedure (i.e., an evolution of the ExAM software, see [3]). Star/galaxy separation is based on the `CLASS_STAR` (star classification) and S/N (signal-to- noise ratio) parameters provided by S-Extractor [1, 6]. From the whole galaxy population extracted within the KIDS DR2 ($\sim$7 million), we have selected those galaxies with r-band signal-to-noise ratio (S/N) >50. This criterion allows one to obtain reliable structural parameters [7]. Using this cut on S/N ratio we ended up with a sample of $\sim$38,000 galaxies. For further details, see the contribution from Tortora et al.

21.3 Structural Parameter Extraction

The structural parameter extraction is performed by 2DPHOT [6], which is an automated software environment which accomplishes several tasks such as catalog extraction (using S-extractor, [1]), star/galaxy separation and surface photometry, for the selected galaxies. Galaxy images were fitted with a PSF convolved Sérsic models [2] having elliptical isophotes plus a local background value.

In Fig. 21.1 we show the 2DPHOT output stamps for few example galaxies. Ninety-three percent of our galaxy sample were fitted by a Sérsic profile with a small reduced χ^2 value (see left panels in Fig. 21.1) without any residuals. However, we found galaxies with $\chi^2 > 1.5$, not perfectly fitted, and leaving in the residual maps signs of substructures, as spiral arms. 2DPHOT model fitting will provide structural parameters like surface brightness μ_e, effective radius R_e, Sérsic index n, model magnitude, axial ratio, position angle, etc.

Because our sample consists of both early-type (ETGs) and late-type (LTGs) galaxies, we have selected ETGs on the base of the following criteria: (1) Sérsic index n $>$ 2.5, since LTGs are found to have lower Sérsic indices (e.g. [8]), (2) $R_e > 0.2$ arcsec (since the VST pixel scale is 0.21 arcsec/pixel), and (3) $\chi^2 < 1.5$, to take the galaxies best-fitted by the Sérsic profile and remove those systems with a clear sign of spiral arms in the fit residuals (see right panels in Fig. 21.1).

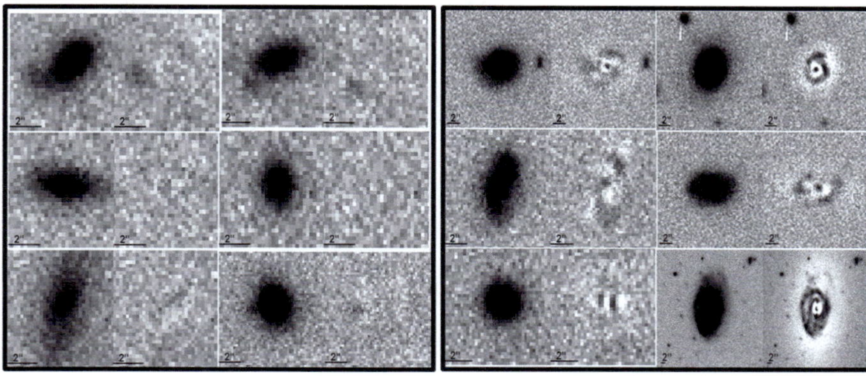

Fig. 21.1 2D fit results for 12 example galaxies in r-band. Each frame shows the galaxy stamp (*left*) and the residual map (*right*) after model subtraction. *Left panels* are for $\chi^2 < 1.5$, while *right panels* for $\chi^2 > 1.5$

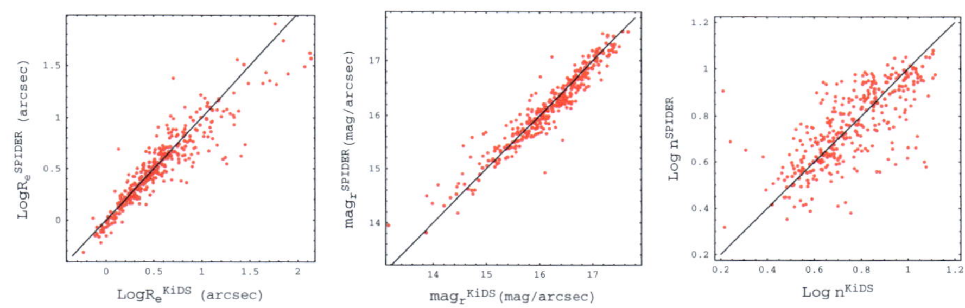

Fig. 21.2 Comparison of KiDS structural parameters with the ones derived within the SPIDER survey. The SPIDER dataset consists of ETGs with redshifts in the range $0.05 < z < 0.095$, selected from SDSS; the structural parameters are derived using 2DPHOT. We show the relationship for effective radius (R_e), model magnitude (mag_r) and Sèrsic index (n) respectively from left to right. Data are shown as *red points*. The *diagonal lines* are the one-to-one relations, corresponding to the perfect agreement

21.4 Comparison with Literature Data

To study the reliability and consistency of the derived structural parameters we made a comparison of our results with an external datasample from the SPIDER survey [7]. This dataset, extracted from SDSS, consists of ~39,000 ETGs with redshifts $0.05 < z < 0.095$, with structural parameters determined by fitting a Sérsic profile and the fitting procedure is performed using 2DPHOT. We found 345 shared sources by cross-matching the two catalogs.

The comparisons of R_e, magnitudes and Sérsic index are shown in Fig. 21.2. There is a good agreement in the parameters from two catalogs, which confirm the reliability of our extracted parameters in the KiDS survey.

21.5 Conclusions

KiDS, being a large scale survey with good image quality, provides a great platform for the study of galaxy evolution. With the extraction of structural parameters we are about to start the analysis of scaling relations, as the relationship between the extracted structural parameters (e.g. R_e and n) as a function of luminosity or stellar mass [9], or the correlation of surface brightness with R_e or magnitude [5]. KiDS will allow to analyze these correlations at different cosmic epochs up to redshift 0.5. Galaxy size and mass evolution with cosmic time provide fundamental details about the physical mechanisms which drive the galaxy evolution, including the effect of mergings or other phenomena related to the galaxy environment.

Acknowledgements CT has received funding from the European Union Seventh Framework Programme (FP7/2007–2013) under grant agreement n. 267251 "Astronomy Fellowships in Italy" (AstroFIt).

References

1. Bertin, E., Arnouts, S.: SExtractor: software for source extraction. Astron. Astrophys. **117**, 393 (1996)
2. Caon, N., Capaccioli, M., D'Onofrio: On the shape of the light profiles of early type galaxies. MNRAS **265**, 1013 (1993)
3. Huang, Z., Radovich, M., Grado, A., et al.: A weak-lensing analysis of the Abell 383 cluster. Astron. Astrophys. **529**, AA93 (2011)
4. Kauffmann, G., Heckman, T.M., White, S.D.M., et al.: The dependence of star formation history and internal structure on stellar mass for 105 low-redshift galaxies. MNRAS **341**, 54 (2003)
5. La Barbera, F., Merluzzi, P., Busarello, G., Massarotti, M., Mercurio, A., et al.: Probing galaxy evolution through the internal colour gradients, the Kormendy relations and the Photometric Plane of cluster galaxies at $z \sim 0.2$. Astron. Astrophys. **425**, 797 (2004)
6. La Barbera, F., de Carvalho, R.R., Kohl-Moreira, J.L., et al.: 2DPHOT: a multi-purpose environment for the two-dimensional analysis of wide-field images. PASP **120**, 681 (2008)
7. La Barbera, F., de Carvalho, R.R., de La Rosa, I.G., Lopes, P.A.A., Kohl-Moreira, J.L., Capelato, H.V.: SPIDER - I. Sample and galaxy parameters in the grizYJHK wavebands. MNRAS **408**, 1313 (2010)
8. Tortora, C., Napolitano, H.V., Cardone, V.F., Capaccioli, M., Jetzer, Ph., Molinaro, R.: Colour and stellar population gradients in galaxies: correlation with mass. MNRAS **407**, 144 (2010)
9. Trujillo, I., Conselice, C.J., Bundy, K., Cooper, M.C., Eisenhardt, P., Ellis, R.S.: Strong size evolution of the most massive galaxies since $z \sim 2$. MNRAS **382**, 109 (2007)
10. Verdoes Kleijn, G., de Jong, J.T.A., Valentijn, E.A., et al.: Astro-WISE for KiDS survey production and quality control (2011). arXiv:1112.0886

Chapter 22
STREGA@VST: Structure and Evolution of the Galaxy

M. Marconi, I. Musella, M. Di Criscienzo, M. Cignoni, M. Dall'Ora, V. Ripepi, G. Bono, E. Brocato, G. Coppola, A. Grado, L. Limatola, M.I. Moretti, G. Raimondo, P.B. Stetson, A. Calamida, M. Cantiello, M. Capaccioli, E. Cappellaro, M.-R.L. Cioni, S. Degl'Innocenti, D. De Martino, A. Di Cecco, I. Ferraro, G. Iannicola, P.G. Prada Moroni, R. Silvotti, R. Buonanno, F. Getman, N.R. Napolitano, L. Pulone, and P. Schipani

Abstract STREGA (STRucture and Evolution of the Galaxy) is an ongoing VLT Survey Telescope Guaranteed Time survey, aimed at investigating the mechanisms of formation and evolution of the Galactic Halo on an area of about 150 square degrees. The core programme searches for the signatures of interaction between

---

M. Marconi (✉)
INAF – Observatory of Capodimonte Naples, Italy
e-mail: marcella.marconi@oacn.inaf.it

I. Musella • M. Dall'Ora • V. Ripepi • G. Coppola • A. Grado • L. Limatola • D. De Martino • F. Getman • N.R. Napolitano • P. Schipani
INAF-Osservatorio Astronomico di Capodimonte, Salita Moiariello 16 I-80131 Napoli, Italy

M. Di Criscienzo • E. Brocato • A. Di Cecco • I. Ferraro • G. Iannicola • L. Pulone
INAF – Osservatorio Astronomico di Roma, Via Frascati 33, I-00044 Monte Porzio Catone, Italy

M. Cignoni • A. Calamida
Space Telescope Science Institute, 3700 San Martin Drive, Baltimore, MD 21218, USA

G. Bono
Dipartimento di Fisica, Università degli Studi di Roma-Tor Vergata, via della Ricerca Scientifica 1, I-00133 Roma, Italy

R. Buonanno
Dipartimento di Fisica, Università degli Studi di Roma-Tor Vergata, via della Ricerca Scientifica 1, I-00133 Roma, Italy

INAF-Osservatorio Astronomico Collurania, via Maggini, I-64100 Teramo, Italy

M.I. Moretti
INAF-Osservatorio Astronomico di Bologna, via Ranzani 1, I-40127 Bologna, Italy

G. Raimondo • M. Cantiello
INAF-Osservatorio Astronomico Collurania, via Maggini, I-64100 Teramo, Italy

M. Capaccioli
Department of Physics "Ettore Pancini", University "Federico II" of Naples, Italy

N.R. Napolitano et al. (eds.), *The Universe of Digital Sky Surveys*, Astrophysics and Space Science Proceedings 42, DOI 10.1007/978-3-319-19330-4_22

selected stellar systems and the Galactic Halo and a complementary part is focussed on exploring the southern portion of the Fornax Stream. The adopted stellar tracers are variable stars and main sequence turn-off stars. We present an overview of the survey, the first results, and the perspectives.

22.1 Introduction

The outer Galactic halo seems to be quite "clumpy", supporting theories based on the hierarchical formation of structures in a Cold Dark Matter cosmological scenario. The most known examples include: (i) the observed merging of Sagittarius with the Galactic halo and its associated stream (see e.g. [2] and references therein); (ii) the stellar over-density in the Canis Major region (e.g. [9] and references therein); (iii) peculiar Galactic Globular Clusters (GGCs), with tidal tails or suspected halos (e.g. [6] and references therein); (iv) ultra-faint satellites of the Milky Way (MW) discovered from analysis of the Sloan Digital Sky Survey (SDSS, e.g. [1, 4] and references therein). The distribution of MW dwarf spheroidal galaxies (Sphs) and GGCs along planar alignments (see e.g. the Vast Polar Structure identified by [10]) suggests the presence of distinct orbital planes that might be the result of the disruption of larger galaxies. Interestingly, similar phenomena are also observed in M31 (e.g. [5]). In this context the STREGA survey uses part of the VST Guaranteed Time Observation (GTO) allocated by ESO to the Italian Istituto Nazionale di Astrofisica (INAF) in return for the building of the telescope, to investigate the Galactic halo formation mechanisms.

22.2 The STREGA Survey

The core programme of STREGA (P.I.: M. Marconi/I. Musella; see also [7, 8]) is aimed at finding and investigating stellar overdensities (tidal tails and/or haloes) around selected dSphs and GGCs up to 2–3 tidal radii. In particular the target

P.B. Stetson
NRC – Herzberg, Dominion Astrophysical Observatory, 5071 West Saanich Road, Victoria, BC V9E 2E7, Canada

E. Cappellaro
INAF-Osservatorio Astronomico di Padova, vicolo dell'Osservatorio 5, I-35122 Padova, Italy

M.-R.L. Cioni
School of Physics, Astronomy and Mathematics, University of Hertfordshire, College Lane, Hatfield AL10 9AB, UK

S. Degl'Innocenti • P.G. Prada Moroni
Dipartimento di Fisica 'Enrico Fermi', Università di Pisa, largo Pontecorvo 3, I-56127 Pisa, Italy

R. Silvotti
INAF – Osservatorio Astrofisico di Torino, via Osservatorio 20, I-10025 Pino Torinese, Italy

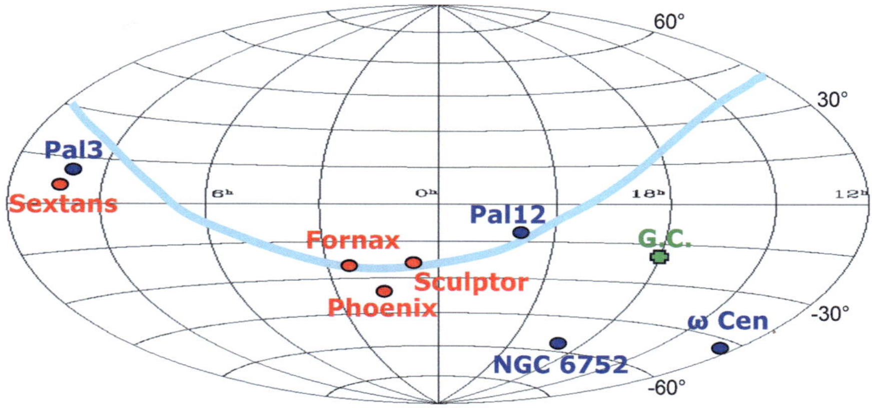

Fig. 22.1 The dSphs (*red symbols*) and GGCs (*blue symbols*) observed by the STREGA survey. The *cyan line* represents the orbit of Fornax in equatorial coordinates. The Galactic Centre is marked with a *green symbol*

systems include: Fornax and Sculptor (38 deg^2), Sextans (13 deg^2), Phoenix (3 deg^2), ω Cen and NGC 6752 (37 and 36 deg^2, respectively), Pal 3 (3 deg^2) and Pal 12 (2 deg^2). These targets are shown in Fig. 22.1 (red and blue symbols for dSphs and GGCs, respectively) in equatorial coordinates. The adopted stellar tarcers of the searched overdensities are mainly RR Lyrae variables stars and Main-Sequence (especially Turn off) stars: the former are bright and easily indentified thanks to their characteristic variability; the latter are fainter by about three magnitudes than RR Lyrae stars but two orders of magnitude more abundant. The multipurpose character of the STREGA survey implies that there are a number of secondary objectives. Some of them are direct byproducts, e.g. the test of Galactic models through the comparisons of observed star counts with the predicted contribution of the various Galactic components. But there are also secondary objectives obtained by slightly changing the strategy, e.g. the investigation of White Dwarf and Interacting Binary Populations as a function of the Galactic latitude in selected fields.

22.3 The Case of ω Cen

In this Section we focus our analysis on the 37 deg^2 that have been observed around the GGC ω Cen. This system is considered to be a tidally disrupted galaxy on the basis of the presence of multi-populations with different chemical compositions, the retrograde orbit coupled with an unusually low inclination and the similarity with the globular cluster M54, currently considered the remnant core of the disrupted Sagittarius galaxy. Our VST observations in the SDSS *gri* bands (see [7, 8] for details) cover 37 deg^2 centred on ω Cen and reach about three tidal radii. Each single exposure reaches at least 2 mag below the turn-off and combining the *g*, *r* and *i* catalogues for all the covered fields, we built the total CMDs and performed star

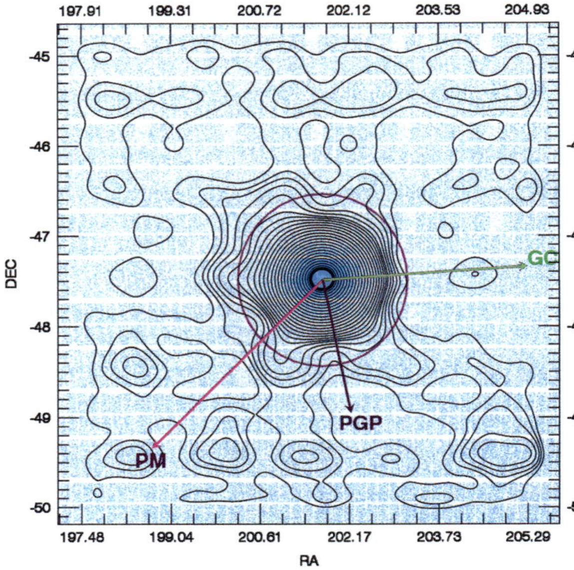

Fig. 22.2 Contour level of the stars around Cen projected on the sky. The *long magenta arrow* shows the direction of proper motion (PM), the *green arrow* shows the Galactic Centre direction (GC) and the *purple arrow* is the projection on the sky of the direction perpendicular to the Galactic Plane (PGP)

counts in selected evolutionary phases (see [7] for details). In particular, we focused on star counts in the most statistically significant phase of the Main Sequence Turn Off (MSTO) and the SubGiant Branch (SGB) both as a function of the radial distance from the cluster center and as a function of the direction.

The contour level map[1] shown in Fig. 22.2 well summarize the obtained results from the star counting procedure in the MSTO + SGB phase. In particular:

- the asymmetry increases when moving towards the outermost cluster regions;
- there is an excess of stars beyond the nominal tidal radius (0.95°, purple circle);
- the orientation of this overdensity, roughly orthogonal to ω Cens proper motion direction (PM, magenta arrow), appears to be intermediate between Galactic Centre direction (GC, green arrow) and the projection on the sky of the direction perpendicular to the Galactic Plane (PGP, purple arrow).

The evidence shown in Fig. 22.2 is consistent with current measurements and predictions for the cluster ellipticity profile and orientation, as well as with the very recent results by [3] who covered a 50 deg area not uniform around the cluster and found no evidence of extratidal stellar populations in the direction perpendicular to the above discussed overdensity orientation.

Acknowledgements This work was supported by PRIN-INAF 2011 Tracing the formation and evolution of the Galactic halo with VST (PI: M. Marconi), PRIN-INAF 2011 Galaxy Evolution with the VLT Surveys Telescope (VST) (PI: A. Grado) and PRIN-MIUR (2010LY5N2T) Chemical and Dynamical Evolution of the Milky Way and Local Group galaxies (PI: F. Matteucci).

[1]Based on 2D number densities.

References

1. Clementini, G., Cignoni, M., Contreras Ramos, R., et al.: Variability and star formation in Leo T, the lowest luminosity star-forming galaxy known today. ApJ **756**, 108 (2012)
2. Deg, N., Widrow, L.: The Sagittarius stream and halo triaxiality. MNRAS **428**, 912 (2013)
3. Fernández-Trincado, J.G., Vivas, A.K., Mateu, C.E., et al.: Searching for tidal tails around ω Centauri using RR Lyrae stars. A&A **574**, AA15 (2015)
4. Garofalo, A., Cusano, F., Clementini, G., et al.: Variable stars in the ultra-faint dwarf spheroidal galaxy Ursa Major I. ApJ **767**, 62 (2013)
5. Ibata, R.A., Lewis, G.F., Conn, A.R., et al.: Does the Sagittarius stream constrain the Milky Way Halo to be triaxial? Nature **493**, 62 (2013)
6. Majewski, S.R., Nidever, D.L., Smith, V.V., et al.: Exploring Halo substructure with Giant stars: substructure in the local Halo as seen in the grid Giant star survey including extended tidal debris from ω Centauri. ApJL **747**, LL37 (2012)
7. Marconi, M., Musella, I., Di Criscienzo, M., et al.: STREGA: STRucture and evolution of the GAlaxy – I. Survey overview and first results. MNRAS **444**, 3809 (2014)
8. Marconi, M., Musella, I., Di Criscienzo, M., et al.: STREGA: STRucture and Evolution of the GAlaxy with the VST. The Messenger **158**, 39 (2014)
9. Mateu, C., Vivas, A.K., Zinn, R., Miller, L.R., Abad, C.: No excess of RR Lyrae stars in the Canis Major Overdensity. AJ **137**, 4412 (2009)
10. Pawlowski, M.S., Pflamm-Altenburg, J., Kroupa, P.: The VPOS: a vast polar structure of satellite galaxies, globular clusters and streams around the Milky Way. MNRAS **423**, 1109 (2012)

Chapter 23
The VST Survey of the SMC and the Magellanic Bridge (STEP): First Results

V. Ripepi, M. Cignoni, M. Tosi, M. Marconi, I. Musella, G. Coppola, A. Grado, L. Limatola, G. Clementini, E. Brocato, M. Cantiello, M. Capaccioli, E. Cappellaro, M.-R.L. Cioni, F. Cusano, M. Dall'Ora, J.S. Gallagher III, E.K. Grebel, A. Nota, F. Palla, D. Romano, G. Raimondo, E. Sabbi, F. Getman, N.R. Napolitano, P. Schipani, and S. Zaggia

V. Ripepi (✉) • I. Musella • G. Coppola • A. Grado • L. Limatola • M. Dall'Ora • F. Getman • N.R. Napolitano • P. Schipani
INAF-Osservatorio Astronomico di Capodimonte, Via Moiariello 16, 80131, Naples, Italy
e-mail: ripepi@oacn.inaf.it

M. Marconi
INAF – Observatory of Capodimonte Naples, Italy

M. Cignoni • A. Nota • E. Sabbi
Space Telescope Science Institute, 3700 San Martin Drive, Baltimore, MD, USA

M. Tosi • G. Clementini • F. Cusano • D. Romano
INAF-Osservatorio Astronomico di Bologna, via Ranzani 1, Bologna, Italy

E. Brocato
INAF-Osservatorio Astronomico di Roma, Via Frascati 33, 00040, Monte Porzio Catone (RM), Italy

M. Cantiello • G. Raimondo
INAF-Osservatorio Astronomico di Teramo, Via M. Maggini, 64100 Teramo, Italy

M. Capaccioli
Department of Physics "Ettore Pancini", University "Federico II" of Naples, Italy

E. Cappellaro • S. Zaggia
INAF-Osservatorio Astronomico di Padova, Vicolo dell'Osservatorio 5, 35122 Padova, Italy

M.-R.L. Cioni
Physics Astronomy and Mathematics, University of Hertfordshire, College daue Hatfield AL10 9AB, UK

J.S. Gallagher III
Department of Astronomy, University of Wisconsin-Madison, 5534 Sterling, 475 North Charter Street, Madison, WI 53706, USA

E.K. Grebel
Astronomisches Rechen-Institut, Zentrum für Astronomie der Universität Heidelberg, Mönchhofstr 12-14, D-69120 Heidelberg, Germany

F. Pallav
INAF-Osservatorio Astrofisico di Arcetri, Largo E. Fermi 5, 50125 Firenze, Italy

N.R. Napolitano et al. (eds.), *The Universe of Digital Sky Surveys*, Astrophysics and Space Science Proceedings 42, DOI 10.1007/978-3-319-19330-4_23

Abstract STEP (Small Magellanic Cloud in Time: Evolution of a Prototype interacting late-type dwarf galaxy) is a Guaranteed Time Observation survey being carried out at the VLT Survey Telescope. STEP will obtain homogeneous photometry in the g-, r-, i- and Hα-bands over an area of 74 deg^2 covering the main body of the Small Magellanic Cloud, the Bridge that connects it to the Large Magellanic Cloud and a small part of the Magellanic Stream (2 deg^2). Our photometry will allow us to detect and measure the magnitudes of individual stars well below the main sequence turnoff of the oldest populations. Here we briefly describe the observing strategy, the photometric techniques, and the upcoming data products of the STEP survey.

23.1 Introduction

STEP (Small Magellanic Cloud in Time: Evolution of a Prototype interacting late-type dwarf galaxy) is a Guaranteed Time Observation survey being carried out at the VLT Survey Telescope. STEP will obtain homogeneous photometry in the g-, r-, i- and $H\alpha$-bands over an area of 74 deg^2 covering the main body of the Small Magellanic Cloud, the Bridge that connects it to the Large Magellanic Cloud and a small part of the Magellanic Stream (2 deg^2). Our photometry will allow us to detect and measure the magnitudes of individual stars well below the main sequence turnoff of the oldest populations. We aim at investigating: (i) trends of Star Formation History (SFH) with position: are there connection with the interaction history of the SMC? (ii) the whole cluster population: same SFH of the field? (iii) Evolution properties of low metallicity stars in the mass range 1–2 $M_\odot$; (iv) the properties of the intra-cloud (Bridge) population: how did the stellar component of the Bridge form? (v) Impact of metallicity on PMS accretion and on the global properties of star formation (star formation rate and efficiency, and initial mass function). A thorough presentation of the survey have been presented in [6] and [7], where we have also presented preliminary results for two fields, namely 3_7 and 4_6.

23.2 Survey Strategy, Observations and Data Reduction

Our observations are carried out in the (SDSS) g, i bands, with a limiting magnitude $g \sim 24$ (AB mag) with S/N$\sim$10. Additionally, we observe in $r, H\alpha$ (with VPHAS+ consortium permission) to $r \sim 23$ mag with S/N = 5. Time-series (TS) photometry (24 epochs) on selected fields of the Bridge was carried out to $g \sim 19.0$ mag (to reach the RR Lyrae variables) with S/N = 100. The sum of all epochs in these fields will eventually provide $g \sim 24$ (AB mag) with S/N$\sim$ 10. Only eight tiles were observed with TS because of low efficiency of these observations (also, the area is covered by OGLE IV). The survey is organised into tiles of 1 deg^2 each, partially

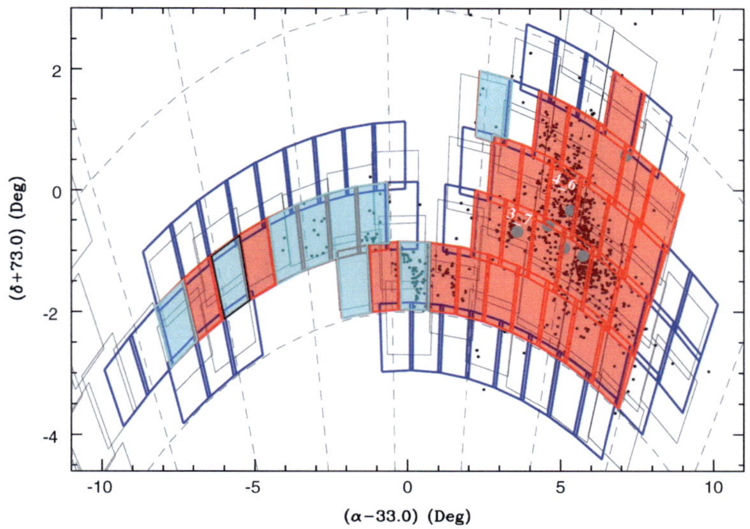

Fig. 23.1 Map of STEP tiles (the two tiles centred in the direction of the Magellanic Stream are outside of the figure). To highlight the location of the SMC body and of part of the Bridge, *black dots* indicate the position of known star clusters and associations [1]. The *thick boxes* correspond to the 1 deg^2 FoV of the VST tiles. *Red boxes* represent tiles whose observations are completed, *green boxes* those with completed time series photometry and *blue boxes* the remaining ones. For comparison, thin *grey boxes* show the VMC tiles, whereas the HST fields are the small *cyan-filled circles* (note that the true size of HST fields is significantly smaller). The two tiles analysed in this work (tiles 3_7 and 4_6) are highlighted with filled *red boxes*

overlapping with each other to allow a homogeneous calibration (see Fig. 23.1). STEP tiles are placed so as to maximise the overlap with the VMC survey [3]. The STEP observing strategy is reported in detail in [6] and will not be repeated here.

The observations are being carried out with the VST 2.6-m wide field optical survey telescope [2]. The telescope is equipped with OmegaCAM, a 1 deg^2 camera (16×16 k detector mosaic with 0.214 arcsec per pixel scale). Observations started in late 2011 during ESO Period 88 and are currently going on. Usually, the observations are carried out in service mode. The different colours of the tiles shown in Fig. 23.1 illustrate the execution status as of Period 94. The total number of hours of observation allocated up to now to STEP during periods 88–94 is 217 h; the hours of actual observations in the same period were about 117.4 h, with an efficiency (hour observed/allocated) of about 54 %. As a consequence of this rather low observing efficiency, the current percentage of completion of the entire survey is about 50 % in g, i and of 7 % in $r, H\alpha$. The data reduction was carried out by means of the VSTTube package [4], specifically developed to handle OmegaCAM data (see [6] for details). The PSF photometry on the final mosaics was carried out with DAOPHOT/ALLSTAR. It can be seen (see [6] for details) that 50 % completeness (limiting magnitude) is achieved for $g \sim 23.5$ mag in tile 4_6, that is much more crowded than tile 3_7, where we reach the same completeness level at g ∼24–24.5 mag.

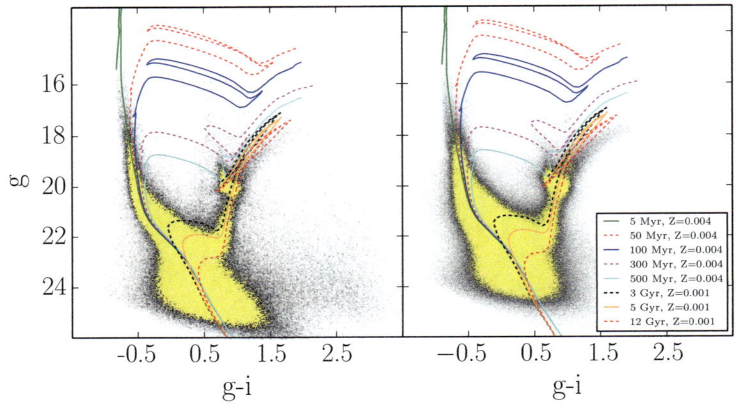

Fig. 23.2 CMD of tile 3_7 (*left panel*) and 4_6 (*right panel*) with overlaid stellar isochrones from [5]: for the labelled ages and abundances. The assumed distance modulus and reddening E(B − V) are 18.9 and 0.08 mag for tile 3_7 (*left panel*), 18.9 and 0.04 mag for tile 4_6 (*right panel*) respectively

23.2.1 First Results and Future Prospects

As an illustration of the results that can be obtained in the the STEP context, Fig. 23.2 shows the CMDs of the stars measured in tile 3_7 (left panel) and 4_6 (right panel). Since tile 4_6 is located in a populous part of the SMC, the corresponding CMD hosts over four times more stars than tile 3_7, which is placed in the relatively low density and peripheral Wing. A detailed discussion of the stellar populations in these fields can be found in [6].

At present we are working to homogenize the photometry among the different tiles in order to have a flat photometric zero point over the whole surveyed area. Once this task will be finished we plan to publish the stellar catalogues and to calculate the SFH of the observed stars. We shall also analyse systematically the stellar cluster population and search for variability the tiles observed by means of time-series photometry. As for the $r - H\alpha$ survey we are starting to exploit the first completed tiles, by cleaning the candidates sample and planning follow-up spectroscopic observations.

Acknowledgements Partial support to this work was provided by the following projects: PRIN-MIUR 2010 (2010LY5N2T) "Chemical and Dynamical evolution of the Milky Way and Local Group galaxies" (PI F. Matteucci).

References

1. Bica, E., Bonatto, C., Dutra, C.M., Santos, J.F.C.: A general catalogue of extended objects in the Magellanic System. MNRAS **389**, 678 (2008)

2. Capaccioli, M., Schipani, P.: The VLT survey telescope opens to the sky: history of a commissioning. Messenger **146**, 2 (2011)
3. Cioni, M.-R.L., Clementini, G., Girardi, L., et al.: The VMC survey. I. Strategy and first data. A&A **527**, AA116 (2011)
4. Grado, A., Capaccioli, M., Limatola, L., Getman, F.: VST processing facility: first astronomical applications. Memorie della Societa Astronomica Italiana Supplementi **19**, 362 (2012)
5. Marigo, P., Girardi, L., Bressan, A., et al.: Evolution of asymptotic giant branch stars. II. Optical to far-infrared isochrones with improved TP-AGB models. A&A **482**, 883 (2008)
6. Ripepi, V., Cignoni, M., Tosi, M., et al.: STEP: the VST survey of the SMC and the Magellanic Bridge – I. Overview and first results. MNRAS **442**, 1897 (2014)
7. Ripepi, V., Cignoni, M., Tosi, M., et al.: STEP: the VST survey of the SMC and the Magellanic Bridge. The Messenger **157**, 32 (2014)

Chapter 24
White Dwarfs in the Galactic Plane: The Clustered and Dispersed Population

R. Raddi, S. Catalán, B.T. Gänsicke, and the EGAPS consortium

Abstract White dwarfs in the densest parts of the Galactic plane have been traditionally overlooked. The reason being a lack of deep optical surveys and the high stellar density, which make it difficult to identify faint stellar objects. We test the potential offered by the new homogeneous optical photometry from the VLT Survey Telescope (VST) Photometric Hα Survey of the Galactic Plane and Bulge (VPHAS+) to white dwarf searches. We have spectroscopically confirmed 17 candidate white dwarfs towards open clusters, for which we determine physical parameters. We suggest six of them to be cluster members, for which we estimate the progenitor masses from stellar evolutionary tracks. We compare our results with relationships between progenitor and white dwarf masses available in the literature. Our pilot study shows that larger samples of white dwarfs in open clusters can be found, providing the sought improvement for the empirical testing of stellar evolutionary theories.

24.1 Introduction

White dwarfs are the end products of low- to intermediate-mass stars ($0.8\,M_{\odot} \leq M \leq 8$–$10\,M_{\odot}$). They occupy a central role in the Galactic ecology, as they experience intense mass loss during the giant branches and enrich the interstellar medium with *s*-process elements [10, 14]. The connection between the progenitor (initial) mass and the white dwarf (final) mass is known as initial-to-final mass relationship (IFMR; [13]). The shape of the IMFR can be affected by several factors, which are metallicity, stellar rotation, convection, and magnetic fields. The relationship is important also for the study of other galaxies, because of its

R. Raddi (✉) • S. Catalán • B.T. Gänsicke
Department of Phsyics, University of Warwick, Gibbett Hill Road, Coventry, CV4 7AL, UK
e-mail: r.raddi@warwick.ac.uk

EGAPS consortium, www.vphasplus.org

N.R. Napolitano et al. (eds.), *The Universe of Digital Sky Surveys*, Astrophysics and Space Science Proceedings 42, DOI 10.1007/978-3-319-19330-4_24

influence on the mass-to-light ratio determination [8, 9]. The IMFR is studied semi-empirically via observations of white dwarfs both in open clusters and in binary systems [1, 2]. The mass of white dwarf progenitors is estimated under the assumption that coeval formation took place. Potentially large numbers of white dwarfs in open clusters can allow the study of the IFMR of homogeneous stellar populations. However, it is a matter of fact that the number of white dwarfs confirmed members of open clusters is still relatively small (about 50 stars in 11 clusters [12]).

We present here a pilot study targeting white darfs candidates towards open clusters, using VPHAS+ photometry ([4], and N. Wright contribution in this volume). Deep optical photometry ($g = 20$, at 10σ) allow us to find white dwarfs in open clusters with distance modulus of $\approx$9.5 mag, which are old enough to host white dwarf members (age $\geq$ 100 Myr).

24.2 Observations and Spectral Analysis

Candidate white dwarfs were selected via colour cuts to VPHAS+ photometry, based on the intrinsic colours of white dwarfs in the *ugr* plane (Fig. 24.1).

We observed 17 white dwarfs with FORS2 at the VLT. The optical spectra cover the 3,500–5,600 Å range, with an average $R = 1{,}000$ and signal-to-noise ratio ≈ 20. We determined $T_{\rm eff}$ and $\log g$ for 13 H-dominated and one He-dominated white dwarfs, via a line-fitting procedure that uses the FITSB2 programme [11] and [7] model atmospheres (Fig. 24.2). Three other white dwarfs present either Zeeman spliting of the Balmer lines, compatible with weak magnetic fields, or no visible lines. Therefore we estimated their $T_{\rm eff}$ via a fit to the VPHAS+ photometry.

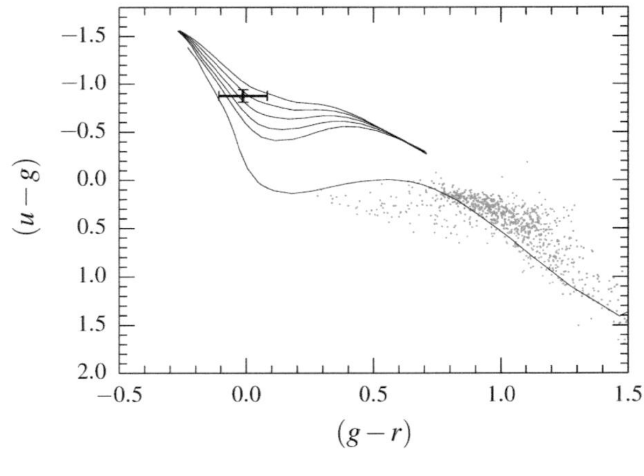

Fig. 24.1 VPHAS+ colour-colour diagram in the direction of NGC 2527. The white dwarf candidate is plotted with error bars. White dwarf cooling tracks and the main sequence are plotted

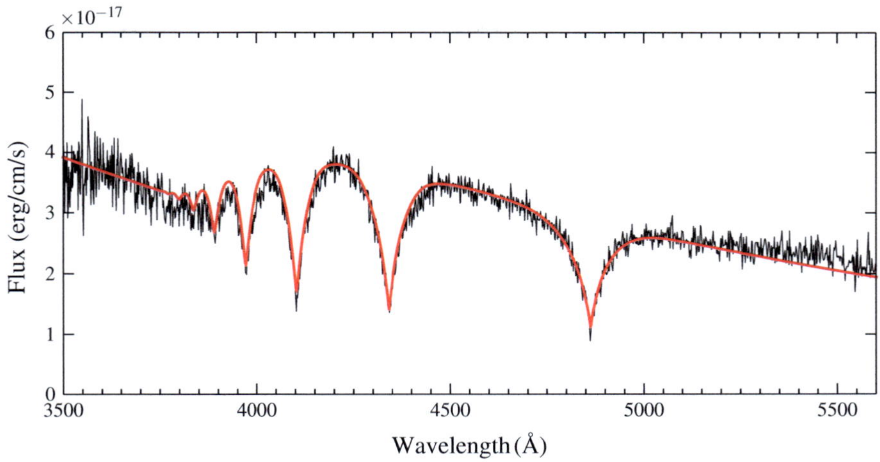

Fig. 24.2 Best fitting model and observed spectrum of VPHAS J1025-5841

24.3 Stellar Parameters

Masses and ages of white dwarfs are estimated via a grid of cooling models [5]. The absolute magnitudes are estimated convolving the model atmospheres [7] with the VPHAS+ filter profiles. Thus, spectroscopic parallaxes are determined using VPHAS+ photometry. Six white dwarfs are found to have distances compatible with those of the open clusters along their sightlines [3, 6]. The progenitor ages are determined as $t_{prog} = t_{cluster} - t_{WD}$, where $t_{cluster}$ are also taken from [3, 6]. Progenitor masses are inferred via interpolation with evolutionary tracks [12].

24.4 Results

The six white dwarfs for which we confirmed membership are compared to the [1] IFMR (Fig. 24.3). We find one massive (1.15 $M_\odot$) white dwarf, probably with an oxygen-neon core, whose progenitor could have been more massive than 5 $M_\odot$. This stars is very interesting, as it experienced the most intense mass loss, and can give insight on the minimum mass of core-collapse supernovae progenitors. In Fig. 24.3, errors due to the cluster-age uncertainty, are the dominat error on the initial mass (up to 1–2 $M_\odot$) for the youngest white dwarfs. The other new white dwarfs are found in the most populated range of initial masses (3–5 $M_\odot$). The scatter of initial/final masses is thought to be a real feature of the IFMR, influenced by stellar parameters such as metallicity and convection. Additional scatter can be due to incorrect association of white dwarfs to an open cluster.

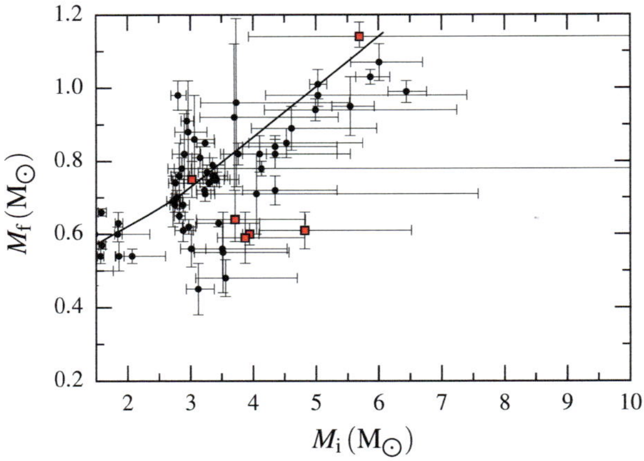

Fig. 24.3 IFMR and data from [1]. The new confirmed cluster members are plotted in *red*

24.5 Summary

We have demonstrated an efficient selection of white dwarfs via colour-cuts applied to VPHAS+ photometry. Our pilot studt delivers six likely members of open clusters, expanding the present sample of 10 %. While the atmospheric parameters of white dwarfs are determined with satisfying precision, the more uncertain cluster-ages still dominate the errors on the progenitor masses. We envisage a neat improvement of the field coming from larger samples cluster white dwarfs, and improved cluster parameters obtained with the *Gaia* mission.

References

1. Catalán, S., Isern, J., García-Berro, E., Ribas, I.: The initial-final mass relationship of white dwarfs revisited: effect on the luminosity function and mass distribution. MNRAS **387**, 1693 (2008)
2. Catalán, S., Isern, J., García-Berro, E., Ribas, I., Allende Prieto, C., Bonanos, A.Z.: The initial-final mass relationship from white dwarfs in common proper motion pairs. A&A **477**, 213 (2008)
3. Dias, W.S., Alessi, B.S., Moitinho, A., Lépine, J.R.D.: New catalogue of optically visible open clusters and candidates. A&A **389**, 871 (2002)
4. Drew, J.E., et al.: The VST Photometric Ha survey of the Southern Galactic Plane and Bulge (VPHAS+). MNRAS **440**, 2036 (2014)
5. Fontaine, G., Brassard, P., Bergeron, P.: The potential of White Dwarf cosmochronology, PASP **113**, 409 (2001)
6. Kharchenko, N.V., Piskunov, A.E., Schilbach, E., Röser, S., Scholz, R.-D.: Global survey of star clusters in the Milky Way. II. The catalogue of basic parameters. A&A **558**, AA53 (2013)
7. Koester, D.: White dwarf spectra and atmosphere models. MmSAI **81**, 921 (2010)
8. Maraston, C.: Evolutionary synthesis of stellar populations: a modular tool. MNRAS **300**, 872 (1998)

9. Maraston, C.: Evolutionary population synthesis: models, analysis of the ingredients and application to high-z galaxies. MNRAS **362**, 799 (2005)
10. Marigo, P.: Chemical yields from low- and intermediate-mass stars: model predictions and basic observational constraints. A&A **370**, 194 (2001)
11. Napiwotzki, R., Yungelson, L., Nelemans, G., et al.: Double degenerates and progenitors of supernovae type Ia. ASPCS **318**, 402 (2004)
12. Salaris, M., Serenelli, A., Weiss, A., Miller Bertolami, M.: Semi-empirical white dwarf initial-final mass relationships: a thorough analysis of systematic uncertainties due to stellar evolution models. ApJ **692**, 1013 (2009)
13. Weidemann, V.: Mass loss towards the white dwarf stage. A&A **59**, 411 (1977)
14. Williams, K.A., Bolte, M., Koester, D.: Probing the lower mass limit for supernova progenitors and the high-mass end of the initial-final mass relation from white dwarfs in the open cluster M35 (NGC 2168). ApJ **693**, 355 (2009)

Chapter 25
VEGAS-SSS: A VST Early-Type GAlaxy Survey: Analysis of Small Stellar System

M. Cantiello on behalf of the VEGAS team*

Abstract VEGAS-SSS is a program devoted to study the properties of small stellar systems (SSSs) around bright galaxies, built on the VEGAS survey. At completion, the survey will have collected detailed photometric information of $\sim$100 bright early-type galaxies to study the properties of diffuse light (surface brightness, colours, SBF, etc.) and the clustered light (compact stellar systems) out to previously unreached projected galactocentric radii. VEGAS-SSS will define an accurate and homogeneous dataset that will have an important legacy value for studies of the evolution and transformation processes taking place in galaxies through the fossil information provided by SSSs.

25.1 Motivation of the Study, Data and Data-Analysis

The surroundings of massive galaxies are populated by a zoo of dynamically hot SSSs: globular clusters (GCs), ultra-compact dwarfs (UCDs), dwarf spheroidals (dSphs), ultra-faint dwarfs (UFD), dwarf ellipticals (dEs), compact ellipticals (cE), etc. (e.g. [9, 14]).

Characterizing the properties of SSSs in the potential well of the host galaxy is fundamental, not only because of the key role in understanding the basic processes of formation and evolution of the structures in the Universe from star clusters to galaxies, but also because (i) the stellar populations of SSSs are less complex than those of massive galaxies, allowing for a more accurate comparison of their ages and metallicities with stellar population models; (ii) SSSs are relatively easy to detect out to large distances, which offers the unique opportunity to scrutinize various environments beyond the Local Group, and thus to study how SSS properties vary

*VEGAS team members: M. Capaccioli (P.I.), E. Brocato, M. Cantiello, D. A. Forbes, A. Grado, E. Iodice, L. Limatola, N. Napolitano, M. Paolillo, T. H. Puzia, G. Raimondo, A. J. Romanowsky, P. Schipani & M. Spavone

M. Cantiello (✉)
INAF-O.A. Teramo, via Maggini snc, 64100 Teramo, Italy
e-mail: cantiello@oa-teramo.inaf.it

N.R. Napolitano et al. (eds.), *The Universe of Digital Sky Surveys*, Astrophysics and Space Science Proceedings 42, DOI 10.1007/978-3-319-19330-4_25

across space and time. SSSs are then ideal tracers of the host galaxy's gravitational potential and of the assembly history of the galaxy's stellar populations.

With the VEGAS-SSS program, by taking advantage of the large field of view of the VLT Survey Telescope, VST [7], we intend to dig into the zoo of SSSs host by bright galaxies in different environments observed as part of the VEGAS survey, and comprehensively study their properties out to very large galactocentric radii R_{gal} on a homogeneous and self-consistent basis.

VEGAS is a deep multiband *gri* imaging survey of early-type galaxies, ETGs, carried out with the 1 deg^2 field-of-view camera OmegaCAM, mounted at the VST. The survey plan is to analyze the photometry of $\sim$100 galaxies with $V_{rad} \leq$ 4,000 km/s in different environments and covering the parameter space. More details on the survey are provided by M. Spavone in this volume, and in [8].

We started the VEGAS-SSS project analyzing the *g* and *i*-band data of NGC 3115, an isolated S0 galaxy at $d \sim 10$ Mpc. The choice of NGC 3115 was also motivated by the many photometric and spectroscopic studies available in the literature, which were particularly useful for testing the procedures and the methods that will be used for the future targets of the survey.

In [6] we provided a step-by-step description of the procedures adopted to (*a*) derive photometry of compact and extended sources, (*b*) model and subtract the galaxy, (*c*) obtain and apply the extinction and aperture corrections, (*d*) match the *g* and *i*-band catalogs and analyze colours, (*e*) obtain a separation between compact objects in the galaxy and fore/background contaminants.

We compared our photometry against the various catalogs existing in the literature, and obtained in all cases satisfactory agreement.

Thanks to the large area covered, we succeeded in inspecting the properties of SSSs in NGC 3115 out to $\sim$23′, more than 20 times the galaxy effective radius, the largest projected distance ever reached to inspect the SSSs in the galaxy. This allowed us to adopt a statistical approach to study the properties of the GCs population versus R_{gal}, as follows. First, we estimated the surface density of objects satisfying some well-defined criteria (colour, magnitude, size, etc.) at different R_{gal}, and analyzed the properties of the residual over-density of the central galaxy regions with respect to the background regions.

Second, we analyzed the half-light radii R_h of the objects in the field of the galaxy as a further attribute to discriminate between SSSs in the potential well of NGC 3115 and contaminating objects. Overall, we obtained a catalog of $\sim$30,000 sources with both photometry and R_h estimates, including $\sim$6,000 objects with $R_h \leq 1.5$ pc as well as very extended sources having $R_h \geq 500$ pc, i.e. likely foreground stars and background galaxies.[1] For the few candidates with R_h available in the literature the matching with our measurement is on average satisfactory.

Some of the results of our study, based on the statistic decontamination are described in the following section. More details can be found in [6].

[1]The full catalog is available at vizier.cfa.harvard.edu/viz-bin/VizieR?-source=J/A%2BA/576/A14.

25.2 Results

GC Luminosity function We analyzed the luminosity function, LF, of sources in the field of NGC 3115, with the specific purpose of inspecting the GCs LF to (i) estimate the galaxy distance modulus μ_0 using the GCLF Turn-Over Magnitude (TOM; e.g. [11] and references therein), and (ii) to study how, and if, the TOM changes as a function of R_{gal}.

The left panel of Fig. 25.1 shows the histogram of the residual density Σ_{Host}[2] versus magnitude.

The density distribution has a peak at ~22.8 mag. Using the TOM calibration from the ACSVCS team we estimate $\mu_0 = 29.9 \pm 0.3$ mag, or 9.8±1.4 Mpc.

For what concerns the study of the properties of the TOM versus R_{gal}, to date the scarce literature on the topic shows both galaxies with and galaxies without a radial dependence of the TOM [11, 12]. For this galaxy we find no obvious variation of with R_{gal} (gray region in Fig. 25.1, left panel). The peak of luminosity of the GCLF is influenced by various physical factors. To first approximation, the simplest interpretation of a gradient in the peak luminosity of the GCLF is a correlation of the mean GCs mass with R_{gal}.

We further inspected how the TOM differs between red and blue GCs. After dividing the blue/red GCs adopting a sharp colour separation at $g{-}i = 0.9$ mag, we found a ~0.2 mag offset exists between the red and blue GCLF, with the red system being fainter. Indeed, from a stellar population viewpoint, such behavior is expected if the GCs mass function is universal across metallicity.

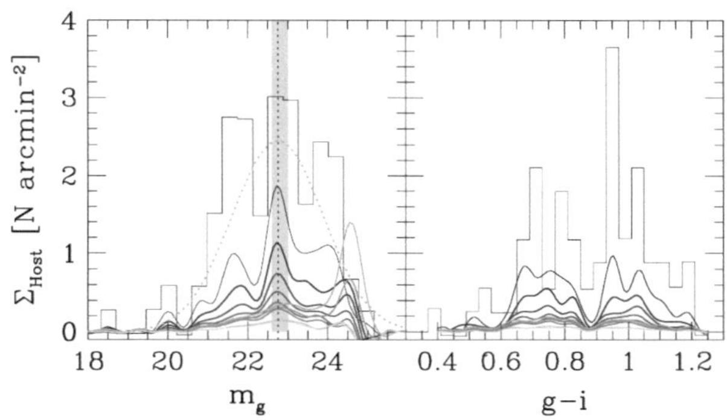

Fig. 25.1 Residual surface density versus magnitude (*left*) and colour (*right*). Darker colour refers to areas with smaller R_{gal}, starting from 2′ out to $R_{gal} = R_{bg} = 23'$. In the *left panel* the *thin dotted line*, *gray shaded area* and *green dotted gaussian curve* show the position of the TOM, the 1-σ error for the TOM, and the GCLF, respectively

[2]The residual density at various radii is obtained from the total density at fixed radius R_{gal}, after subtracting the density background sources at $R_{gal} \geq 23' \equiv R_{bg}$.

With the future VEGAS-SSS studies, we will further explore the dependence of TOM with the projected radii and the possible systematic differences between the blue and red GCLFs, for galaxies in different environments and with different masses.

GCs colour distribution The choice of NGC 3115 as first VEGAS-SSS target was also motivated by the fact that the galaxy is the first one beyond the Local Group with confirmed bimodal metallicity distribution [3, 5]. In the last decade, the nearly universal presence of two well-separated peaks in the optical colour distribution of GCs in ETGs, has inspired a vigorous and prolific debate. Historically, the bimodal GCs colour distribution in optical bands has been equated to metallicity [Fe/H] bimodality, implying a fundamental constraint on GCs and galaxy formation scenarios. This assumption became the subject of debate when various authors demonstrated that the presence of non-negligible nonlinearities in the colour-metallicity relations of GCs naturally produces bimodal colour histograms from non-bimodal [Fe/H] distributions [2, 4, 15].

Using the same approach adopted for the GCLF (i.e. by subtracting the background surface density to the total density at various radii), we analyzed the colour distribution of GC candidates. The results are shown in the right panel of Fig. 25.1. The plot shows (i) the presence of a dip at $g{-}i \sim 0.9$ mag with two well-defined peaks at $\sim$0.75 and 1.00 mag, and (ii) the dip and the peaks are seen at all inspected radii.

To investigate the properties of the GC color distributions at each given radius, we used the Gaussian mixture modelling, GMM, code [13]. GMM uses the likelihood-ratio test to compare the goodness of fit for double-Gaussians versus a single-Gaussian. It also provides the positions, relative widths, and fraction of objects associated with each peak.

Figure 25.2 shows positions (blue/red points), widths (bars), and fractions (given by symbol sizes) of the blue and red colour distributions obtained with GMM, as a function of R_{gal}. Various features appear: (i) a colour-R_{gal} correlation for the blue GC component, and no or only a very weak correlation for the red GCs (Pearson correlation coefficients $r_{xy}^{blue} \sim -0.8$, $r_{xy}^{red} \sim -0.25$); (ii) the fraction of red GCs has a slight but significant decrease with respect to the blue GCs at large radii; (iii) the width of the two sequences is relatively stable with radius, with the blue distribution being broader at all radii.

In the figure we also added data from the SLUGGS survey [1]. There is a good matching between the VEGAS-SSS and SLUGGS results. In particular, we highlight the very good matching with the colours of spectroscopically confirmed GCs (gray dots). The agreement appears even more striking if one takes into account that the SLUGGS data were obtained by coupling photometry and spectroscopy from the 8–10 m class telescopes Subaru and Keck-II.

Finally, we analyzed the radial profiles of the projected surface density of GC candidates inspecting the full sample, and the red and blue GCs separately. The results, shown in Fig. 25.3, highlight that the total, red and blue GCs follow very closely a de Vaucouleurs profile, and that the profile of red GCs is steeper than

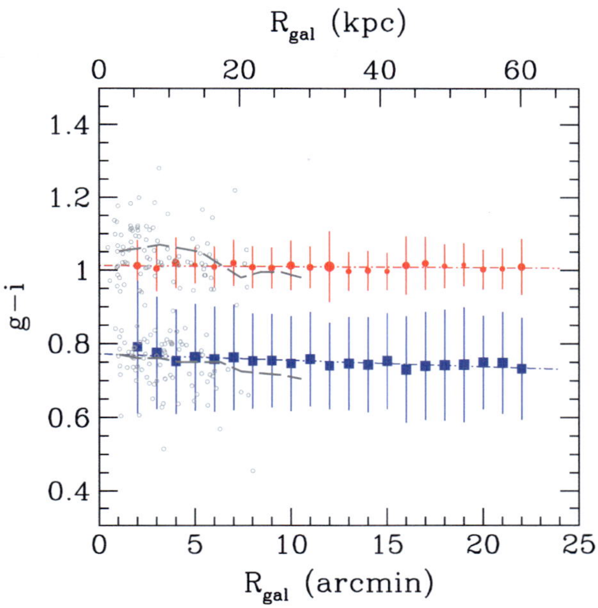

Fig. 25.2 Position and width of the *blue* and *red* GCs at different R_{gal}. Symbol size is proportional to the fraction of objects associated to each peak. A fit to the data is shown with *dot-dashed lines*. *Grey dots* show spectroscopic confirmed GCs, *gray long-dashed lines* mark the rolling fits of the *blue* and *red* GC peaks obtained from a combination of spectroscopically and photometrically selected GCs [1]

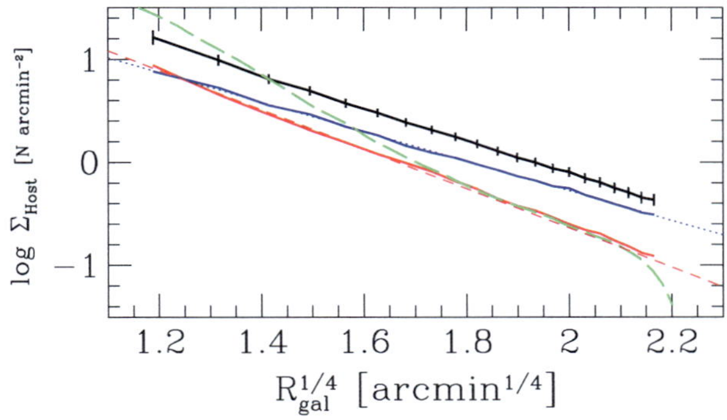

Fig. 25.3 Surface density profiles of *blue*, *red*, and total GC population (*blue*, *red* and *black line*, respectively). The *green long-dashed* shows the *g*-band surface brightness profile from [8] with arbitrary scale. The linear fit to the surface density is shown with *dotted lines*

that for the blue GCs while both are shallower than the galaxy profile. Moreover, the red GCs and the galaxy profile match at $R_{gal} \geq 7.5'$, while the GCs density is lower at smaller radii. Such depletion, already observed in other galaxies, has been interpreted with an higher efficiency of GC-disruption mechanisms in the inner galaxy regions by dynamical friction, two-body relaxation, and GC tidal shocking mechanisms [10].

Overall, the observed properties support a scenario where blue GCs are associated with the galaxy halo, while red ones are more centrally concentrated and associated with the bulge stellar component in the galaxy, suggesting that NGC 3115 has undergone a relatively quiescent evolution, without major star-forming events.

25.3 Future Perspectives

The final catalogs from VEGAS-SSS will contain accurate photometry, position and sizes for thousands of SSS candidates in the field of bright ETGs. The program will have a twofold legacy value. First, for preparing future observational studies, like near-IR imaging and spectroscopic follow-ups. Second, future analysis with the other VEGAS targets, and extended to other SSS classes, will have great value in studies of the evolution and transformation processes taking place in galaxies in a range of environments, as suggested by the results on the GCs population in NGC 3115 summarized here.

Acknowledgements M. Cantiello acknowledges support PO-FSE Abruzzo "Sistema Sapere e Crescita"

References

1. Arnold, J.A., Romanowsky, A.J., Brodie, J.P., et al.: The fossil record of two-phase galaxy assembly: kinematics and metallicities in the nearest S0 galaxy. ApJL **736**, L26 (2011)
2. Blakeslee, J.P., Cantiello, M., Peng, E.W.: The mass-metallicity relation of globular clusters in the context of nonlinear color-metallicty relations. ApJ **710**, 51 (2010)
3. Brodie, J.P., Usher, C., Conroy, C., et al.: The SLUGGS survey: NGC 3115, a critical test case for metallicity bimodality in globular cluster systems. ApJL **759**, L33 (2012). doi:10.1088/2041-8205/759/2/L33
4. Cantiello, M., Blakeslee, J.P.: On the Metallicity-color relations and bimodal color distributions in extragalactic globular cluster systems. ApJ **669**, 982 (2007)
5. Cantiello, M., Blakeslee, J.P., Raimondo, G., et al.: Globular clusters of NGC 3115 in the near-infrared. Demonstrating the correctness of two opposing scenarios. A&A **564**, L3 (2014)
6. Cantiello, M., Capaccioli, M., Napolitano, N., et al.: VEGAS-SSS. A VST early-type galaxy survey: analysis of small stellar systems. Testing the methodology on the globular cluster system in NGC 3115. A&A **576**, A14 (2015)
7. Capaccioli, M., Schipani, P.: The VLT survey telescope opens to the sky: history of a commissioning. The Messenger **146**, 2 (2011)
8. Capaccioli, M., Spavone, M., Grado, A., et al.: VEGAS: a VST early-type GAlaxy survey. I. Presentation, wide-field surface photometry, and substructures in NGC 4472. A&A **581**, A10 (2015)
9. Forbes, D., Pota, V., Usher, C., et al.: Filling the gap: a new class of old star cluster? MNRAS **435**, L6 (2013)

10. Goudfrooij, P., Schweizer, F., Gilmore, D., Whitmore, B.C.: Dynamical evolution of globular cluster systems formed in galaxy mergers: deep hubble space telescope advanced camera for surveys imaging of old and intermediate-age globular clusters in NGC 3610. AJ **133**, 2737 (2007)
11. Harris, W.E., Morningstar, W., Gnedin, O.Y., et al.: Globular cluster systems in brightest cluster galaxies: a near-universal luminosity function? ApJ **797**, 128 (2014)
12. Jordán, A., McLaughlin, D.E., Côté, P., et al.: The ACS Virgo Cluster Survey. XII. The luminosity function of globular clusters in early-type galaxies. ApJS **171**, 101 (2007)
13. Muratov, A.L., Gnedin, O.Y.: Modeling the metallicity distribution of globular clusters. ApJ **718**, 1266 (2010)
14. Tolstoy, E., Hill, V., Tosi, M.: Star-formation histories, abundances, and kinematics of dwarf galaxies in the local group. ARA&A **47**, 371 (2009)
15. Yoon, S.J., Yi, S.K., Lee, Y.W.: Explaining the color distributions of globular cluster systems in elliptical galaxies. Science **311**, 1129 (2006)

Chapter 26
Deep Photometry of Galaxies in the VEGAS Survey: The Case of NGC 4472

M. Spavone on behalf of the VEGAS team*

Abstract The VST-VEGAS project is aimed at observing and studying a rich sample of nearby early-type galaxies in order to systematically characterize their properties over a wide baseline of sizes and out to the faint outskirts where data are rather scarce so far. The external regions of galaxies more easily retain signatures about the formation and evolution mechanisms which shaped them, as their relaxation time are longer, and they are more weakly influenced by processes such as mergers, secular evolution, central black hole activity, and supernova feedback on the ISM, which tend to level age and metallicity gradients. The collection of a wide photometric dataset of a large number of galaxies in various environmental conditions, may help to shed light on these questions. To this end VEGAS exploits the potential of the VLT Survey Telescope (VST) which provides high quality images of 1 deg^2 field of view in order to satisfy both the requirement of high resolution data and the need of studying nearby, and thus large, objects. We present a detailed study of the surface photometry of the elliptical galaxy NGC4472 and of smaller ETGs in its field, performed by using new g and i bands images to constrain the formation history of this nearby giant galaxy, and to investigate the presence of very faint substructures in its surroundings.

26.1 The VEGAS Survey

The VST Elliptical GAlaxies Survey (VEGAS) is a deep multi-band (g, r, i) imaging survey of early-type galaxies in the Southern hemisphere carried out with VST at the ESO Cerro Paranal Observatory (Chile). The survey goal is to map the surface brightness of galaxies with $V_{rad} < 4{,}000$ km/s, sampling all environmental

*VEGAS team: M. Capaccioli (P.I.), Michele Cantiello, D. A. Forbes, A. Grado, E. Iodice, L. Limatola, N. Napolitano, M. Paolillo, T. H. Puzia, G. Raimondo, A. J. Romanowsky, P. Schipani & M. Spavone

M. Spavone (✉)
INAF-Astronomical Observatory of Capodimonte, Salita Moiariello 16, 80133, Naples, Italy
e-mail: spavone@na.astro.it

N.R. Napolitano et al. (eds.), *The Universe of Digital Sky Surveys*, Astrophysics and Space Science Proceedings 42, DOI 10.1007/978-3-319-19330-4_26

conditions and the whole parameter space. The expected depths at S/N $>$ 3 in the g, r and i bands are 27.3, 26.8, and 26 mag arcsec^{-2} respectively, enough to detect signatures of diffuse star components around galaxies (see e.g. [11]) and the dynamical interactions of ETGs with the intergalactic medium. The main aspects that the VEGAS survey will investigate are: (1) 2D light distribution out to 8–10 R_e: galaxy structural parameters and diffuse light component, inner substructures as a signature of recent cannibalism events, inner disks and bars fueling active nuclei present in almost all the objects of our sample; (2) radially averaged surface brightness profiles and isophote shapes up to 10 R_e; (3) color gradients and the connection with galaxy formation theories; (4) detection of external low-surface brightness structures of the galaxies and the connection with the environment; (5) census of small stellar systems (SSS: GCs, ultra compact dwarfs and galaxy satellites) out to $\sim$200 kpc from the main galaxy center, and their photometric properties (e.g. GC luminosity function and colors and their radial changes out to several R_e allowing to study the properties of GCs in the outermost "fossils" regions of the host galaxy. VEGAS will provide a volume limited survey in the South, complementary to the Next generation Virgo Cluster Survey (NGVS), with comparable depth but no environmental restrictions.

26.2 The NGC 4472 Field: A Test Case

This first VEGAS case [2] deals with a deep photometric analysis of the ETGs in the VST field of the galaxy NGC 4472 (M49), the brightest member of the Virgo cluster. This field has been chosen for the following reasons:

- it is well-studied, with an ample scientific photometric literature [5, 7–10];
- it offers a wide range of cases where to investigate the ability of VEGAS to map the faint galaxy outskirts. In fact, together with a nearby supergiant object filling almost the entire OmegaCAM field, there are smaller ETGs either embedded in the light of NGC 4472 or close to the edges of the frame.

26.2.1 Light and Color Distribution

We used the standard IRAF[1] task ELLIPSE to perform the isophotal analysis of the VEGAS galaxies on the final mosaic in each band, after proper background subtraction. We have also derived the azimuthally averaged surface brightness profiles in isophotal annuli of specified thickness. The azimuthally averaged profiles

[1]IRAF is distributed by the National Optical Astronomy Observatories, which is operated by the Associated Universities for Research in Astronomy, Inc. under cooperative agreement with the National Science Foundation.(*Image Reduction and Analysis Facility*) environment.

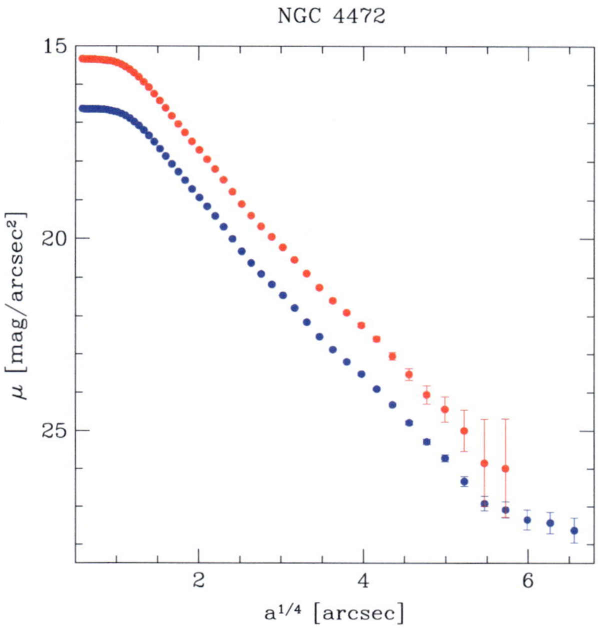

Fig. 26.1 Azimuthally averaged light profiles for NGC 4472 in the *g* (*blue*) and *i* (*red*) bands

for NGC 4472 in the *g* and *i* bands are shown in Fig. 26.1 as a function of the isophote semi major axis *a*.

The light profile in the *g* band, shows a neat change in the slope at $a_e^{1/4} \simeq 5.5$, where $\mu_g \sim 27\,\mathrm{mag\,arcsec}^{-2}$. The level at which the break occurs is compatible with the typical values at which [11] have observed a change of slope induced by the ICL in a series of stacked galaxy clusters. Moreover we note the presence of an outer and more elliptical component with a significant gradient in the P.A. which, as suggested by [6], is likely due to a population of some ICL.

Our azimuthally averaged *g* band profile is compared with the available literature in Fig. 26.2. The residuals with respect to $r^{1/4}$ fits show a spectacular agreement with the literature, and from the comparison with NGC3379 ([4], blue line) we note clear similarities due to the presence of diffuse shells in both systems.

We have also studied the fainter ETGs in the 1 deg^2 of the OmegaCAM field: NGC 4434, NGC 4464, NGC 4467, and VCC 1199, including the dwarf irregular, UGC 7636 in the proximity of the giant NGC 4472, reaching an even larger depth for these systems.

The (g-i) color profiles show an indication that for $r > 3r_e$ a very negative colour gradient develops in some galaxies, which apparently vanishes at $r \simeq 8r_e$ (see Fig. 26.3).

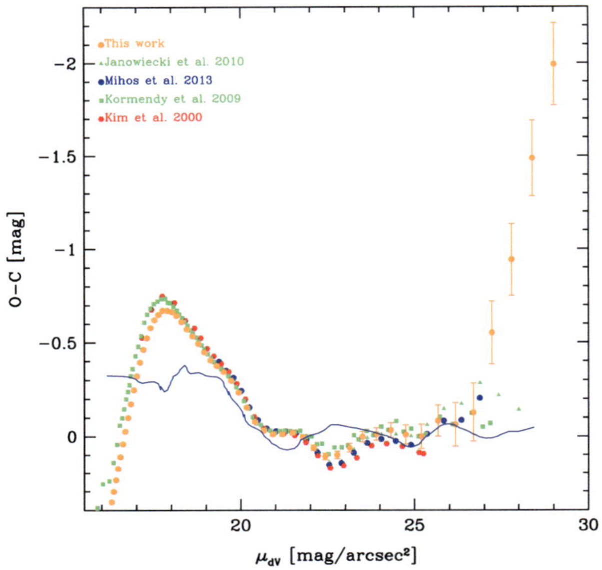

Fig. 26.2 (O-C) residuals of mean profiles from a best fitting $r^{1/4}$ model used only to remove the main gradient and make comparisons easier. The *blue solid line* plots the O-C residuals for the East-West photometric cross-section of the standard elliptical galaxy NGC3379 from [4]

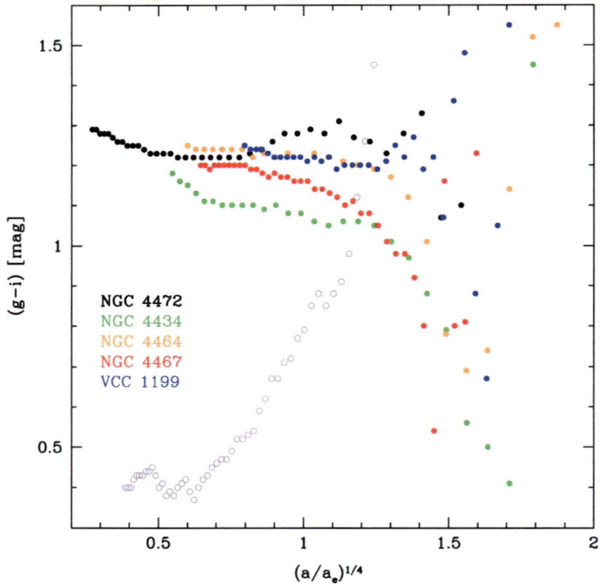

Fig. 26.3 *Left*-Azimuthally averaged light profiles in the *g* band for the five ETGs of this paper scaled to their effective parameters. *Right*-Assembly of the *(g-i)* color profiles for the five ETGs and for the interacting system UGC 7636

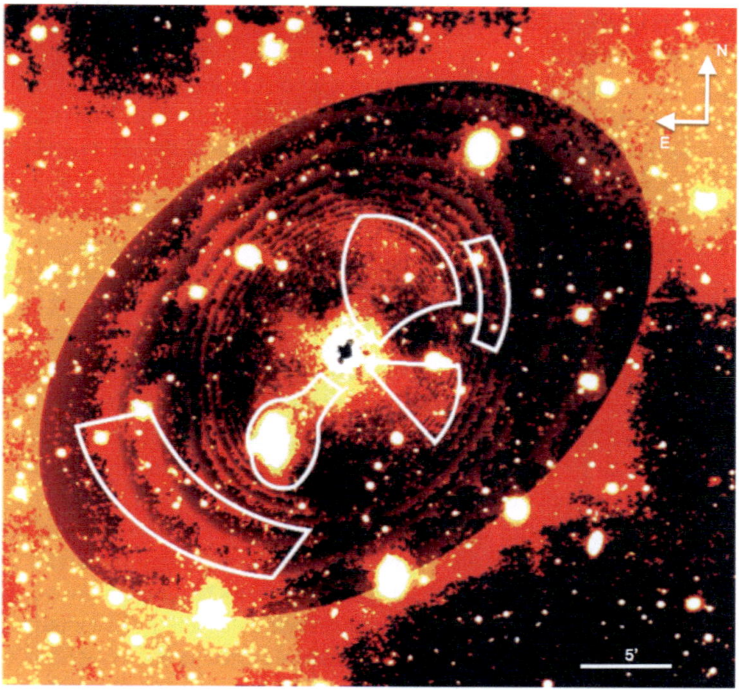

Fig. 26.4 NGC 4472. Zoom (37 × 35 arcmin) of the median-smoothed residual image. Superimposed white contours are (1) the tail connecting UGC 7636 to the giant ETG, and (2) shells and fan of material identified by [1] and visible in our residual image

26.2.2 2-Dimensional Model

In order to enlighten possible larger substructures, a 2-dimensional elliptical model of NGC4472 best fitting the azimuthally averaged isophotes has been produced using the IRAF task BMODEL. Figure 26.4 plot the difference between the original *g* band image and the model. This residual map shows a clear asymmetry in the nuclear region and some diffuse features such as a tail associated with the dwarf irregular galaxy UGC 7636 interacting with NGC 4472, as well as the presence of concentric shells and fans of material (white contours), also identified by [1].

26.3 Conclusions

We have presented the VST Early-type Galaxy Survey (VEGAS), currently ongoing with VST/OmegaCAM (PI: M. Capaccioli). In particular, we present the deep observations in two bands (*g* and *i*), collected with the VST/OmegaCAM, for NGC4472.

The surface brightness profiles of NGC 4472 reach a depth of 27.5 mag/arcsec^2 in g band and 26 mag/arcsec^2 in i band, comparably to previous deep studies (see Fig. 26.2). This depth allowed us to spot deviations from a simple de Vaucouleurs profile and in particular a change of slope at $a \sim 14'.2$ (see Fig. 26.1) that we have associated to the presence of a decoupled ICL component which was not detected in previous analyses.

Here we stress that the simple inspection of the deep surface brightness profile of NGC 4472, clearly shows the presence of a diffuse component starting to dominate at $\mu_g \sim 26.5$ mag/arcsec^2 (see Fig. 26.1), which is compatible with the typical values at which [11] have observed change of slope induced by the ICL in a series of stacked galaxy clusters.

We notice that the trend of the residuals of the luminosity profiles of NGC 4472 with respect to an $r^{1/4}$ best fitting model has some striking analogies with the similar curve for NGC 3379 [4]. Besides a bright extended core, we find evidence for a wavy pattern possibly associated with shells of diffuse material. The presence of such shells also stand out very clear from the 2-D residual map.

We have also studied the fainter ETGs in the 1 deg^2 of the OmegaCAM field: NGC 4434, NGC 4464, NGC 4467, and VCC 1199, including the dwarf irregular, UGC 7636 in the proximity of the giant galaxy NGC 4472. For all these systems we have highlighted the presence of some substructures defined as deviations from a simple de Vaucouleurs [3] best fit profile. These deviations are associated to strong varying values of the ellipticity and P.A. as well as a_4 and b_4 parameters, suggesting the presence of some substructures.

The color profiles show an indication that for $r > 3r_e$ a very negative colour gradient develops in some galaxies, which apparently vanishes at $r \simeq 8r_e$. In a forthcoming paper [2] we will show that our results are basically unaffected in their qualitatively conclusions by the extended wings of the PSF (scattered light).

To conclude, this paper illustrates the performance and the accuracy achieved with the VST/OmegaCAM to produce surface photometry of early-type galaxies, also in very extreme conditions. For the case of NGC 4472 the presence of an extended halo around the giant galaxy, reaching the edge of the 1 deg^2 field of view, has allowed us to fully test the procedure for data reduction and background subtraction. The results obtained with our observations are comparable for accuracy to the collection of observations gathered from different telescopes (see [9]).

In the future we expect to implement a more variegate surface analysis including a wider set of photometric law in order to characterize the SB measurements in a larger sample of galaxies and thus discuss results in the context of galaxy formation theories.

References

1. Arrigoni Battaia, F., Gavazzi, G., Fumagalli, M., Boselli, A., Boissier, S., Cortese, L., et al.: Stripped gas as fuel for newly formed H ii regions in the encounter between VCC 1249 and M 49: a unified picture from NGVS and GUViCS. A&A **543**, AA112 (2012)
2. Capaccioli, M., Spavone, M., Grado, A., et al.: VEGAS: a VST Early-type GAlaxy Survey. I. Presentation, wide-field surface photometry, and substructures in NGC 4472. A&A **581**, A10 (2015)
3. de Vaucouleurs, G.: Recherches sur les Nebuleuses Extragalactiques. Annales d'Astrophysique **11**, 247 (1948)
4. de Vaucouleurs, G., Capaccioli, M.: Luminosity distribution in galaxies. I – the elliptical galaxy NGC 3379 as a luminosity distribution standard. ApJS **40**, 699 (1979)
5. Ferrarese, L., Côté, P., Jordán, A., Peng, E.W., Blakeslee, J.P., Piatek, S., et al.: The ACS Virgo cluster survey. VI. Isophotal analysis and the structure of early-type galaxies. ApJS **164**, 334 (2006)
6. Gonzalez, A.H., Zabludoff, A.I., Zaritsky, D.: Intracluster light in nearby galaxy clusters: relationship to the halos of brightest cluster galaxies. ApJ **618**, 195 (2005)
7. Janowiecki, S., Mihos, J.C., Harding, P., Feldmeier, J.J., Rudick, C., Morrison, H.: Diffuse tidal structures in the halos of Virgo ellipticals. ApJ **715**, 972 (2010)
8. Kim, E., Lee, M.G., Geisler, D.: Wide-field CCD surface photometry of the giant elliptical galaxy NGC 4472 in the Virgo cluster. MNRAS **314**, 307 (2000)
9. Kormendy, J., Fisher, D.B., Cornell, M.E., Bender, R.: Structure and formation of elliptical and spheroidal galaxies. ApJS **182**, 216 (2009)
10. Mihos, J.C., Harding, P., Rudick, C.S., Feldmeier, J.J.: Stellar populations in the outer halo of the massive elliptical M49. ApJL **764**, L20 (2013)
11. Zibetti, S., White, S.D.M., Schneider, D.P., Brinkmann, J.: Intergalactic stars in z$\sim$ 0.25 galaxy clusters: systematic properties from stacking of sloan digital sky survey imaging data. MNRAS **358**, 949 (2005)

Chapter 27
Shapley Supercluster Survey

P. Merluzzi, G. Busarello, C.P. Haines, A. Mercurio, A. Grado, L. Limatola, and K.J. Pimbblett

Abstract Our multi-wavelength survey of the Shapley supercluster ($z \sim 0.05$) covers a contiguous area of $260\,h_{70}^{-2}\,\mathrm{Mpc}^2$ including the supercluster core. The project's main aim is to quantify the influence of cluster-scale mass assembly on galaxy evolution in one of the most massive structures in the local Universe. The Shapley supercluster survey (ShaSS) includes nine Abell clusters (A 3552, A 3554, A 3556, A 3558, A 3559, A 3560, A 3562, AS 0724, AS 0726) and two poor clusters (SC 1327-312, SC 1329-313) showing evidence of cluster-cluster interactions. Optical (*ugri*) and near-infrared (*K*) imaging acquired with VST and VISTA allow us to study the galaxy population down to $\mathrm{m}^{\star}+6$ at the supercluster redshift. A dedicated spectroscopic survey with AAOmega on the Anglo-Australian Telescope provides a magnitude-limited sample of supercluster members with 80 % completeness at $\sim\mathrm{m}^{\star}+3$.

27.1 Survey Objectives and Target Choice

In a Lambda-CDM universe in which structures assemble hierarchically, the galaxies evolve but also move, tending towards denser regions with time, while the environments change too, thus what we actually observe is *galaxy evolution in an evolving environment.*

The prototypes of evolving environments are the superclusters, where cluster-cluster collisions and group-cluster mergers occur, and different environments from

P. Merluzzi (✉) • G. Busarello • A. Mercurio • A. Grado • L. Limatola
INAF – Osservatorio Astronomico di Capodimonte, Salita Moiariello 16, I-80131 Napoli, Italy
e-mail: merluzzi@na.astro.it

C.P. Haines
Departamento de Astronomía, Universidad de Chile, Casilla 36-D, Correo Central, Santiago, Chile
e-mail: cphaines@das.uchile.cl

K.J. Pimbblet
Department of Physics and Mathematics, University of Hull, Cottingham Road, Kingston-upon-Hull, HU6 7RX, UK
e-mail: k.pimbblet@hull.ac.uk

N.R. Napolitano et al. (eds.), *The Universe of Digital Sky Surveys*, Astrophysics and Space Science Proceedings 42, DOI 10.1007/978-3-319-19330-4_27

cluster cores to filaments and the field coexist. Furthermore, within a dynamically active and locally dense structure the probability of observing evidence of environmental effects on galaxy evolution is dramatically enhanced, making these systems a sort of *magnifying glass* to identify the different physical mechanisms which transform the properties of galaxies. Finally, in order to study in detail the galaxy properties a resolution $\lesssim$1 kpc is required, across a wide range of galaxy masses, extending down to the dwarf regime where such galaxies are not quenched by internal processes, but are more susceptible to environmental transformations. With all this in mind, we have undertaken a study of the Shapley supercluster which is the largest conglomeration of Abell clusters in the local Universe.

The Shapley Supercluster Survey (ShaSS) will map a 23 deg^2 region ($\sim$260 Mpc^2) of the Shapley supercluster at $z = 0.048$, containing filaments and embedded galaxy groups, in order to identify the primary locations (groups, filaments, clusters) and mechanisms for the transformation of spirals into the S0s and dEs.

Our study has the following main objectives.

- To investigate the role of cluster-scale mass assembly on the evolution of galaxies, mapping its effects in the cluster outskirts and along the filaments with the aim of identifying the very first interactions between galaxies and their environment.
- To identify and measure signs of ongoing transformation in galaxies belonging to a complex structure with the goal of improving our comprehension of what drives their star-formation quenching and structural modification.
- To obtain detailed maps of the dark matter and baryonic matter distributions (galaxies, intra-cluster medium, ICM), combining weak lensing, X-ray and dynamical analyses.
- To quantify the variation in the stellar mass fractions going from cluster cores to groups, by comparing the near-infrared light distribution with the dark matter maps and dynamical masses.
- To build up a multi-band homogeneous data-set on this area of the sky, combining sub-kiloparsec resolution imaging and magnitude-limited spectroscopy, thus providing the community with a solid background for studies of the Shapley supercluster.

To address the above objectives we will explore the global and internal properties of galaxy populations extending outside the cluster/group virial radius and aim for an accurate characterization of the environment. This will be defined through galaxy density, dark matter distribution, dynamical substructure, and ICM properties. The different quantifications of the environment will allow us to disentangle the effects of local and large-scale density, cluster and group merging, dynamical state and mass of the host systems on the properties of galaxies in different ranges of mass (for details see [1] and references therein).

We will map a region of $\sim$260 Mpc^2 including the Shapley supercluster core (SSCC) and other six galaxy clusters, see Fig. 27.1). The supercluster region is chosen to ensure to map the structures directly connected to the SSCC.

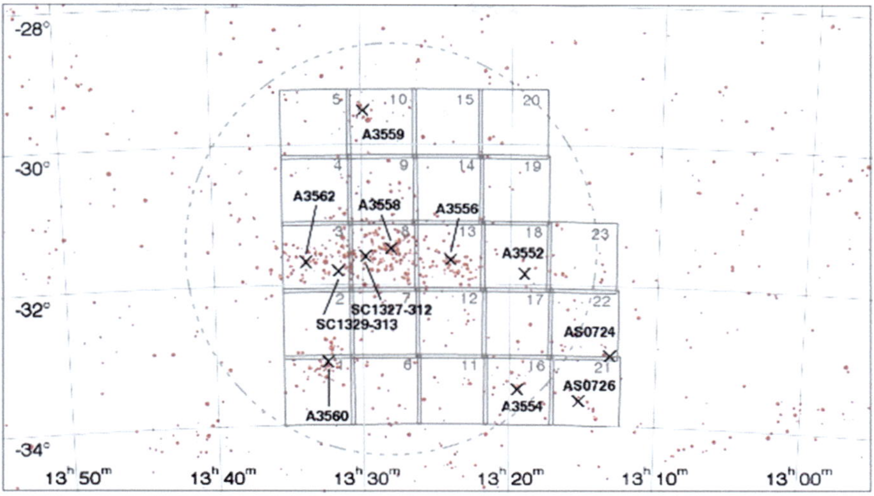

Fig. 27.1 The 23 1 deg^2 VST fields mapping the ShaSS region. *Red dots* indicate the supercluster members in the range $V_h = 11{,}300$–$17{,}000\,\mathrm{km\,s^{-1}}$ taken from literature. The size of the dots are proportional to the K-band flux. *Black crosses* show the cluster centres. The 10 Mpc radius *dotted circle* encloses the supercluster region believed to be dynamically bound. The SSCC corresponds to fields #3,8,13. The positions of all structures present in the plotted area are indicated

The data-set includes optical (*ugri*) and NIR (K) imaging acquired with VST and VISTA respectively, and optical spectroscopy with AAOmega.

The optical and near-IR data allow us to study the global and internal properties of the supercluster galaxies down to $m^\star + 6$. For the morphological analysis we will use the r-band imaging collected with seeing ~0.6 arcsec (0.75 kpc at $z \sim 0.05$). Near- and mid-IR (3.4–22 μm) imaging from the WISE satellite is also available for the whole area reaching a depth corresponding to $\sim \mathrm{m}^\star_{3.4\,\mu\mathrm{m}} + 5$ for supercluster galaxies. The spectroscopic survey carried out with AAOmega spectrograph at the 3.9 m Anglo Australian Telescope provided us with a final sample which is 80 % complete down to $i = 17.6$ mag and a stellar mass $\mathscr{M}_\star \sim 8.7 \times 10^9\,\mathrm{M}_\odot$.

27.2 Galaxy Density

To map the structure of the supercluster, and determine its extent in redshift space and across the plane of the sky, we take advantage of our redshift survey which allows us to demarcate the supercluster in redshift space as lying within the recession velocities 11,300–17,000 $\mathrm{km\,s^{-1}}$ selecting a supercluster sample of 2,281 galaxies across the ShaSS area. Figure 27.2 shows the resulting density map in which each galaxy is further weighted by its 3.4 μm flux as a proxy for its stellar mass (for details see [1]).

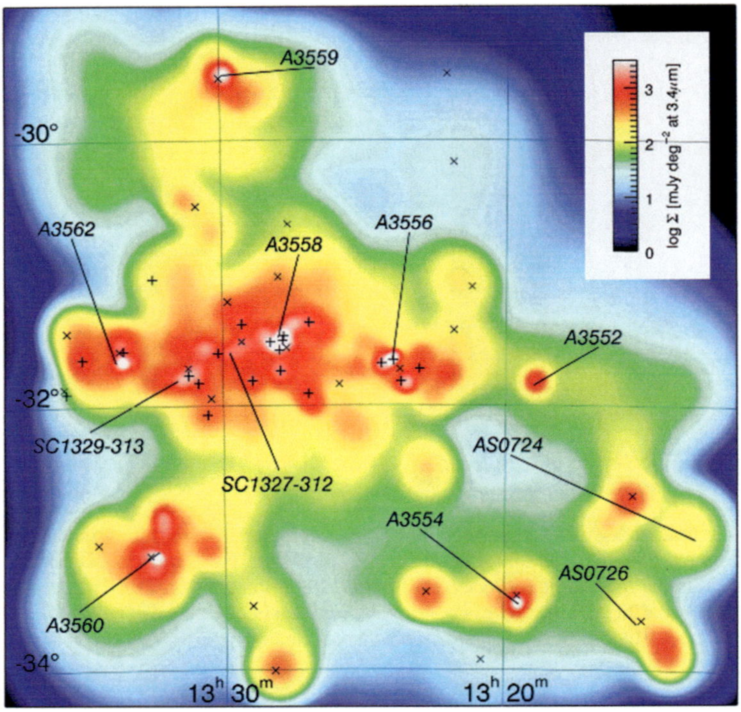

Fig. 27.2 ShaSS density map in units of mJy per square degree at 3.4 μm. Abell clusters and groups are labelled, the *black straight lines* pointing to the X-ray centre for all the systems except AS 0726 and A 3559. Cluster substructures and groups identified by previous studies (+ and × symbols) are shown in the map. The *upper right corner* is not covered by ShaSS

The map shows a clumpy structure both in the SSCC and the surrounding clusters with several substructures, most of them already identified in previous works. All the clusters in the ShaSS area are embedded in a common network. We estimate the mean overdensity across the SSCC being ∼40× with respect to the cosmic total stellar mass density for galaxies in the local Universe ($z < 0.06$).

Some new substructures with respect to previous works have been identified in the ShaSS density map. The most important new feature is however the 7 Mpc filament connecting the SSCC and the cluster A 3559 as well as the less pronounced overdensity extending from the SSCC towards A 3560. Both of them support a scenario of dynamical interaction among the clusters.

Reference

1. Merluzzi, P., Busarello, G., Haines, C.P., et al.: Shapley supercluster survey: galaxy evolution from filaments to cluster cores. MNRAS **446**, 803–822 (2015)

Chapter 28
The Wide-Field Nearby Galaxy-Cluster Survey (WINGS) and Its Extension OMEGAWINGS

B.M. Poggianti, G. Fasano, D. Bettoni, A. Cava, W. Couch, M. D'Onofrio, A. Dressler, J. Fritz, P. Kjaergaard, M. Gullieuszik, M. Moles, A. Moretti, A. Omizzolo, A. Paccagnella, J. Varela, and B. Vulcani

Abstract WINGS is a wide-field multi-wavelength survey of 76 X-ray selected clusters at low redshift. The WINGS database has been used for a variety of cluster and cluster galaxy studies, investigating galaxy star formation, morphologies, structure, stellar mass functions and other properties. We present the recent wider-

---

B.M. Poggianti (✉)
INAF-Astronomical Observatory of Padova, Padova, Italy
e-mail: bianca.poggianti@oapd.inaf.it

G. Fasano • D. Bettoni • M. Guilleuszik • A. Paccagnella • A. Omizzolo
INAF-Astronomical Observatory of Padova, Padova, Italy

A. Moretti
INAF-Astronomical Observatory of Padova, Padova, Italy

Department of Physics and Astronomy, University of Padova, Padova, Italy

M. D'Onofrio • A. Paccagnella
Department of Physics and Astronomy, University of Padova, Padova, Italy

A. Omizzolo
Specola Vaticana, Vatican State, Vatican

A. Cava
Observatoire de Geneve, Versoix, Switzerland

W. Couch
Australian Astronomical Observatory, North Ryde, Australia

A. Dressler
Carnegie Institute, Pasadena, CA, USA

J. Fritz
CRyA, UNAM, Morelos, Mich., Mexico

P. Kjaergaard
Niels Bohr Institute, Copenhagen, Denmark

M. Moles • J. Varela
CEFCA, Teruel, Spain

B. Vulcani
IPMU, Tokyo, Japan

N.R. Napolitano et al. (eds.), *The Universe of Digital Sky Surveys*, Astrophysics and Space Science Proceedings 42, DOI 10.1007/978-3-319-19330-4_28

field extension of WINGS, OMEGAWINGS, conducted with OmegaCAM@VST and AAOmega@AAT. We show two of our latest results regarding jellyfish galaxiesİ and galaxy sizes. OMEGAWINGS has allowed the first systematic search of galaxies with signs of ongoing ram pressure stripping (jellyfishes), yielding a catalog of ∼240 galaxies in 41 clusters. We discuss the first results obtained from this sample and the prospects for integral field data. Finally, we summarize our results regarding the discovery of compact massive galaxies at low redshift, their properties, dependence on environment and the implications for the evolution of galaxy sizes from high- to low-z.

28.1 Introduction

WINGS is a large survey targeting 76 clusters of galaxies selected on the basis of their X-ray luminosity [2–4], covering a wide range in cluster masses (velocity dispersion 500–1,200 + km s^{-1} and X-ray luminosity log $L_X = 43.3$–$44.7\,\mathrm{erg\,s^{-1}}$). The original WINGS dataset [6] consisted of B and V deep photometry on a $34' \times 34'$ field-of-view with the WFT@INT and the WFC@2.2mMPG/ESO, with spectroscopic follow-ups for a subset of clusters done with 2dF@AAT and WYFFOS@WHT, J and K imaging with WFC@UKIRT and some U-band taken with several telescopes. This database is presented in Moretti et al. [11], together with all the catalogs of data products made publicly available, which include galaxy morphology and structural parameters, star formation histories, galaxy masses and spectral measurements. A list of the WINGS publications can be found at http://web.oapd.inaf.it/wings/pub_wings.html.

OMEGAWINGS is a recent extention of this project, that quadruples the area covered by the optical imaging and allows to reach up to 2.5 cluster virial radii. The core of this project are two GTO programs with OmegaCAM@VST for 45 WINGS clusters: a B and V Omegacam GTO campaign completed in P93 and an ongoing u-band VST GTO programme. The B and V data, the data reduction and the photometric catalogs are presented in Gullieuszik et al. [9]. Spectra are being obtained with AAOmega@AAT on the $1 \times 1\,\mathrm{deg}^2$ OmegaCAM field. With the 18 nights of observations obtained so far we have secured high quality spectra for about 30 OMEGAWINGS clusters, reaching very high levels of spectroscopic completeness (typically >90 %) for all galaxies brighter than V $=$ 20 from the cluster cores to their periphery. Star formation histories, galaxy stellar masses and other stellar population properties have been obtained from the spectra with our full spectral fitting tool Fritz et al. [7]. The new dataset is being employed for a number of studies, and in this contribution we present two of our most recent results.

28.2 Jellyfish Galaxies

The so-called "jellyfishes galaxies" are galaxies that exhibit "tentacles" of material that appear to be stripped from the galaxy, most likely due to ram pressure stripping (e.g. [5, 8, 10]). We visually inspected OMEGAWINGS B images and identified ~240 jellyfish candidates in 41 clusters (two examples are shown in Fig. 28.1, Poggianti et al. [14]). This is the first systematic search for such signatures in clusters at low redshift, and the first large sample that can be used for follow-up studies of the gas and for a detailed analysis of the cluster properties and the location within clusters where such episodes of gas stripping occur.

An analysis of the star formation activity in these galaxies reveals that the majority are vigorously forming stars, at a rate that is typically enhanced by a factor of 2 compared to non-jellyfish galaxies of similar mass. The few post-starburst and passive jellyfishes identified have weak jellyfish signatures (weak trails/tentacles), indicating that our selection preferentially singles out galaxies in their initial stripping phases.

As jellyfishes in nature are found in every ocean and every sea, jellyfish galaxies are found in essentially all clusters we have inspected and their number does not

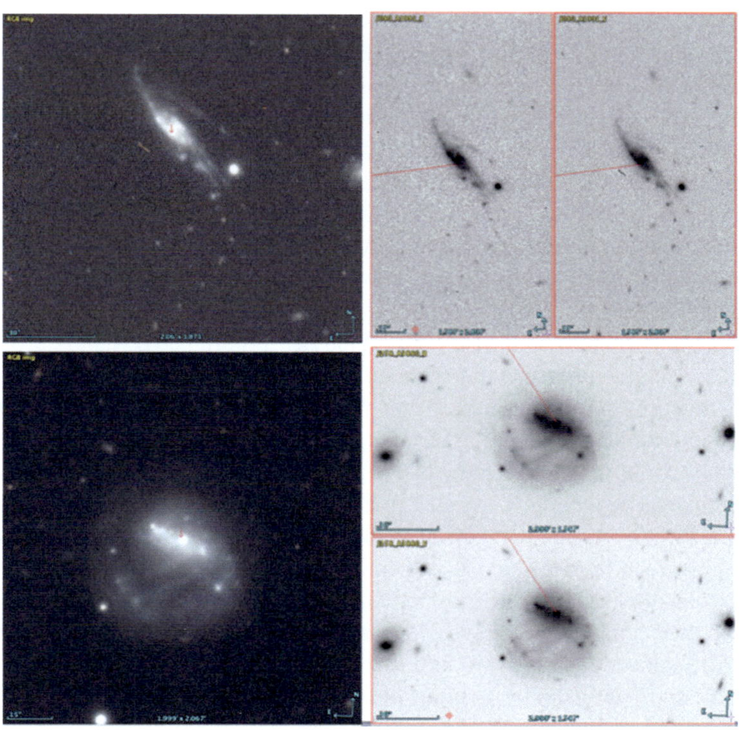

Fig. 28.1 Examples of jellyfish galaxies in WINGS clusters [14]

seem to correlate with cluster global properties, such as velocity dispersion and X-ray luminosity. They can be numerous even in 500 km s^{-1} clusters.

MUSE data of two of our jellyfishes recently acquired beautifully shows the emission lines (NII, Hα and SII) associated with the ionized gas in the trails, out to several tens of kpc from the galaxy. This gas is ionized from the massive stars formed in the stripped gas that are well visible in the optical images (Fig. 28.1). The star-formation knots in the tentacles resemble those observed in ram pressure simulations (e.g. [16]). The ongoing analysis of these data will clarify the amount of gas stripped, the amount of stars formed and injected into the intracluster medium, as well as the detailed spatially resolved map of the parent galaxy.

28.3 Galaxy Sizes and Compact Massive Galaxies at Low Redshift

Several authors have found a population of high-z ($z = 1$–3), massive (stellar masses $M > 10^{10}$–10^{11} $M_{\odot}$) galaxies with small sizes, thus "compact" (effective radii of a few kpc or less), usually selected to be already devoid of star formation (passive) at the redshift they are observed. Searches for similarly massive and compact galaxies in the SDSS found very few such objects, raising the question whether any of the high-z compact massive galaxies have survived until today and suggesting a strong evolution in size, due for example to minor merging.

In the WINGS clusters we found a significant population of massive compact galaxies, whose masses and sizes resemble those of the high-z population [17, 18]. In the field at the WINGS redshifts, compact galaxies are three times less common than in clusters, pointing to a strong environmental variation [13]. Both in clusters and in the field, compact galaxies are mostly S0s and have old stellar population ages, compatible with being the undisturbed descendants ("relics") of the high-z compact galaxies.

These works of ours have also highlighted the importance of taking progenitors biases into account when trying to measure the evolution of individual galaxy sizes. In fact, at a given mass, galaxies that have stopped forming stars sooner have on average smaller sizes. As time goes by, more and more galaxies are added to the population of passive galaxies. Since these have average larger sizes than older galaxies, the median mass-size relation of the passive population appears to evolve, without necessarily implying an evolution in size of those galaxies that were already passive at high-z [13, 17]. When this effect is correctly taken into account, we find that the evolution of the median-mass size relation is very mild, only a factor 1.6 between high- and low-z. In addition, comparing our field results with the CANDELS data at z = 1–3, we find that also the evolution of the number density of massive compact galaxies is mild: *at most* 50 % of the high-z compact galaxies have evolved in size by z = 0 (Fig. 28.2, [14]).

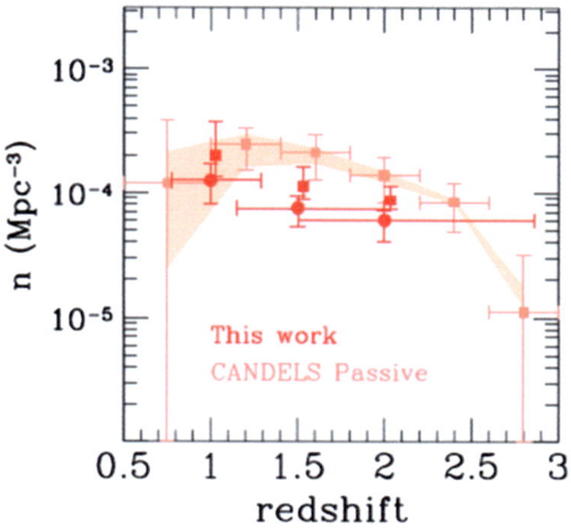

Fig. 28.2 Number density of compact quiescent galaxies [14]. CANDELS data (*pink*, [1]) are compared with the number of low-z PM2GC compact galaxies with luminosity-weighted ages compatible with being passive at z = 1, 1.5 and 2 (*red*, *circles* for Bruzual and Charlot 2003 masses and *rectangles* for Maraston 2005 masses). The number density evolves by at most a factor of 2 between high- and low-z

Moreover, our simulations (Millennium Simulation + semianalytic models) predict that 60 % of all galaxies that at z = 2 are already massive ($M > 10^{11}\,M_{\odot}$) and passive end up in haloes with masses larger than $10^{14}\,M_{\odot}$ at low-z, thus in WINGS-like clusters, and in 2 out of 3 cases become the Brightest Cluster Galaxy (BCG) [13]. Thus, studying massive galaxies at high-z, especially if passive, means to a large extent studying the progenitors of today's cluster galaxies. It is thus no surprise to find a difference with environment in the incidence of compacts at low-z. Furthermore, the evolution of BCGs seems to be markedly different from that of the general massive galaxy population. BCGs do evolve strongly, both in size and in mass, as found by us and several other authors (e.g. [18]).

Finally, an IFU study of 18 of our compact massive WINGS galaxies with GMOS@Gemini and VIMOS@VLT finds large dynamical masses, compact sizes and old ages confirming the compact nature of our cluster relics [12].

Acknowledgements I would like to thank the organizers for the great job they have done, and Massimo Capaccioli and the whole team that made VST a success and, therefore, our research possible.

References

1. Barro, G., et al.: CANDELS: the progenitors of compact quiescent galaxies at z ∼ 2. ApJ **765**, 104 (2013)
2. Ebeling, H., et al.: Properties of the X-ray-brightest Abell-type clusters of galaxies (XBACs) from ROSAT All-Sky survey data – I. The sample. MNRAS **281**, 799 (1996)
3. Ebeling, H., et al.: The ROSAT brightest cluster sample – I. The compilation of the sample and the cluster log N-log S distribution. MNRAS **301**, 881 (1998)
4. Ebeling, H., et al.: The ROSAT brightest cluster sample – IV. The extended sample. MNRAS **318**, 333 (2000)

5. Ebeling, H., Stephenson, L.N., Edge, A.C.: Jellyfish: evidence of extreme Ram-pressure stripping in massive galaxy clusters. ApJL **781**, L40 (2014)
6. Fasano, G., et al.: WINGS: a WIde-field nearby galaxy-cluster survey. I. Optical imaging. A&A **445**, 805 (2006)
7. Fritz, J., et al.: WINGS-SPE. III. Equivalent width measurements, spectral properties, and evolution of local cluster galaxies. A&A **566**, 32 (2014)
8. Fumagalli, M., Fossati, M., Hau, G., Gavazzi, G., Bower, R., Sun, M., Boselli, A.: MUSE sneaks a peek at extreme ram-pressure stripping events – I. A kinematic study of the archetypal galaxy ESO137-001. MNRAS **445**, 4335 (2014)
9. Gullieuszik, M., et al.: OmegaWINGS: OmegaCAM-VST observations of WINGS galaxy clusters. A&A **581**, 41 (2015)
10. Merluzzi, P., et al.: ACCESS – V. Dissecting ram-pressure stripping through integral-field spectroscopy and multiband imaging. MNRAS **429**, 1747 (2013)
11. Moretti, A., et al.: WINGS data release: a database of galaxies in nearby clusters. A&A **564**, 138 (2014)
12. Moretti, A., et al.: (2016, in prep.)
13. Poggianti, B.M., et al.: Superdense galaxies and the mass-size relation at low redshift. ApJ **762**, 77 (2013)
14. Poggianti, B.M., et al.: The evolution of the number density of compact galaxies. ApJ **777**, 125 (2013)
15. Poggianti, B.M., et al.: AJ, Jellyfish galaxy candidates at low redshift. AnJ (2015, in press). arXiv 1504.07105
16. Roediger, E., Bruggen, M., Owers, M.S., Ebeling, H., Sun M.: Star formation in shocked cluster spirals and their tails. MNRAS **443**, L114 (2014)
17. Valentinuzzi, T., et al.: Superdense massive galaxies in wings local clusters. ApJ **712**, 226 (2010)
18. Valentinuzzi, T., et al.: Superdense massive galaxies in the ESO distant cluster survey (EDisCS). ApJL **721**, L19 (2010)

Chapter 29
The Properties of Faint Galaxies in Nearby Clusters of the WINGS Sample

D. Bettoni, P. Kjærgaard, B. Milvan-Jensen, M. D'Onofrio, A. Moretti, B.M. Poggianti, G. Fasano, A. Cava, W. Couch, J. Fritz, and M. Moles

Abstract We present the results of our X-shooter observations for a sample of dwarf ($-17 < M_B < -15$) galaxies in nearby ($0.040 < z < 0.068$) galaxy clusters of the WINGS sample. The study of galaxies in this faint luminosity range is fundamental to trace the evolution of high-z star-forming cluster galaxies down to the present day. We measure the velocity dispersion of 22 galaxies in this range of luminosity and we explore their scaling relations. We found that the Fundamental Plane has a peculiar feature, suggesting the existence of some kind of warping at low luminosities.

D. Bettoni (✉)
INAF – Osservatorio Astronomico di Padova, Padova, Italy
e-mail: daniela.bettoni@oapd.inaf.it

P. Kjærgaard
Niels Bohr Institute, Univeristy of Copenhagen, Copenhagen, Denmark

B. Milvan-Jensen
Dark Cosmology Centre, Niels Bohr Institute, University of Copenhagen, Copenhagen, Denmark

M. D'Onofrio • A. Moretti
Dipartimento di Fisica e Astronomia Universita di Padova, Padova, Italy

B.M. Poggianti • G. Fasano
INAF – Osservatorio Astronomico di Padova, Padova, Italy

A. Cava
Observatoire de Geneve, Universite de Geneve, Geneva, Switzerland

W. Couch
Australian Astronomical Observatory, North Ryde, NSW, Australia

J. Fritz
Centro de Radioastronomia y Astrofisica, UNAM, Morelia, Mexico

M. Moles
Centro de Estudios de Fisica del Cosmos de Aragon, Teruel, Spain

N.R. Napolitano et al. (eds.), *The Universe of Digital Sky Surveys*, Astrophysics and Space Science Proceedings 42, DOI 10.1007/978-3-319-19330-4_29

29.1 Introduction

Faint galaxies in nearby clusters have been scarcely studied in terms of their luminosity weighted ages and metallicities. Faint cluster galaxies with low luminosity weighted ages (<3 Gyr) showed evidence of a bimodal metallicity distribution [18]. If confirmed, this could indicate that dwarf galaxies are a diverse population, with the metal-poor dwarfs being the primordial population, and the metal rich dwarfs perhaps being remnants of more massive, stripped galaxies. Moreover, a significant fraction of the faint galaxies in the Coma cluster have been discovered to be post-starburst galaxies in which the star formation activity has been abruptly truncated during the last Gyr [19].

A complementary approach to study the evolution of galaxies in clusters is to analyze the scaling relations of galaxies of different masses. The Fundamental Plane (FP) [7, 11, 14] is a remarkably tight relation among size, surface brightness and velocity dispersion of early-type galaxies. Although the FP itself seems to be quite flat and tight for early-type galaxies over a wide range of luminosities [1], there are a number of indications that the slopes of the FP projections [12, 15] could change in both their bright and faint tails [2, 6, 17]. In particular, the face-on view of the FP shows that dwarfs Ellipticals (dE) are located in a disjoint area from bright E galaxies. Recently [9] studied the FP of the non-dwarfs in the WINGS (Wide-field Nearby Galaxy-cluster Survey) survey [13]. Although the luminosity of this galaxy sample does not extend below $M_V \sim -17.5$, they confirm the indication of an increasing slope both in the low mass region of the correlation between dynamical mass and M/L ratio and in the faint tail of the FP. To extend the previous study beyond $M_V \sim -17.5$, (a critical luminosity as far as the evolution and the star formation history of early-type galaxies are concerned) we have undertaken a high resolution spectroscopic study with X-shooter of a sample of faint galaxies selected from WINGS and here we present our results.

29.2 Observations

The sample of targets was derived from the whole WINGS database. From the galaxies with measured redshift [4] (to ensure membership) and a surface brightness μ_v <19.5 mag/arcsec^2 we select objects with $-16 < M_B < -15$ and $0.040 < z < 0.060$, or $-17 < M_B < -16$ and $0.06 < z < 0.068$. The final list is of 45 objects of which we could observe 22.

Observations were performed in visitor mode using the XSHOOTER spectrograph [8] on the ESO VLT-UT2 (Kueyen). Exposure time was of ~1 h per galaxy, which allows us to reach a signal-to-noise ratio, for resolution element, in the continuum from ~7 (UVB-arm) to ~30 (VIS-arm) depending on the echelle order. The targets were observed with the $11'' \times 1''.0$ slit in the UVB-arm and with $11'' \times 0''.9$ slit in the VIS and NIR-arms. Some standard stars of spectral types

G9III-K0III, for the measure of the velocity dispersion, were also observed. The spectral resolution R is ~5,100 over the UVB and NIR-arms, and ~8,800 in the VIS-arm. The data were reduced using the X-SHOOTER ESO pipeline following the standard steps and analyzed with the Penalized Pixel Fitting (PPXF) software [3] to extract velocity and velocity dispersion.

29.3 Results

Our aim is the study of the FP and of their main projections i.e. the Kormendy [15] and the Faber and Jackson (FJ) [12] relations.

The Kormendy relation is shown in Fig. 29.1 (upper panel) with superimposed the fit to all the galaxies of the WINGS sample [9, 10], the red points are the data for our sample of dwarfs. As expected they are in a different region of the plane, the one where we expect the faint early-type population lies, in agreement with [9]. Figure 29.1 shows that for faint galaxies the $\log(R_e) - < \mu_e >$ linear trend is steeper than that for the bright ones. The Faber and Jackson relation is expressed as $L \propto \sigma^{\alpha}$; the classical exponent is $\alpha \sim 4$, while for fainter objects the slope is shallower [5, 16]. In Fig. 29.1 (lower panel) the classical FJ relation is superimposed as a continuous line, the fit to the old WINGS data is the dashed line for which we derive $L \propto \sigma^{1.69}$ [9]. The faint galaxy sample is found to lie in between the two fits. It is not consistent with the classical FJ relation and also deviates from the relation we derived for the bright sample of galaxies in WINGS.

In our previous works [9, 10] we derived the FP of galaxies in 59 nearby clusters ($0.04 < z < 0.06$) using data from the WINGS survey. In those works we speculate that, rather than a plane, the so called FP is actually a curved surface, which is approximated by different planes depending on the different regions of the FP space occupied by the galaxy samples under analysis.

In Fig. 29.2 we show the edge-on FP of the whole WINGS sample [9] with superimposed (red dots) our 22 faint galaxies as it appears along the direction of luminosity. The dwarf galaxies are located at more than 3σ from the best fit for the whole WINGS sample. This particular projection highlights a peculiar feature of the FP, suggesting the existence of some kind of warping at low luminosities, as pointed out also by [9]. In particular the faint end of the FP seems to depart from the FP of bright ellipticals at $M_B \sim -19.0$. This fact has been noted also by [6] and [16] in the low-luminosity region and may actually highlight a first hint of the connection between the FP of giant and dwarf ellipticals.

Acknowledgements We acknowledge partial financial support by contract PRIN/MIUR 2009: "Dynamics and Stellar Populations of Superdense Galaxies" (Code: 2009L2J4MN); by INAF/PRIN 2011: "Galaxy Evolution with the VLT Survey Telescope (VST)" and INAF/PRIN2014:"Galaxy evolution from cluster cores to filaments".

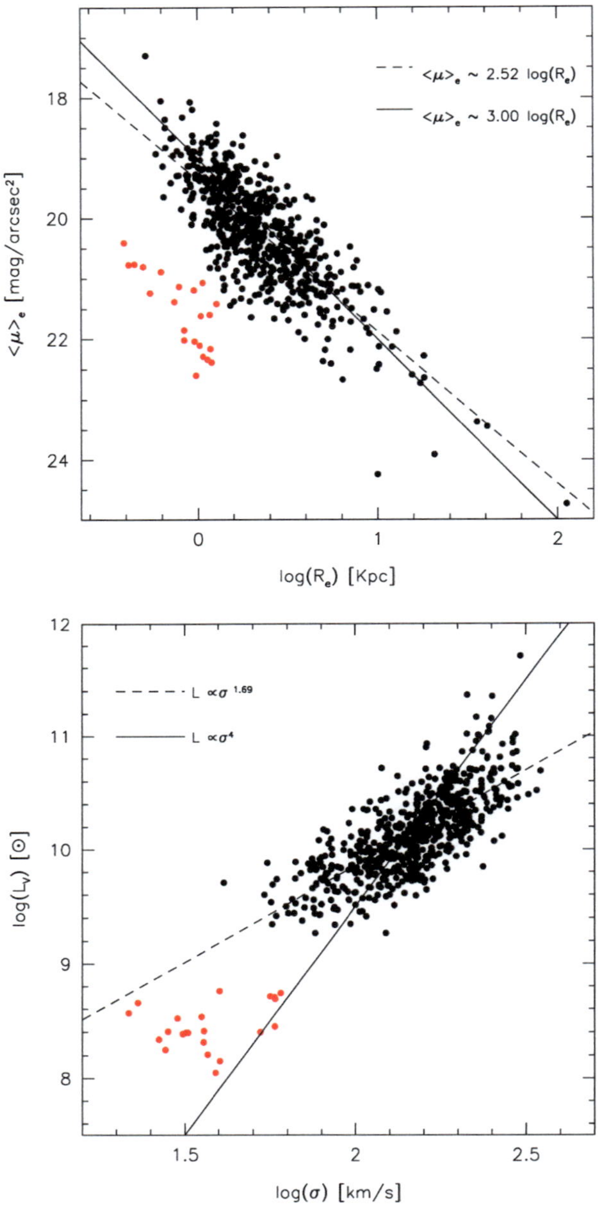

Fig. 29.1 The Kormendy relation (*upper panel*). The *red points* are our 22 galaxies and *black dots* are from the whole WINGS sample. Our sample of dwarfs is clearly offset from the relation defined by the bright early-type galaxies. The Faber and Jackson relation $\mu_e - r_e$ (*lower panel*). In both panels the *solid line* are the classical $\mu_e \propto 3 * log(r_e)$ and L$\propto$ σ^4 fits and the *dashed line* is the fit to our whole sample of bright WINGS galaxies

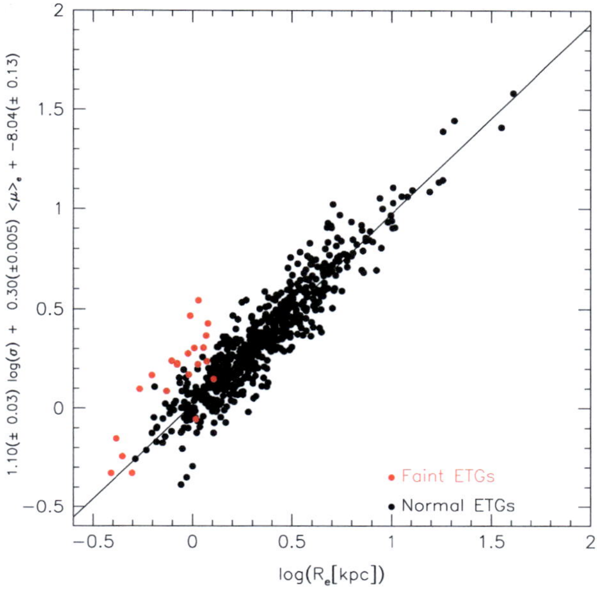

Fig. 29.2 Edge-on FP of the WINGS galaxy sample as it appears along the direction of luminosity. The *black line* is the edge-on view of the best-fitting FP of bright ellipticals from [9]. *Red* and *black points* are dwarfs and bright early-type galaxies, respectively

References

1. Bernardi, M., Sheth, R.K., Annis, J., et al.: Early-type galaxies in the Sloan digital sky survey. IV. Colors and chemical evolution. AJ **125**, 1866 (2003)
2. Bernardi, M., Nichol, R.C., Sheth, R.K., Miller, C.J., Brinkmann, J.: A search for the most massive galaxies: double trouble? AJ **131**, 1288 (2006)
3. Cappellari, M., Emsellem, E.: Parametric recovery of line-of-sight velocity distributions from absorption-line spectra of galaxies via penalized likelihood. PASP **116**, 138 (2004)
4. Cava, A., Bettoni, D., Poggianti, B.M., et al.: WINGS-SPE spectroscopy in the WIde-field nearby galaxy-cluster survey. A&A **495**, 707 (2009)
5. Davies, R.L., Efstathiou, G., Fall, S.M., Illingworth, G., Schechter, P.L.: The kinematic properties of faint elliptical galaxies. ApJ **266**, 41 (1983)
6. Desroches, L.B., Quataert, E., Ma, C.P., West, A.A.: Luminosity dependence in the Fundamental Plane projections of elliptical galaxies. MNRAS **377**, 402 (2007)
7. Djorgovski, S., Davis, M.: Fundamental properties of elliptical galaxies. ApJ **313**, 59 (1987)
8. D'Odorico, S., Dekker, H., Mazzoleni, R., et al.: X-shooter UV- to K-band intermediate-resolution high-efficiency spectrograph for the VLT: status report at the final design review. Proc. SPIE **6269**, 98 (2006)
9. D'Onofrio, M., Fasano, G., Varela, J., et al.: The fundamental plane of early-type galaxies in nearby clusters from the WINGS database. ApJ **685**, 875 (2008)
10. D'Onofrio, M., Fasano, G., Moretti, A., et al.: The hybrid solution for the fundamental plane. MNRAS **435**, 45 (2013)
11. Dressler, A., Lynden-Bell, D., Burstein, D., et al.: Spectroscopy and photometry of elliptical galaxies. I – a new distance estimator. ApJ **313**, 42 (1987)

12. Faber, S.M., Jackson, J.: Velocity dispersions and mass-to-light ratios for elliptical galaxies. ApJ **204**, 668 (1976)
13. Fasano, G., Marmo, C., Varela, J., et al.: WINGS: a WIde-field Nearby Galaxy-cluster Survey. I. Optical imaging. A&A **445**, 805 (2006)
14. Jørgensen, I., Franx, M., Kjæergaard, P.: The Fundamental Plane for cluster E and S0 galaxies. MNRAS **280**, 197 (1996)
15. Kormendy, J.: Brightness distributions in compact and normal galaxies. II – structure parameters of the spheroidal component. ApJ **218**, 333 (1977)
16. Kourkchi, E., Khosroshahi, H.G., Carter, D., et al.: Dwarf galaxies in the Coma cluster – I. Velocity dispersion measurements. MNRAS **420**, 2819 (2012)
17. Matković, A., Guzmań, R.: Kinematic properties and stellar populations of faint early-type galaxies – I. Velocity dispersion measurements of central Coma galaxies. MNRAS **362**, 289 (2005)
18. Poggianti, B.M., Bridges, T.J., Mobasher, B., et al.: A Photometric and spectroscopic study of dwarf and giant galaxies in the coma cluster. III. Spectral ages and metallicities. ApJ **562**, 689 (2001)
19. Poggianti, B.M., Kashikawa, N., Bridges, T., et al.: Two formation paths for cluster dwarf galaxies? ApJ **217**, 562 (2004)

Chapter 30
Searching for Galaxy Clusters in the VST-KiDS Survey

M. Radovich, E. Puddu, F. Bellagamba, L. Moscardini, M. Roncarelli, F. Getman, A. Grado, and the KiDS collaboration

Abstract We present the methods and first results of the search for galaxy clusters in the Kilo Degree Survey (KiDS). The adopted algorithm and the criterium for selecting the member galaxies are illustrated. Here we report the preliminary results obtained over a small area (7 $\deg^2$), and the comparison of our cluster candidates with those found in the RedMapper and SZ Planck catalogues; the analysis to a larger area (148 $\deg^2$) is currently in progress. By the KiDS cluster search, we expect to increase the completeness of the clusters catalogue to $z = 0.6$–0.7 compared to RedMapper.

30.1 Introduction

Clusters of galaxies are described as the most massive gravitationally bound quasi-equilibrium structures in the Universe. They represent a powerful tool for cosmology [2]: a large and complete catalogue of galaxy clusters spanning a broad range of redshifts would be important to understand the history of cosmic structure formation, and to provide accurate measurements of cosmological parameters, in particular σ_8 and Ω_M. In this perspective, it becomes crucial the role of extragalactic surveys that can supply a statistically significant sample for the detection of clusters. Among the already available surveys of different sizes and depths, one of the landmarks is the Sloan Digital Sky Survey (SDSS; [16]), which allows to probe the low-redshift Universe over an area of $20,000 \deg^2$ up to $z \sim 0.45$. The Kilo Degree Survey (KiDS; http://kids.strw.leidenuniv.nl/) and the Dark Energy Survey (DES; http://www.darkenergysurvey.org/) will allow to obtain samples of clusters

M. Radovich (✉)
INAF – OAPD, Padova, Italy
e-mail: mario.radovich@oapd.inaf.it

E. Puddu • F. Getman • A. Grado
INAF – OACN, Napoli, Italy

F. Bellagamba • L. Moscardini • M. Roncarelli
Department of Physics and Astronomy (DIFA), University of Bologna, Bologna, Italy

KiDS collaboration: http://kids.strw.leidenuniv.nl/team.php

N.R. Napolitano et al. (eds.), *The Universe of Digital Sky Surveys*, Astrophysics and Space Science Proceedings 42, DOI 10.1007/978-3-319-19330-4_30

spanning a wider range of redshift and mass, thanks to their superior depth and image quality. When completed, KiDS will cover 1,500 deg^2 in the *ugri* bands, with optimal seeing conditions in the r–band; DES will cover a larger area (5,000 deg^2 in *gri*), with an image quality between SDSS and KiDS. Here we show the very first results obtained from the comparison of clusters detected in KiDS with the RedMapper catalogue based on the SDSS [12], containing $\sim$25, 000 clusters with masses $>10^{14} M_{\odot}$ in a volume limited up to $z < 0.35$.

30.2 The Data

The Kilo Degree Survey [6] is one of the ESO VST public surveys and is designed to observe an area of 1,500 deg^2 in the *ugri* bands (with depths at 10 σ respectively of 23, 24, 24 and 23 mags in the given bands). At present, a total of 148 deg^2, complete in *ugri*, have been published in the KiDS ESO Data Release DR2 [7]. Data reduction was performed by the AstroWISE system [15] within the KiDS consortium. Photometric redshifts were computed by the KiDS lensing team using the BPZ code [5] and benefit from the usage of the GAMA [8] spectroscopy as a training set (for details and photometric redshift accuracy, see [10]). An initial guess of the number of clusters vs. redshift expected in KiDS was derived using the mock catalogues by [9] from the Millennium Simulation [13] assuming KiDS photometric depths. As it can be seen from Fig. 30.1, KiDS should detect many more clusters at intermediate redshifts ($0.4 < z < 0.6$) with respect to SDSS (red inset). However, as it is described e.g. in [3], the semi-analytic galaxy formation models are not fully representing the photometric properties of galaxies in clusters; therefore, we are

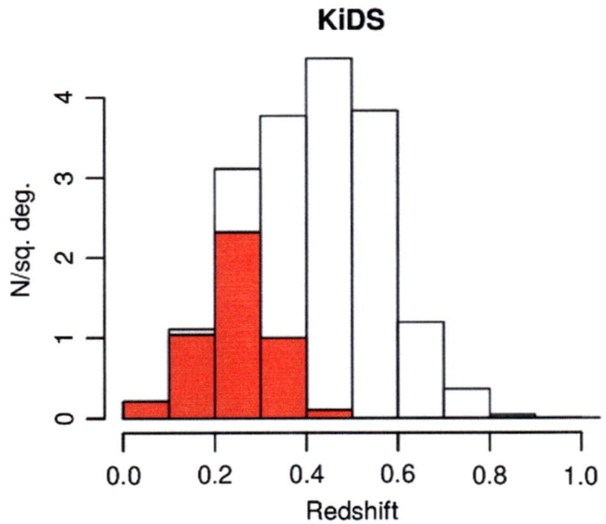

Fig. 30.1 Results from the Millennium Simulation derived with the KiDS photometric depths (*black*) and SDSS depths (*red inset*)

preparing more appropriate simulations built on KiDS data in order to estimate the completeness and purity of our cluster catalogue.

30.3 The Cluster Search

Cluster detection is performed with the Optimal Filtering technique, described in [4]. This method aims to maximise the signal-to-noise ratio of cluster detection by convolving the observed data with a suitable filter, which is proportional to the expected galaxy distribution in the clusters over the random fluctuations of the field distribution. This approach is very flexible, as it adapts to the properties of the different galaxy populations (cluster and field) in the available dataset, without selecting a priori a single signature of clusters. In particular, in the analysis of the KiDS data, for each galaxy we use the position in the sky, the magnitude in the r-band, and the full redshift probability distribution. For a set of redshift values, the cluster model is built from the mean properties of clusters in the MaxBCG sample derived from SDSS observations, appropriately modified to account for the evolution of the cluster galaxy population. The background is instead estimated from the mean density of galaxies in the field as a function of magnitude. The filter is then highlighting positions in the space where there are very luminous galaxies, which are more common in the clusters than in the field, and where the density is significantly higher than the mean one.

30.4 Membership, BCGs and Richness

The algorithm provides for each cluster the center, the estimated redshift (z_{CL}) and the signal to noise ratio of the detection. In order to have a direct comparison with the RedMapper cluster catalogue (which is based on the selection of Early-Type galaxies), we need to define a similar galaxy membership criterium and apply it to our detections: to this purpose we rerun the photo-z code (BPZ) for all galaxies within 2 Mpc from the center with the redshift fixed to z_{CL} and selected galaxies with a best-fit template consistent with Early-Type galaxies. Richness was, then, computed as the number of red galaxies placed within 1 Mpc from the center and with $L > 0.2L_*$ (see Fig. 30.3, red dots). We want to stress that these preliminary results depend on this specific data-set comprising a small area; the procedure will be tuned as extensive simulations and a more significant area data-set will be available. We initially run the cluster search on 7 KiDS fields (a total area of 7 deg^2) to check the performances of the cluster search algorithm and to validate it. There are 74 cluster candidates detected by our algorithm in this area. For comparison, in the same field there are 17 candidate clusters in the RedMapper catalogue, of which 13 (76 %) are common detections. Furthermore, we find all the 3 ACOs [1] and the 2 XMMs [11] clusters and recover 3 of the 5 Planck clusters [14]. Missing detections are located close to the masked regions of the fields, corresponding to the presence of bright foreground stars; clusters detected in KiDS and not in RedMapper are all

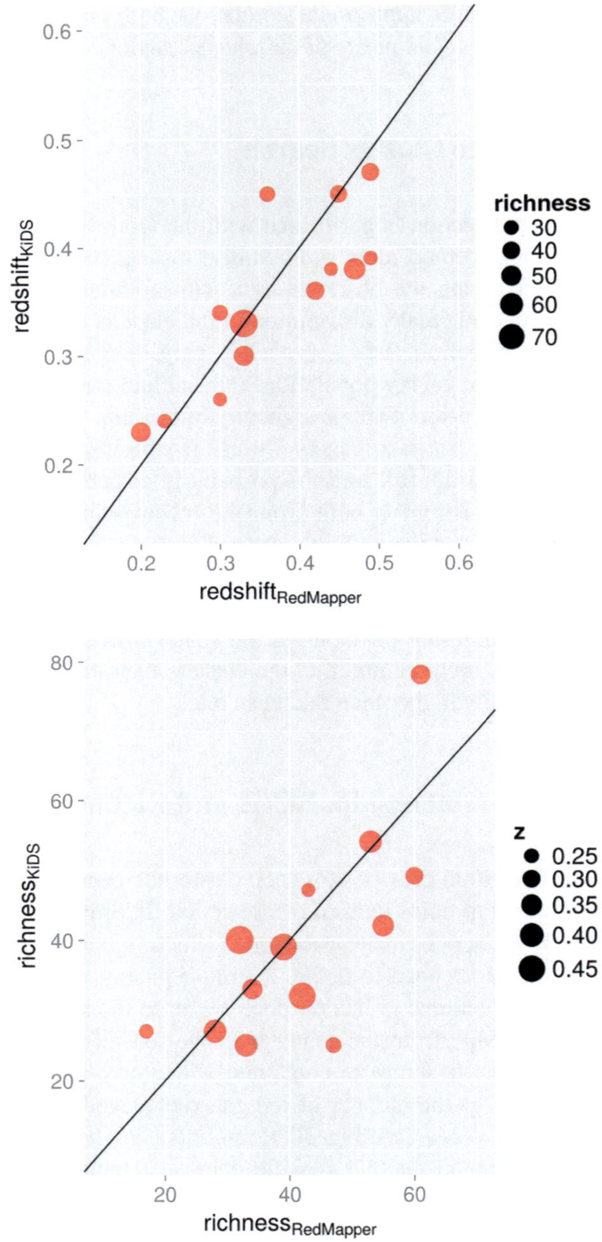

Fig. 30.2 Comparison of the photo-z and richness for the common KiDS/RedMapper clusters

at redshift ≥0.35–0.4, at the detection limit of RedMapper. In Fig. 30.2 we show the comparison of redshift and richness estimates for the KiDS candidate clusters with galaxy clusters found in the RedMapper survey. In Fig. 30.3, we show the *gri* KiDS image of a $3' \times 3'$ region around the cluster center for two RedMapper clusters and one SZ Planck cluster, recovered from our algorithm on KiDS data. Figure 30.4

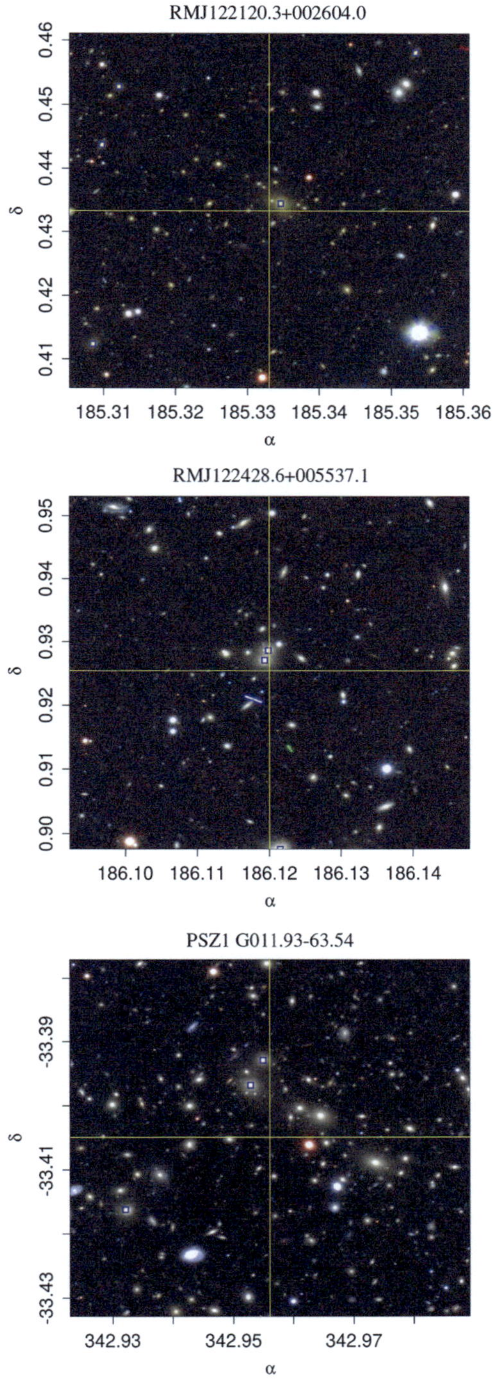

Fig. 30.3 *gri* KiDS image of three RedMapper/SZ Planck clusters. The five brightest galaxies (if present in this central part of the field) are marked with *blue squares*

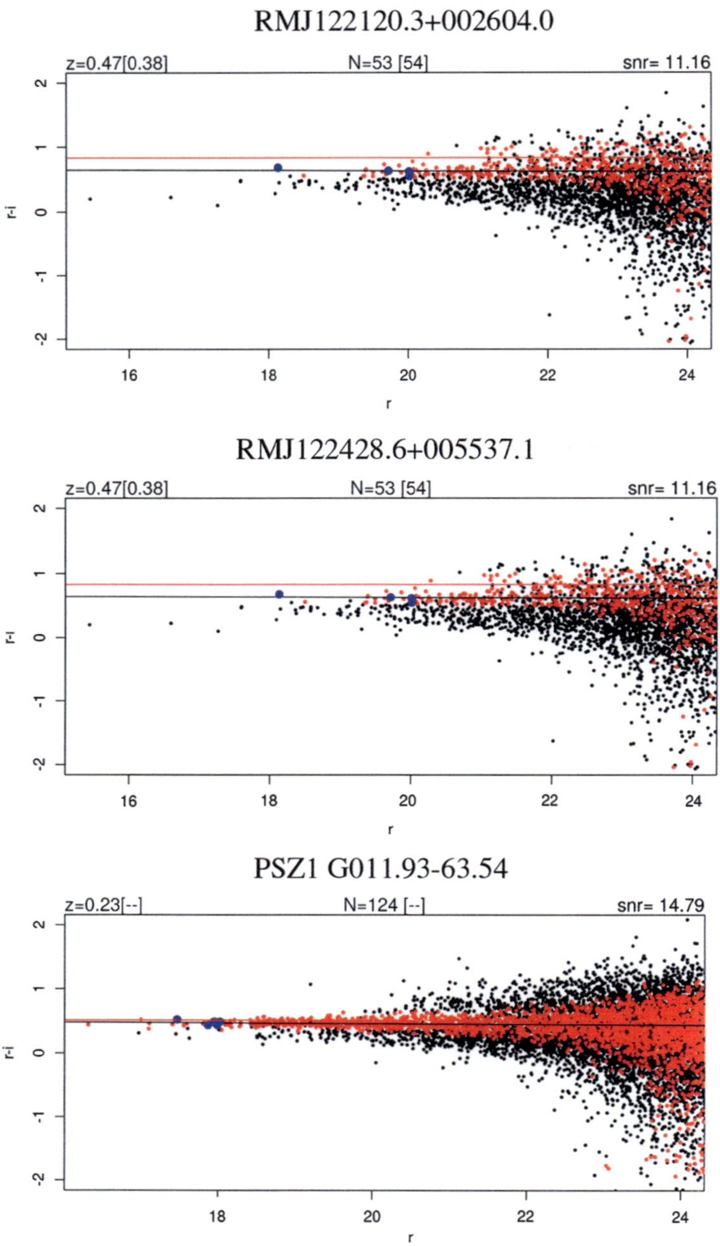

Fig. 30.4 r-i vs. r plot from KiDS data, for the clusters in Fig. 30.2, of field (*black*) and Elliptical cluster member (*red*) galaxies. The five brightest galaxies are marked in *blue* as in Fig. 30.2. Redshift estimate, richness and signal to noise for the detection are reported on each panel; the RM values are also displayed in *square brackets*

displays the related color-magnitude diagrams. Based on these preliminary results, an extension of this analysis to the 148 deg^2 of the KiDS DR2 is currently in progress.

References

1. Abell, G.O., Corwin, H.G., Olowin, R.P.: A catalog of rich clusters of galaxies. ApJS **70**, 1 (1989)
2. Allen, S.W., Evrard, A.E., Mantz, A.B.: Cosmological parameters from observations of galaxy clusters. ARA&A **49**, 409 (2011)
3. Ascaso, B., Mei, S., Benitez, N.: Simulating mock catalogues to provide accurate cluster selection functions. In: Proceedings of the Annual Meeting of the French Society of Astronomy and Astrophysics, Paris, vol. 299 (2014)
4. Bellagamba, F., Maturi, M., Hamana, T., et al.: Optimal filtering of optical and weak lensing data to search for galaxy clusters: application to the COSMOS field. MNRAS **413**, 1145 (2011)
5. Benítez, N.: Bayesian photometric Redshift estimation. ApJ **536**, 571 (2000)
6. De Jong, J., Kuijken, K., Applegate, D., et al.: The Kilo-Degree Survey. The Messenger **154**, 44 (2013)
7. de Jong, J., Verdoes Kleijn, G., Boxhoorn D. et al. 2015, The first and second data releases of the Kilo-Degree Survey, A&A, **582**, 62 (2015)
8. Driver, S.P., Hill, D.T., Kelvin, L.S., et al.: Galaxy and Mass Assembly (GAMA): survey diagnostics and core data release. MNRAS **423**, 971 (2011)
9. Henriques, B.M.B., White, S.D.M., Lemson, G., et al.: Confronting theoretical models with the observed evolution of the galaxy population out to z= 4. MNRAS **421**, 2904 (2012)
10. Kuijken, K., et al.: Gravitational lensing analysis of the Kilo-Degree Survey. MNRAS **454**, 3500 (2015)
11. Mehrtens, N., et al.: The XMM Cluster Survey: optical analysis methodology and the first data release. MNRAS **423**, 1024 (2013)
12. Rykoff, E.S., Rozo, E., Busha, M.T., et al.: redMaPPer. I. Algorithm and SDSS DR8 catalog. ApJ **785**, 104 (2014)
13. Springel, V., White, S.D.M., Jenkins, A., et al.: Simulations of the formation, evolution and clustering of galaxies and quasars. Nature **435**, 629 (2005)
14. The Planck Collaboration: Planck 2013 results. XXIX. The Planck catalogue of Sunyaev-Zeldovich sources. A&A **571**, 29 (2014)
15. Verdoes Kleijn, G., de Jong, J., Valentijn, E.A., et al.: Astro-WISE for KiDS survey production and quality control. In: Astronomical Data Analysis Software and Systems XXI. Proceedings of a Conference held at Marriott Rive Gauche Conference Center, Paris **461**, 237 (2011). astro-ph1112.0886
16. York, D.G., Adelman, J., Anderson, J.E., et al.: The Sloan Digital Sky Survey: technical summary. AJ **120**, 1579 (2000)

Chapter 31
First Results from Supernova Diversity and Rate Evolution (SUDARE) Survey at VST

M.T. Botticella, E. Cappellaro, G. Pignata, A. Grado, L. Limatola, M. Della Valle, M. Vaccari, L. Greggio, S. Spiro, F. Bufano, L. Tomasella, G. Covone, M. Capaccioli, N. Napolitano, L. Marchetti, E. Gonzales-Solares, M. Jarvis, M. Radovich, S. Benetti, A. Pastorello, M. Turatto, M. Paolillo, P. Schipani, A. Baruffolo, and E. Cascone

Abstract Despite the key role played by Supernovae (SNe) in discovering the accelerating expansion of the Universe, there are still fundamental questions to answer about their progenitor systems and explosion mechanisms. Furthermore the discovery of a significant number of both exceptionally bright and extremely faint SNe, as well as peculiar events, suggests the existence of an unexpected diversity. Important clues on the SN progenitors can be derived by examining the rate of type Ia and core collapse SNe. With this goal in mind we started the SUpernova Diversity And Rate Evolution (SUDARE) programme currently running at the VLT

M.T. Botticella (✉) • A. Grado • L. Limatola • M. Della Valle • N. Napolitano • P. Schipani • E. Cascone
INAF-Osservatorio Astronomico di Capodimonte, Salita Moiariello 16, 80131 Napoli, Italy
e-mail: botticela@na.astro.it

E. Cappellaro • L. Greggio • S. Spiro • L. Tomasella • M. Radovich • S. Benetti • A. Pastorello • M. Turatto • A. Baruffolo
INAF-Osservatorio Astronomico di Padova, Vicolo dell'Osservatorio 5 , 35122 Padova, Italy

G. Pignata • F. Bufano
Departemento de Ciencias Fisicas, Universidad Andres Bello, Santiago, Chile

M. Vaccari
Astrophysics Group, Physics Department, University of the Western Cape, Cape Town, South Africa

G. Covone • M. Capaccioli • M. Paolillo
Department of Physics "Ettore Pancini", University "Federico II" of Naples, Italy

L. Marchetti
Department of Physical Sciences, The Open University, Milton Keynes, MK7 6AA, UK

E. Gonzales-Solares
Institute of Astronomy, University of Cambridge, Cambridge, CB3 0HA, UK

M. Jarvis
CAR, University of Hertfordshire, Hertfordshire, AL10 9AB, UK

N.R. Napolitano et al. (eds.), *The Universe of Digital Sky Surveys*, Astrophysics and Space Science Proceedings 42, DOI 10.1007/978-3-319-19330-4_31

Survey Telescope (VST). We present a measurement of the volumetric SN rates as a function of redshift for the first 2 years of data from SUDARE.

31.1 Supernovae

SNe are energetic explosions involving the formation of neutron stars, black holes, and gamma-ray bursts and the main producers of heavy elements in galaxies. We recognise two different classes of SNe: core collapse-induced explosions of massive stars (CCSNe) and thermonuclear explosions of low mass stars in binary systems (SNeIa). SNeIa are one of the more promising tools to probe the nature of dark energy in the Universe, but the nature of their progenitor system is still debated and the assumption that SNIa class is highly homogeneous has been revised [e.g. 9]. The evolutionary paths leading to the various types of CCSNe are also unsettled [e.g. 12]. The simultaneous analysis of the cosmic evolution of SN rates and the dependence of SN rates on the average properties of the stellar population from which SNe were formed can help for constraining the progenitor scenarios and for understanding the SN diversity.

31.2 SUDARE

The wide field of view and excellent quality of VST optics [3] gave us the opportunity of carrying out a SN survey in the range $0.1 \leq z \leq 0.7$, that is crucial to connect measurements in the nearby Universe and future measurements at high redshifts. The goal of SUDARE is to measure both SNIa and CCSN rate as a function of cosmic time, galaxy age, mass, star formation rate (SFR) and in different environments. The observing strategy consists in a frequent (every 3–4 nights in r band and every 7–10 days in g, i bands), long term (4 years) monitoring of selected sky fields. The repeated imaging allows not only the construction of multiband light curves and photometric typing for each transient source, but also deep images of the survey fields to retrieve fundamental properties of the surveyed galaxies. We are monitoring two sky fields with an extraordinary amount of ancillary data from Chandra Deep Field South (CDFS) and Cosmic Evolution Survey (COSMOS). Mosaic stacks for each epoch and deep stacks were generated for each filter with the VSTTube pipeline [5].

31.3 SN Sample

The presence of SNe and other variable sources (AGNs, variable stars, moving objects) in a "search" image is revealed through the subtraction of a "reference" image acquired at a different epoch. The subtracted images were searched to build a catalog of SN candidates and apply a series of tunable culling procedures. Each catalog of detections was cleaned of objects with very low S/N ratio or a significant fraction of bad pixels, as well as objects with irregular profiles. The resulting list of variable sources in each epoch was then visually reviewed to select SN candidates. SN flux measurements were calculated for each epoch using a PSF fitting routine on both search and difference images. The multi-colour light curves were then fit to SN templates that were selected to take into account the diversity in the photometric evolution for SNe of the same type. SN detection and classification were performed with a pipeline implemented for SUDARE and its possible inefficiencies are accounted for performing detailed simulations. In its first 2 years of full operation, we discovered hundreds of variable sources and classified 117 SNe (57 % Ia up to $z = 0.8$, 43 % CC up to $z = 0.4$).

31.4 Galaxy Sample

A photometric catalog[1] based on 30 band photometry of the sources in the COSMOS field has been produced by [10]. To collect multi-band photometry in CDFS we exploited: (i) deep stacks in u, g, r, i bands from SUDARE and VOICE surveys, (ii) deep stacks in J, H, K_s from the VISTA Deep Extragalactic Observations [VIDEO, 6], (iii) UV data from *GALEX*, (iv) *Spitzer* IRAC data from the SERVS and SIMPLE surveys. Source detection and photometry are performed using SExtractor routine in dual image mode with the K_s stack as detection image. By fitting the SED of a series of template galaxies to the broadband fluxes available for each galaxy, we estimate the photometric redshifts with EAZY code [2]. The stellar masses and SFRs have been estimated, for each galaxy, using FAST code [7].

31.5 Preliminary Results

The rate for each SN class is computed as the number of observed SNe divided by the control time, defined as the time interval during which a SN occurring at a given redshift is detectable from the survey [1]. The rates are then converted to a rest frame value and normalised to the cosmic volume. A comparison of our results with all measurements in literature is shown in Fig. 31.1.

Concerning the SNIa rate, our results are in broad agreement with literature values and show a steady rise of the SNIa rate up to $z = 0.7$, with a steeper trend than indicated by measurements from the Supernova Legacy Survey [11]. The CC SN rate at redshift 0.25 is in agreement within 1σ with other measurements and confirm an evolutionary trend similar to that of SFR. Our approach allows us to measure the ratio between type Ia and CC rates and analyse its evolution with redshift. We found

[1] http://www.strw.leidenuniv.nl/galaxyevolution/ULTRAVISTA

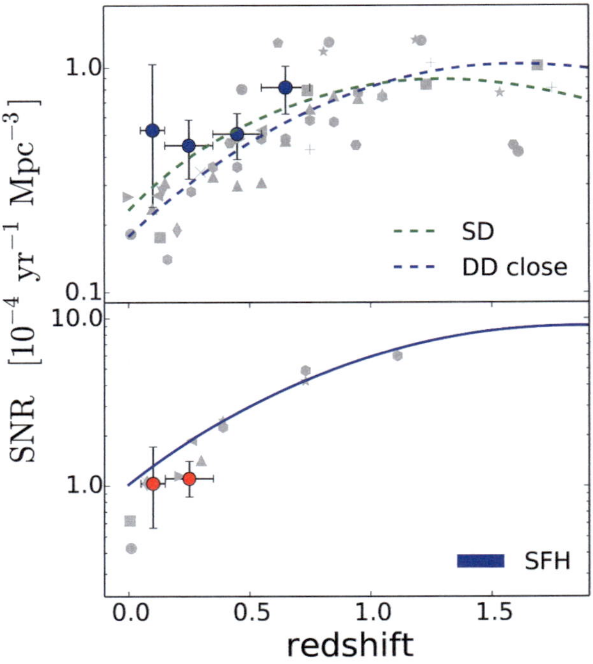

Fig. 31.1 *Top panel*: SN Ia rates per unit volume. *Blue points* show SUDARE measurements. As a comparison, the lines show the evolution of SN rates predicted by two models (*SD* single degenerate and *DD* double degenerate) for SNe Ia progenitors by Greggio [13] and the SFH by Madau and Dickinson [14]. *Bottom panel*: CC SNe rates per unit volume. *Red points* show SUDARE measurements. The *blue line* shows the predicted CC SN rate assuming the SFH by Madau and Dickinson [14]

a ratio r_{CC}/r_{Ia} of about 2.5 at $z = 0.25$ in agreement with the value of 2.4 measured in the local Universe [4, 8].

References

1. Botticella, M.T., Riello, M., Cappellaro, E., et al.: Supernova rates from the Southern inTermediate Redshift ESO Supernova Search (STRESS). Astr. Astrophys. **479**, 49 (2008)
2. Brammer, G.B., van Dokkum, P.G., Coppi, P.: EAZY: a fast, public photometric redshift code. ApJ **686**, 1503 (2008)
3. Capaccioli, M., Schipani, P.: The VLT survey telescope opens to the sky: history of a commissioning. The Messenger **146**, 2 (2011)
4. Cappellaro, E., Evans, R., Turatto, M.: A new determination of supernova rates and a comparison with indicators for galactic star formation. Astr. Astrophys. **351**, 459 (1999)
5. Grado, A., Capaccioli, M., Limatola, L., Getman, F.: VST processing facility: first astronomical applications. MSAIS **19**, 362 (2012)
6. Jarvis, M.J., Bonfield, D.G., Bruce, V.A., et al.: The VISTA deep extragalactic observations (VIDEO) survey. MNRAS **428**, 1281 (2013)
7. Kriek, M., van Dokkum, P.G., Labbé, I., et al.: An ultra-deep near-infrared spectrum of a compact quiescent galaxy at z = 2.2. ApJ **700**, 221 (2009)
8. Li, W., Chornock, R., Leaman, J., et al.: Nearby supernova rates from the lick observatory supernova search - III. The rate-size relation, and the rates as a function of galaxy Hubble type and colour. MNRAS **412**, 1473 (2011)
9. Maoz, D., Mannucci, F., Nelemans, G.: Observational clues to the progenitors of type Ia supernovae. ARA&A **52**, 107 (2014)

10. Muzzin, A., Marchesini, D., Stefanon, M., et al.: A public Ks-selected catalog in the COSMOS/ULTRAVISTA field: photometry, photometric redshifts, and stellar population parameters. ApJs **206**, 8 (2013)
11. Perrett, K., Sullivan, M., Conley, A., et al.: Evolution in the volumetric type Ia supernova rate from the supernova legacy survey. Astr. J. **144**, 59 (2012)
12. Smartt, S.J.: Progenitors of core-collapse supernovae. ARA&A **47**, 63 (2009)
13. Greggio, L.: The rates of type Ia supernovae. I. Analytical formulations. A&A **441**, 1055 (2005)
14. Madau, P., Dickinson, M.: Cosmic star-formation history. Annu. Rev. Astron. Astrophys. **52**, 415 (2014)

Part III
The Universe of Spectroscopic Surveys

Chapter 32
The Wide Area VISTA Extra-Galactic Survey (WAVES)

S.P. Driver, L.J. Davies, M. Meyer, C. Power, A.S.G. Robotham, I.K. Baldry, J. Liske, and P. Norberg

Abstract The "Wide Area VISTA Extra-galactic Survey" (WAVES) is a 4MOST Consortium Design Reference Survey which will use the VISTA/4MOST facility to spectroscopically survey ~2 million galaxies to $r_{AB} < 22$ mag. WAVES consists of two interlocking galaxy surveys ("WAVES-Deep" and "WAVES-Wide"), providing the next two steps beyond the highly successful 1M galaxy Sloan Digital Sky Survey and the 250k Galaxy And Mass Assembly survey. WAVES will enable an unprecedented study of the distribution and evolution of mass, energy, and structures extending from 1-kpc dwarf galaxies in the local void to the morphologies of 200-Mpc filaments at $z \sim 1$. A key aim of both surveys will be to compare comprehensive empirical observations of the spatial properties of galaxies, groups, and filaments, against state-of-the-art numerical simulations to distinguish between various Dark Matter models.

32.1 Introduction

Since the pioneering days of the 2dFGRS and SDSS, extra-galactic spectroscopic surveys have come in two flavours: those optimised for cosmology, and those optimised for galaxy evolution. The distinction is important. Cosmology surveys (e.g., WiggleZ, BOSS, DESI) use specific tracers to probe the underlying large scale

S.P. Driver (✉) • L.J. Davies • M. Meyer • C. Power • A.S.G. Robotham
International Centre for Radio Astronomy Research (ICRAR), School of Physics, University of Western Australia, M468, 35 Stirling Highway, Crawley, WA 6009, Australia
e-mail: simon.driver@uwa.edu.au

I.K. Baldry
Astrophysics Research Institute, Liverpool John Moores University, IC2, Liverpool Science Park, 146 Brownlow Hill Liverpool, L3 5RF, UK

J. Liske
European Southern Observatory, Karl-Schwarzschild-Str. 2, 85748, Garching, Germany

P. Norberg
International Cosmology Centre, Durham University, Durham, DH1 3LE, UK

N.R. Napolitano et al. (eds.), *The Universe of Digital Sky Surveys*, Astrophysics and Space Science Proceedings 42, DOI 10.1007/978-3-319-19330-4_32

structure. These surveys advance cosmology but provide a biased cross-section of the galaxy population. Conversely galaxy-evolution studies (e.g., MGC, GAMA, zCOSMOS, DEEP2, MOONS) uniformally sample the full galaxy population, but only cover modest areas. These samples are ideal for studying galaxy evolution and its interplay with environment, but lack the area coverage for cosmological studies. WAVES, in the era of dedicated cosmology experiments, represents the next step in galaxy evolution studies, bridging the gap between the very near ($z < 0.3$; SDSS, GAMA) and the very far ($z > 0.8$; MOONS, HST, JWST), as well as probing the intrinsically faint (low mass) and dim (low surface density) populations within the nearby Universe ($z < 0.2$).

The WAVES target catalogue will be based on the VST KiDS South sub-arcsecond optical imaging and is intended to complement Euclid, LSST, and the SKA. The key science drivers and initial survey concept design are outlined in the sections which follow. Here we briefly introduce the two WAVES components (Deep and Wide), outline the primary science motivations, and describe the preliminary design concept.

WAVES-Deep will cover 100 deg^2 to $r_{\rm AB} \sim 22$ mag and extend the power of SDSS and GAMA like population statistics out to $z \sim 0.7$, tracing the rarest structures even up to and slightly beyond $z \sim 1$. This ~1.2 million galaxy redshifts survey is key in providing the largest and most representative galaxy group and filament catalogue ever constructed.

WAVES-Wide will cover 750 deg^2 to $r_{\rm AB} \sim 22$ mag with additional photo-z pre-selection ($z_{\rm photo} < 0.2$). This will target ~0.9 million galaxies allowing a detailed study of the occupancy of $\sim 10^{11}$–$10^{14} M_{\odot}$ halos to a stellar mass limit of $10^7 M_{\odot}$, and providing a dwarf galaxy sample over a representative volume of 10^6 Mpc^3.

32.2 WAVES Science Drivers

32.2.1 Ensemble of Milky-Way Sized Systems to Test CDM

The nature of dark matter is one of the key questions in modern day cosmology. The currently favoured Cold Dark Matter model, ΛCDM, provides a good description of the large scale structure of the Universe (see Figs. 32.1 and 32.2). Comparison of robust model predictions with empirical galaxy clustering measurements on Mpc scales supports the Cold Dark Matter model for the growth of structure (e.g. [26]). On sub-Mpc scale, i.e. on galaxy and group scales, baryons and baryonic physics become critical: the kpc to Mpc range is the key scale over which Dark Matter halos virialize and merge, and baryons decouple, collapse and eventually form complex structures such as galaxies. In this regime, our theoretical understanding is less well-founded, in great part due to the immense complexity of the physics encountered.

Galaxy group samples are now able to probe down to a few $10^{12} M_{\odot}$, with arguably the most complete being the "GAMA Galaxy Group Catalogue" (G^3C;

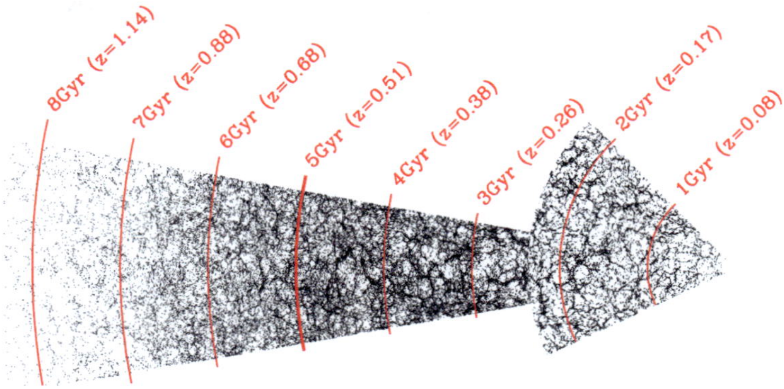

Fig. 32.1 A representation of the RA geometry of the WAVES survey (derived from the Theoretical Astrophysical Observatory), highlighting the complexity of structures that will be sampled

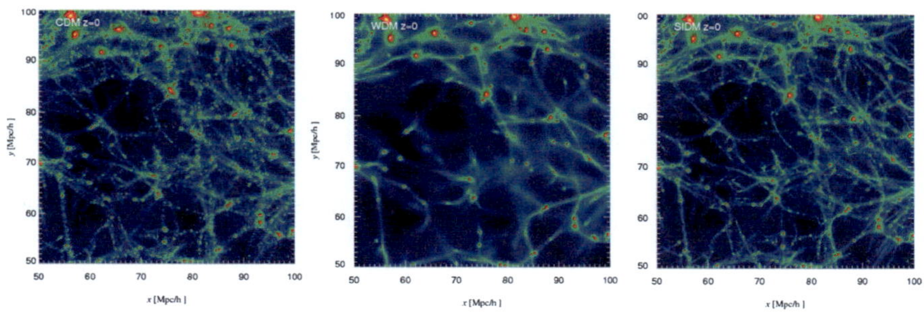

Fig. 32.2 Numerical simulations of the galaxy distribution at z = 0 from (*left*-to-*right*) Cold, Warm, and Self-interacting dark matter. The images show the dark matter density within a cube of $50 \times 50 \times 5$ Mpc

[19]). Properties of low mass galaxy groups are limited by the intrinsic lack of survey depth. This explains why for $10^{12}\,M_{\odot}$ systems ($\mathscr{M}_{h12}$ hereafter), only two really well studied examples exist: the Milky-Way and the Andromeda systems, both in our own neighbourhood. Galaxies in the Local Group are extensively used to probe in detail the Cold Dark Matter paradigm. They are often central to the strongest evidence against the standard Cold Dark Matter model, from "the missing satellite problem" (e.g. [15]) to "the too big too fail problem" (e.g. [5]). via "the co-planar location of satellite galaxies" (e.g. [13]). Given that $\mathscr{M}_{h12}$ halos are the most important ones in terms of galaxy formation (with galaxy formation efficiency peaking just around that mass scale in all standard CDM model) and one of the more critical in terms of testing CDM, it is fundamental to ensure that this limited sample of well studied $\mathscr{M}_{h12}$ systems is representative. To create such statistical sample is a central goal of WAVES.

WAVES-Wide, which survey depth results in LMC like galaxies to be close to volume limited out to $z \sim 0.2$, is specifically designed to deliver a high fidelity group catalogue (∼4k groups with 5 or more members) probing to the very lowest halo masses. This sample size should allow the intrinsic scatter in the sub-halo/stellar mass occupation statistics to be measured. WAVES-wide will result in a proper characterisation of $\mathcal{M}_{h12}$ groups, including assessing how representative our two best studied examples are.

32.2.2 The Low Surface Brightness and Dwarf Domains

The study of the field dwarf galaxy population offers a unique testing ground for galaxy formation and the underlying physics. Cold dark matter (CDM) simulations predict that there are many more low-mass than high-mass halos remaining today [15]. Observationally there appears to be a deficit with respect to this prediction; the observed galaxy stellar mass function is not as steep as the halo mass function. This could be in part because the mass of the dark matter particle is in the keV range (i.e., WDM), which suppresses power on dwarf galaxy scales [4]. However, dwarf galaxy formation is also sensitive to the impact of the photo-ionizing background, supernovae feedback, and environmental effects [2]. Disentangling the baryonic effects from any change in the power spectrum requires large statistically representative samples.

Local surveys such as the SDSS and follow-on campaigns such as GAMA are fundamentally limited by surface brightness sensitivity of the imaging data, and the reality of the luminosity-surface brightness relation [7, 8]. Various studies including [1, 3, 9] clearly demonstrate that the SDSS becomes incomplete for systems with $\mu_e \geq 23.0$ mags/sq. arcsec which become frequent below absolute magnitudes of $M_r = -18$ mag (Fig. 32.3 left). For this reason the low surface brightness and

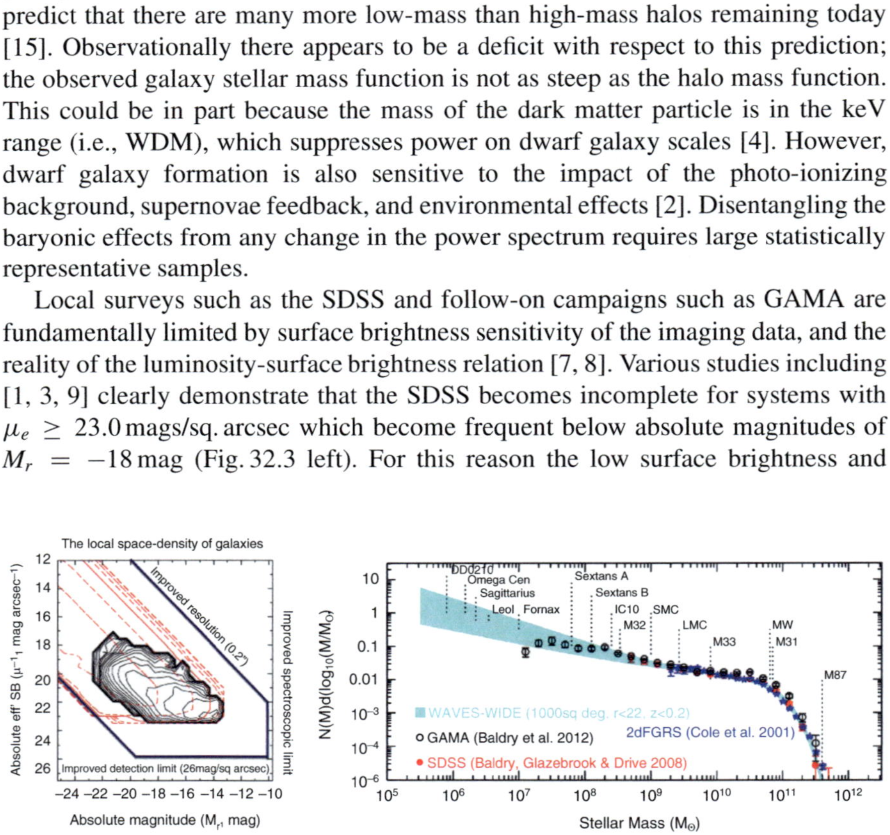

Fig. 32.3 (*left*) The space-density of galaxies in the luminosity-surface brightness plane (derived from GAMA). The logarithmic density contours in *red* show the current robust sampling region bounded by the selection limits of the Sloan Digital Sky Survey. WAVES will extend our census of the nearby galaxy population to the *blue boundary* due to the superb imaging quality and limiting surface brightness sensitivity of the VST KiDS data. (*right*) The improvement in the measurement of the stellar mass function possible with WAVES-Wide

dwarf galaxy domains continue to remain uncharted territory for modern surveys. Only deep surveys of clusters (highly unrepresentative of the average Universe) have entered this dwarf galaxy low-SB regime. The first deep wide-area imaging survey capable of probing into this domain will be VST KiDS covering 750 deg^2 in the Southern and Northern Galactic Caps. WAVES will use the VST KiDS data as its input survey to provide targets to $r_{AB} < 22$ mag with $\mu_e < 26$ mag/sq. arcsec allowing for the construction of a complete sample of galaxies to $M_r = -14$ mag, i.e., 4 mags deeper than SDSS (see the blue boundary in Fig. 32.3 left). The WAVES survey, $r_{AB} < 22$, with high-completeness for low-SB galaxies would enable the galaxy stellar mass function to be measured accurately down to $10^6\,M_\odot$ (see Fig. 32.3 right).

32.2.3 The Evolution of Galaxy Structure (with Euclid)

Studies of the mass-size relation of galaxies in the nearby ($z < 0.1$) and distant Universe ($z > 1$) show a $\times 5$ growth in galaxy sizes at fixed stellar mass (e.g., [23, 24]) and in number-density (e.g., [12]). This extraordinary result, confirmed by numerous groups (e.g., [6]), implies a dramatic physical change occurring in the galaxy population over the redshift range 0.2–1.0. Possible explanations include dynamical relaxation, major mergers, minor mergers, and disc growth [11]. However this result is only clearly established for high stellar mass systems (i.e., $>10^{11}\,M_\odot$), found in extreme dense cluster environments. Whether this growth is endemic or confined to a specific mass or environment remains unclear. A deep spectroscopic survey with HST resolution imaging over a sustained area is needed to extend the measurements to fainter mass limits and to distinguish between the competing hypotheses. WAVES-Deep in combination with Euclid imaging (see Fig. 32.4 left), provides exactly the dataset required to study this extraordinary growth.

A key starting point will be to address whether the fundamental nature of galaxies is its bimodality (red vs. blue), or its evident duality (bulges plus discs). At very low redshift ($z < 0.1$) this issue is clear-cut with multi-component decomposition a routine part of the analysis toolkit. At high redshift ($z > 1.5$) the case is less clear as galaxies no-longer appear to adhere to the simple idea of bulge plus disc systems, but exhibit highly asymmetrical and irregular shapes. As such the language of high-redshift galaxy work is typically focused on turbulence, distortions, and the global colour (red v blue).

To date the largest contiguous survey by HST covers 1.8 deg^2 (COSMOS). Euclid will transform this by imaging upto 8,000 deg^2 of sky at $0.2''$ resolution. Surveying this entire area spectroscopically is unrealistic, however WAVES-Deep will provide over 1-million galaxies in the range $0.2 < z < 0.8$ with imaging resolution sufficient to discern and measure bulge, bar, and disc components to 1 kpc resolution (see Fig. 32.4 right). This will open the door for direct measurements of the mass and size evolution of the distinct structural components (bulges, bars, discs) within a

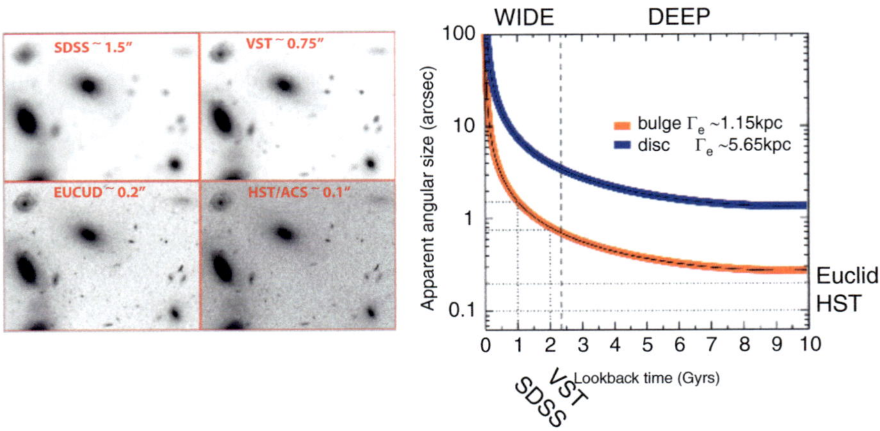

Fig. 32.4 Angular-size versus lookback time with the low-z limitations of the SDSS and VST shown. (*right*) Comparison of various imaging datasets, with median seeing as indicated

sufficiently comprehensive survey to dissect trends by halo mass, star-formation rate or a multitude of other indicators. *If* tracing the duality of galaxies is critical for understanding galaxy formation, as we suspect [11], then WAVES-Deep/Euclid will provide more than an order of magnitude advancement over COSMOS/zCOSMOS, and in doing so firmly bridge the near and far Universe.

32.2.4 The Evolving HI Universe (with ASKAP/SKA)

The SKA and its pathfinders will allow, for the first time, a direct study of the HI content of galaxies and its role in galaxy evolution over cosmologically representative volumes and over significant cosmic time. Understanding the complicated interaction between the gas content of galaxies, their environments, and other major galactic constituents, remains a largely unsolved problem. In anticipation of the next generation of observational programs, current simulations of cosmic evolution now increasingly provide measures of gas content and its phase breakdown (atomic, molecular, ionized; e.g., [14, 16, 17]), and numerous models have been advanced for the fundamental physical processes that underly the observable gas scaling relations. These models must now be tested, and our observational understanding of evolution in the scaling relations improved to advance the field.

The advantages of alignment between HI and optical spectroscopic programs are many-fold: a comparison of gas content with high fidelity measures of environment and halo mass from optical spectroscopy, metrics that are imperfectly traced by HI galaxy redshifts alone due to the strongly anti-biased nature of this population; improved treatment of issues such as galaxy confusion, satellites, and multi-wavelength counterpart identification due to the comparatively higher spatial resolution of optical data; optically motivated source finding to increase the size of HI source catalogues; and the maximal exploitation of deep HI data through the application of statistical techniques, such as HI stacking, that increase the cosmic baseline over which evolutionary measurements of gas content can be made. Optical redshift catalogues further enable the properties of the gas-rich population to be directly compared to the general galaxy population as selected by tracers much more closely linked to stellar and halo mass.

While the current generation of optical redshift surveys are well-suited to the SKA pathfinders, such datasets are absent for SKA phase 1. In particular, SKA1-mid is expected to survey a 100 $\deg^2$ region (i.e., comparable to WAVES-Deep) out to $z \sim 1$ and beyond. No such dataset currently exists, being too expensive in telescope time for current facilities, and a unique opportunity for VISTA that will be met by WAVES. Combined with data from SKA phase 1 and Euclid, Deep-WAVES will enable robust measurement of HI over half the history of the Universe, and unique measurements over cosmologically representative volumes of the complex interplay of gas content with other major galactic constituents.

32.2.5 A Legacy Resource

WAVES is designed as a legacy survey from which numerous science questions can be asked and numerous follow-on campaigns launched. For this reason the survey selection, crucial for establishing legacy value, is kept as simple as possible with only flux selection for WAVES-Deep and flux and photo-z selection for WAVES-Wide. More complex selection ultimately minimises the generic usefulness of a survey. For example the very specific selection algorithms of large cosmology programmes (e.g., BOSS) reduces the usability of such surveys for evolutionary studies of the galaxy population. There is also a direct opportunity to connect the WAVES survey to planned campaigns (Fig. 32.5) with Euclid—capable of providing $0.2''$ images over the entire WAVES region in optical and near-IR bands—SKA Phase I—capable of sampling HI to $z \sim 0.2$ over WAVES-Wide and to $z \sim 1.0$ over the WAVES-Deep region as well as the all-hemisphere/sky surveys to be conducted by LSST and eROSITA. Apart from fulfilling the science cases outlined above this presents a unique opportunity to combine a deep spectroscopic campaign with both Euclid, SKA Phase I, LSST and eROSITA to study the late-time assembly and evolution of galaxies over a broad mass and redshift baseline with robust stellar masses, gas masses, morphologies and structural decompositions.

Galaxy evolution is complex, and the current picture murky. Clarity will come from comprehensive studies based on the highest quality inputs combined with rigorous analysis such as that provided by the Sloan Digital Sky Survey [25] and Galaxy And Mass Assembly [10] teams. WAVES will game-change by extending to flux limits 4 magnitudes fainter than SDSS locally, and outward in redshift ($z \sim 1.0$). The scale of each of the proposed surveys are SDSS-like in scale and the 1-million galaxy ball-park has proven crucial in establishing key trends such as the

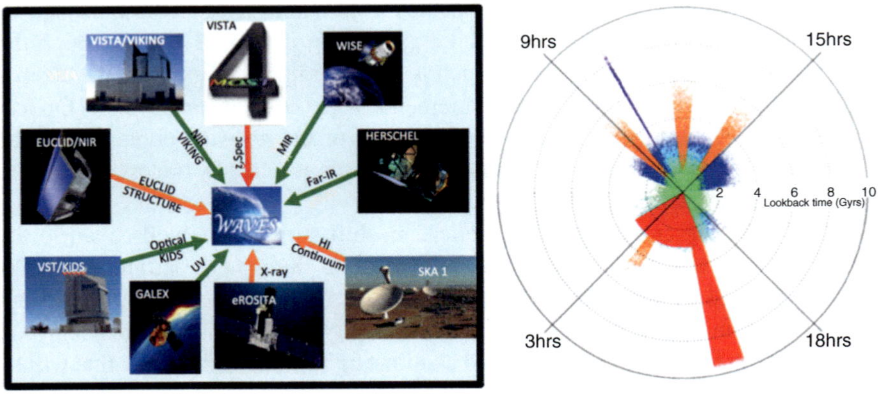

Fig. 32.5 (*left*) The facilities ultimately contributing to the WAVES survey. (*right*) A cylinder-plot showing the region of the Universe probes by WAVES compared to other notable surveys

mass-metallicity relation [22] and other key measurements. In addition conducting a fully sampled contiguous survey provides the opportunity to construct a robust halo mass catalogue to explore the role and influence of the halo on galaxy evolution [20] as well as the intrinsic properties of halos themselves [18].

Hence with WAVES-Deep one can ask questions such as whether the stellar mass-size relation of spheroids and discs evolves in the same manner irrespective of halo mass, or how does the stellar-to-gas mass fraction vary over 4dex in stellar mass from rich clusters, to filaments, to voids. Perhaps most important of all but not discussed in detail here are the possibilities opened up by the $R \sim 5{,}000$ mid-resolution spectrograph which will provide robust metalicity, age, dust and star-formation measurements (e.g., GANDALF; [21]) for the higher signal-to-noise subset (i.e., $r_{AB} < 21$ mag and S/N/Å ~ 20), of the WAVES galaxies. Finally we note that the proposed regions are or will be pre-surveyed by notable facilities including GALEX, VST (KiDs), VISTA (VIKING), WISE, and Herschel (Herschel-Atlas).

32.3 WAVES Survey Design

The survey scope is bounded by the likely capabilities of the VISTA/4MOST facility, the availability of suitable data to define a target catalogue, and consideration of available complementary datasets. Within these bounds the design is then driven by the science drivers which push towards a comprehensive (100–750 deg^2), contiguous, faint flux limited ($r_{AB} < 22$ mag) survey. The tension between studying galaxy evolution over a broad timeline versus probing down to the faintest halo mass and halo occupation limits leads to a survey split into WAVES-Deep and WAVES-Wide. For efficiency and legacy purposes these surveys are interleaved with identical

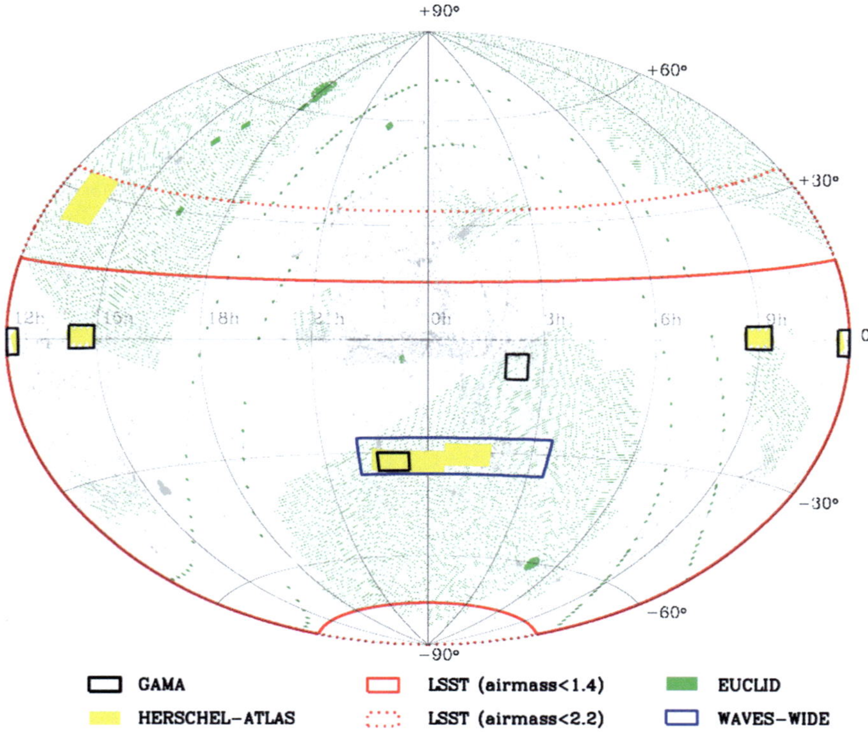

Fig. 32.6 An Aitoff projection showing the location on the sky of various surveys including GAMA, Herschel-Atlas, Euclid and WAVES-Wide

initial target selection but with an additional photo-z selection applied to WAVES-Wide (derived from *ugriZYJHK* VST/VISTA matched aperture photometry where $\Delta z/(1+z) \sim 0.03$ is realistic).

The final WAVES catalogue is expected to contain ~2 million galaxies, 140k groups, and 14k filaments from within the VST KiDS South footprint (see Fig. 32.6). This region of sky passes directly overhead at Paranal and would represent the first major spectroscopic campaign South of the Galactic equator since 2dFGRS over a decade ago.

Both WAVES surveys are designed to operate to the limiting sensitivity of the 4MOST facility, taken here as $r_{\mathrm{AB}} < 22$ mag for a 2 h (5×24 min) integrations with the Mid-Res spectrograph ($R \sim 5{,}000$, $\Delta\lambda 0.4$–$0.9\,\mu$m, S/N/Å~ 3). All 1,800 low-res fibres would be engaged, and the survey density (1,000 and 9,800 gals/deg^2 for WAVES-Wide and WAVES-Deep) is such that the WAVES-Wide footprint will be sampled only once while the WAVES-Deep footprint will be sampled ~6× (ideal for sampling the cores of dense groups). Pre-selection will consist solely of flux and colours consistent with galaxies for WAVES-Deep while WAVES-Wide will include an additional photo-z selected ($z < 0.2$) using *ugriZYJHK* photometry provided by VST KiDs and VISTA VIKING.

32.4 Summary

As of November 2014 WAVES has been adopted as one of the eight key Design Reference Surveys for VISTA/4MOST and should commence circa 2021. The WAVES team remains committed to four overriding principles: (i) simplicity of survey design, (ii) high contiguous coverage, (iii) maximising the synergy with complementary facilities, and (iv) the professional release of high quality data products to the community. Updates on the progress of the project will be broadcast via the website: http://www.wave-survey.org/

We would like to finish by thanking the organisers for the invitation to participate in "The Universe of Digital Sky Surveys" meeting, our thanks in particular to Massimo Capaccioli for his vision and resolve in delivering VST (upon which WAVES builds), and the editors for kindly allowing us to exceed our page allocation in describing the WAVES concept.

References

1. Baldry, I.K. Driver, S.P., Loveday, J., et al.: MNRAS **421**, 621 (2012)
2. Benson, A.J. Bower, R.G., Frenk, C.S., Lacey, C.G., Baugh, C.M.: ApJ **599**, 38 (2003)
3. Blanton, M.R. Lupton, R.H., Schlegel, D.J., et al.: ApJ **631**, 208 (2005)
4. Bode, P., Ostriker, J., Turok, N.: ApJ **556**, 93 (2001)
5. Boylan-Kolchin, M., Bullock, J.S., Kaplinghat, M.: MNRAS **415**, 40 (2011)
6. Bruce, V.A., Dunlop, J.S., Cirasuolo, M., et al.: MNRAS **427**, 1666 (2012)
7. Choloniewski, J.: MNRAS **214**, 197 (1985)
8. de Jong, R.S., Lacey, C.: ApJ **545**, 781 (2000)
9. Driver, S.P., Allen, P.D., Liske, J., Graham, A.W.: MNRAS **360**, 81 (2005)
10. Driver, S.P., Hill, D.T., Kelvin, L.S., et al.: MNRAS **413**, 971 (2011)
11. Driver, S.P., Robotham, A.S.G., Bland-Hawthorn, J., et al.: MNRAS **430**, 2622 (2013)
12. Faber, S.M., Willmer, C.N., Wolf, C., et al.: ApJ **665**, 265 (2007)
13. Ibata, R.A., Lewis, G.F., Conn, A.R., et al.: Nature **493**, 62 (2013)
14. Lagos, C., Del, P., Baugh, C.M., Lacey, C.G., Benson, A.J., Kim, H.-S., Power, C.: MNRAS **418**, 1649 (2011)
15. Moore, B., Ghigna, S., Governato, F., et al.: MNRAS **524**, 19 (1999)
16. Obreschkow, D., Croton, D., De Lucia, G., Khochfar, S., Rawlings, S.: ApJ **698**, 1467 (2009)
17. Power, C., Baugh, C.M., Lacey, C.G.: MNRAS **406**, 43 (2010)
18. Robotham, A.S.G., Phillipps, S., de Propris, R.: ApJ **672**, 834 (2008)
19. Robotham, A.S.G., Norberg, P., Driver, S.P., et al.: MNRAS **416**, 2640 (2011)
20. Robotham, A.S.G., Liske, J., Driver, S.P., et al.: MNRAS **431**, 167 (2013)
21. Sarzi, M., Falcón-Barroso, J., Davies, R.L., et al.: MNRAS **366**, 1151 (2006)
22. Tremonti, C.A., Heckman, T.M., Kauffmann, G., et al.: ApJ **613**, 898 (2004)
23. Trujillo, I., Forster, S.N.M, Rudnick, G.: ApJ **650**, 18 (2006)
24. van Dokkum, P.G., Franx, M.M., Holden, B., et al.: ApJ **677**, 5 (2008)
25. York, D.G., Adelman, J., Anderson, J.E. Jr., et al.: **120**, 1579 (2000)
26. Zehavi, I., Zheng, Z., Weinberg, D.H., et al.: ApJ **736**, 59 (2011)

Chapter 33
Systematic Variation of Central Mass Density Slope in Early-Type Galaxies

C. Tortora, F. La Barbera, N.R. Napolitano, A.J. Romanowsky, I. Ferreras, and R.R. de Carvalho

Abstract We study the total density distribution in the central regions ($\lesssim$1 effective radius, R_e) of early-type galaxies (ETGs), using data from the SPIDER survey (La Barbera et al., MNRAS 408:1313, 2010). We model each galaxy with two components (dark matter halo + stars), exploring different assumptions for the dark matter (DM) halo profile, and leaving stellar mass-to-light ($M_\star/L$) ratios as free fitting parameters to the data. For a Navarro et al. (ApJ 462:563, 1996) profile, the slope of the total mass profile is non-universal. For the most massive and largest ETGs, the profile is isothermal in the central regions ($\sim R_e/2$), while for the low-mass and smallest systems, the profile is steeper than isothermal, with slopes similar to those for a constant-M/L profile. For a concentration-mass relation steeper than that expected from simulations, the correlation of density slope with mass tends to flatten. Our results clearly point to a "non-homology" in the total mass distribution of ETGs, which simulations of galaxy formation suggest may be related to a varying role of dissipation with galaxy mass.

We use a sample of ~4,300 local ETGs from the SPIDER survey [6] in the redshift range 0.05–0.1. Our analysis extends the range of galaxy stellar mass ($M_\star$) probed by gravitational lensing, down to $\sim 10^{10}\,M_\odot$. We model the observed velocity dispersion of each individual galaxy using spherical isotropic Jeans equations. The

C. Tortora (✉) • F. La Barbera • N.R. Napolitano
INAF – Osservatorio Astronomico di Capodimonte, Salita Moiariello, 16, 80131 – Napoli, Italy
e-mail: ctortora@na.astro.it

A.J. Romanowsky
Department of Physics and Astronomy, San José State University, San Jose, CA 95192, USA

University of California Observatories, 1156 High Street, Santa Cruz, CA 95064, USA

I. Ferreras
Mullard Space Science Laboratory, University College London, Holmbury St Mary, Dorking, Surrey RH5 6NT, UK

R.R. de Carvalho
Instituto Nacional de Pesquisas Espaciais/MCTI Av. dos Astronautas 1758, Jd. Granja São José dos Campos – 12227-010 SP, Brazil

N.R. Napolitano et al. (eds.), *The Universe of Digital Sky Surveys*, Astrophysics and Space Science Proceedings 42, DOI 10.1007/978-3-319-19330-4_33

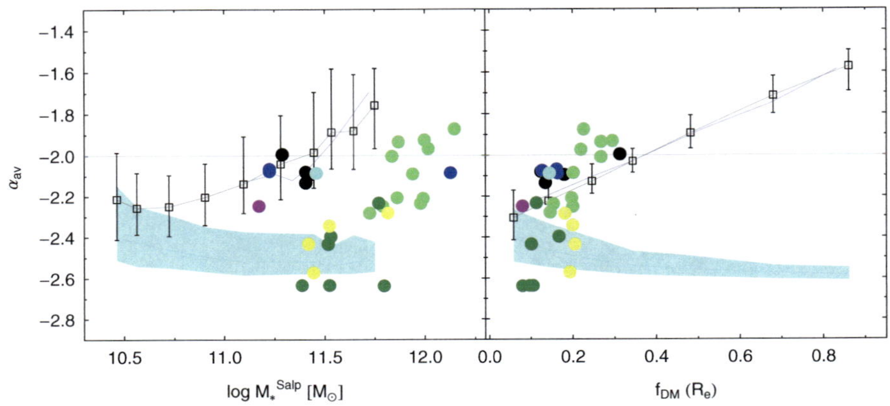

Fig. 33.1 Comparison of total mass density slope trends with predictions for simulated galaxies from [12]. Open squares and error bars are median and 16–84th percentile trends for our reference NFW halo model and a variable IMF, [12] adopt a [13] IMF. *Left* and *right panels* plot the average slope (according to the definition in [12]) vs. Salpeter-based stellar mass (derived from SPS modelling) and vs. DM fraction within 1 R_e (derived from the variable IMF models), respectively. *Purple lines* are the observed trends for the subsample of SPIDER ETGs with Salpeter-like IMF normalization. The *cyan line* and *shaded regions* mark median and 16–84th percentile slopes for the stellar mass distribution only. Simulations consist of binary (spiral-spiral and spiral-elliptical) mergers of different mass ratios and remnants of multiple mergings from cosmological simulations. Dots with different colours are simulated galaxies from [12]: *black*, *blue*, *cyan* and *pink dots* correspond to idealized single binary mergers, while *light-green*, *dark-green*, and *yellow dots* are for mergings systems drawn from cosmological simulations (See [14] for further details about colour coding)

stellar density is provided by a Sérsic fit of the photometric data from SDSS-DR6 and UKIDSS-LAS-DR4 (see [6] for further details). DM mass is modelled with a suite of profiles. Further details are provided in [14].

Independently of the DM halo adopted, we find that the Initial Mass Function (IMF) becomes bottom-heavier than a "standard" [3] distribution in high- relative to low-mass ETGs [7, 14, 15]. One of the most important results of our work is that total mass density slope in ETGs is not universal, assuming a NFW halo model [11] with standard recipes for the virial to stellar mass relation [10] and c_{vir}–M_{vir} relation [9]. Our results suggest that total mass density slope gets shallower with galaxy mass, galaxy size and central DM fraction (see Fig. 33.1). In more detail, we find that low-mass (small) ETGs have slopes consistent with those for constant-M/L profiles (see cyan shaded region in Fig. 33.1), while massive (large R_e) systems have a nearly isothermal density slope ($= -2$) or shallower, consistent with gravitational lensing results [1]. The trends of mass density slope are consistent with independent results from the literature [4, 5]. The trends of mass density slope are the same for NFW and contracted-NFW models, and do not change when assuming a fixed virial mass (and concentration) for all galaxies. When adopting a [2] profile, the slope tends to be more constant as a function of all galaxy parameters explored. However, for the most massive ETGs, the "light" haloes described by Burkert models seem to be rejected by lensing results [1]. Using $c_{\rm vir} - M_{\rm vir}$ relation from observations [8] rather than simulations affects significantly some trends of density slope with galaxy parameters. In particular, while the slope keeps increasing with galaxy radius also for such "high-concentration" models, the trends with mass become flatter in this case.

Our results draw a picture whereby *the total mass density profile in the central regions of ETGs is "non-homologous"*, approaching a constant-M/L distribution at low mass – where stars dominate the total mass budget in the ETGs center –, and an isothermal profile in the most massive ETGs, whose central regions are more DM dominated. To understand the implications of our findings in the framework of galaxy assembly, in Fig. 33.1 we have also compared our results with predictions from N-body simulations [12]. We find that both observations and simulations predict an increase of the total mass density slope with galaxy mass. This comparison indicates that black hole growth and feedback are fundamental ingredients during the formation of ETGs, as only simulations including these processes are able to reproduce the mass density slopes, DM fractions, and stellar masses in the central regions of ETGs. We argue that this trend is because gas dissipation has been more important during the formation of low-, relative to high-, mass ETGs. Thus, a steep profile is due to the formation of new stars inwards, as the gas, dissipating its kinetic energy, falls into the galaxy central regions, while gas-poor mergers tend to make the slopes isothermal.

Acknowledgements CT has received funding from the European Union Seventh Framework Programme (FP7/2007–2013) under grant agreement n. 267251 "Astronomy Fellowships in Italy" (AstroFIt)

References

1. Auger, M.W., et al.: The sloan lens ACS survey. X. Stellar, dynamical, and total mass correlations of massive early-type galaxies. ApJ **724**, 511 (2010)
2. Burkert, A.: The structure of dark matter halos in dwarf galaxies. ApJl **447**, L25 (1995)
3. Chabrier, G.: The galactic disk mass budget. I. Stellar mass function and density. ApJ **554**, 1274 (2001)
4. Dutton, A.A., Treu, T.: The bulge-halo conspiracy in massive elliptical galaxies: implications for the stellar initial mass function and halo response to baryonic processes. MNRAS **438**, 3594 (2014)
5. Humphrey, P.J., Buote, D.A.: The slope of the mass profile and the tilt of the fundamental plane in early-type galaxies. MNRAS **403**, 2143 (2010)
6. La Barbera, F., et al.: SPIDER – I. Sample and galaxy parameters in the grizYJHK wavebands. MNRAS **408**, 1313 (2010)
7. La Barbera, F., et al.: SPIDER VIII – constraints on the stellar initial mass function of early-type galaxies from a variety of spectral features. MNRAS **433**, 3017 (2013)
8. Leier, D., Ferreras, I., Saha, P.: Diagnostics of baryonic cooling in lensing galaxies. MNRAS **424**, 104 (2012)

9. Macciò, A.V., Dutton, A.A., van den Bosch, F.C.: Concentration, spin and shape of dark matter haloes as a function of the cosmological model: WMAP1, WMAP3 and WMAP5 results. MNRAS **391**, 1940 (2008)
10. Moster, B.P., et al.: Constraints on the relationship between stellar mass and halo mass at low and high redshift. ApJ **710**, 903 (2010)
11. Navarro, J.F., Frenk, C.S., White, S.D.M.: The structure of cold dark matter halos. ApJ **462**, 563 (1996)
12. Remus, R.-S., et al.: The dark halo-spheroid conspiracy and the origin of elliptical galaxies. ApJ **766**, 71 (2013)
13. Salpeter, E.E.: The luminosity function and stellar evolution. ApJ **121**, 161 (1955)
14. Tortora, C., et al.: Systematic variations of central mass density slopes in early-type galaxies. MNRAS **445**, 115 (2014)
15. Tortora, C., Romanowsky, A.J., Napolitano, N.R.: An inventory of the stellar initial mass function in early-type galaxies. ApJ **765**, 8 (2013)

Chapter 34
The Low-Mass End of the Initial Mass Function in Massive Early-Type-Galaxies

C. Spiniello

Abstract I present a spectroscopic method to constrain the low-mass end of the Initial Mass Function (IMF) from integrated optical spectra of massive early-type galaxies. I use the new set of non-degenerate optical spectroscopic indices defined in Spiniello et al. (MNRAS 438:1483, 2014) which are strong in cool giants and dwarfs and almost absent in main sequence stars. The use of these spectroscopic indicators together with an extended version of the simple stellar population models by Conroy and van Dokkum (ApJ 747:69, 2012) allows me to decouple the IMF from stellar population age, metallicity and elemental abundance, and thus to constrain the low-mass end of the IMF slope. I present evidence that the low-mass end of the IMF is not universal. A clear trend of steepening IMF slope with increasing velocity dispersion from 150 to 310 km s^{-1} is found in stacked Sloan Digital Sky Survey (SDSS) galaxies spectra with varying velocity dispersions.

34.1 Introduction

In early-type galaxies (ETGs) the stellar mass-to-light ratio ($\Upsilon_\star$) scales with the luminous mass of the system (e.g. [3, 4]). This has been explained with the observational evidence that the dark matter fraction (f_{DM}) in the internal region of these systems increases with the mass of the galaxy (e.g. [2, 18]) under the crucial assumption that the IMF is universal. However, very recently, evidence has emerged that the IMF might evolve [7] and may depend on the (stellar) mass of the system [4, 8–12, 15]. This could imply that the increase in the $\Upsilon_\star$ with galaxy mass is due to a steeper low-mass end (i.e. a larger number of low-mass stars) of the IMF rather than an increasing fraction of internal f_{DM}. To disentangle between these two scenarios a quantitative study of the luminous unresolved stellar content of galaxies is necessary.

C. Spiniello (✉)
Max-Planck Institute for Astrophysics, Karl-Schwarzschild-Strasse 1, 85740 Garching, Germany
e-mail: spini@mpa-garching.mpg.de

N.R. Napolitano et al. (eds.), *The Universe of Digital Sky Surveys*, Astrophysics and Space Science Proceedings 42, DOI 10.1007/978-3-319-19330-4_34

34.2 Optical IMF-Sensitive Indices and Stellar Population Models

Stellar features that show different strengths in M-dwarfs and cool giants can potentially reveal a galaxy's low-mass stellar content ($\leq 0.3\,M_{\odot}$) in spectra of unresolved stellar populations [6, 9, 12]. However, a large set of indicators (i.e. a large spectral coverage) and flexible simple stellar population (SSP) models, exploring a wide parameter-space, are necessary to break degeneracies between IMF variations, age, abundance pattern and effective temperature of the cool red giant (RGB) population and to study possible correlation of IMF slope variation with galaxy masses (e.g. [5, 17]).

In Spiniello et al. [12] (hereafter S14), we search for optical indices that are sensitive to cool dwarf stars and that do not or only weakly depend on age and metallicity. We found several promising indices of molecular TiO and CaH lines (see Table 1 in S14) that allowed us to determine the low-mass end of the IMF slope from integrated-light measurements when combined with age- and metallicity-dependent features such as Hβ, Mg*b*, Fe5270 and Fe5335. We studied the behaviour of these newly defined IMF-sensitive features in new SSP models presented by Conroy & van Dokkum ([6], hereafter CvD12), with variable age, $\alpha-$enhancement, single elements abundance patterns and stellar IMFs suitable for studying the integrated-light spectra of red galaxies. All of the CvD12 models use *solar* metallicity isochrones, even when synthesizing with different abundance patterns or different [α/Fe]. This is restrictive if one aims to disentangle elemental enhancements, metallicity changes and IMF variations, especially in the case of very massive ETGs, which are known to be α-enriched and to have slightly super-solar metallicity, due to star formation histories different from the solar neighborhood [1, 14]. To resolve this, at least on a qualitative level, in Spiniello et al. [13] (hereafter S15), we extended the parameter space of CvD12 (i.e. covering super-solar Z) multipling spectra for each given age and IMF with a metallicity-response function built from the MIUSCAT models [16], which cover six metallicity bins but do not allow to vary the [α/Fe]. I refer to S15 for more details.

34.3 Results: The Low-Mass End of the IMF in ETGs

In this work I compare the extended-CvD12 models to ETGs, for which they are specifically designed. In particular I use the sample of galaxies presented in Spiniello et al. [11] and analyzed in S14 and S15: ETGs from the SDSS DR8 binned in five velocity-dispersion bins spread over 150–310 km s^{-1}.

To investigate the IMF variation with galaxy mass bin and concurrently break the degeneracies between age, metallicity, abundance ratio and temperature, I use the following indices: Hβ, Mg*b*, Fe5270, Fe5335, bTiO, aTiO, TiO1, TiO2, CaH1, and CaH2, CaT1, CaT2, CaT3. The first four indicators are almost IMF-independent and they are mainly used to constrain age and [α/Fe], while the combination of CaH1 with TiO features is used to count dwarfs stars and to break the degeneracy with effective temperature. I do not use the sodium lines (NaD, NaI) because they are highly sensitive to both IMF variation and [Na/Fe] variation, and are therefore complicated to interpret (see S15 for more details). Using this new set of spectral indices defined in S14 and extending the CvD12 SSP models, I clearly demonstrate in S15 that it is possible to constrain the low-mass end of the IMF slope from optical spectra.

Following the same approach as in S14, I compare each stacked SDSS spectrum with grids of interpolated and extended SSPs, spanning a range of ages (log(age) = [0.8 − 1.15] Gyr), [α/Fe] ([−0.2 − +0.4]), total metallicity ([M/H] =[−0.4 − +0.4]), and changes in the effective temperature of the RGB ($\Delta T_{\mathrm{eff,RGB}}$ =[−200 − 50 K]) for different values of the IMF (single power-law with slope between 1.8 [MW-like] and 3.5). Indices in both the galaxy and the model spectra are measured at the same effective velocity dispersion with the same definitions and methods.

For each galaxy spectrum I compute the χ^2 values and then I obtain a probability density function (PDF) for each model via the likelihood function $L \propto \exp(-\chi^2/2)$. I marginalize over age, metallicity, ΔT_{eff} and [α/Fe] to obtain a best-fitting slope of the IMF and its uncertainty (1σ error on the cumulative probability distribution) for each velocity dispersion bin, assuming flat priors on all parameters. The result of my analysis is presented in Fig. 34.1 where I plot the IMF slope inferred from SSP analysis against the stellar velocity dispersion. Data points are SDSS ETGs spectra stacked in velocity dispersion bins, from 150 to 310 km s^{-1} and horizontal lines show the case of a Milky Way-like IMF normalization (x = 1.8) and a Salpeter IMF normalization (x = 2.35). Finally, the green solid line shows a linear regression

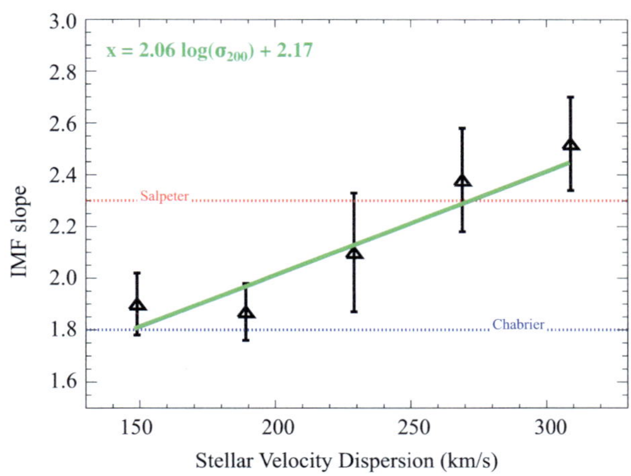

Fig. 34.1 IMF slope as a function of stellar velocity dispersion. Points are SDSS ETGs stacked by σ The Chabrier (x = 1.8) and Salpeter (x = 2.35) cases are shown as *horizontal dotted lines*. *Green line* shows a linear regression model fitted to the data

model ($x = a \times \log \sigma_{200} + b$) where x is the IMF slope and σ_{200} is the stellar velocity dispersion normalized to 200 km s^{-1}. For the stacked SDSS spectra I find:

$$x = (2.06 \pm 0.14) \times \log \sigma_{200} = (2.17 \pm 0.12) \quad (34.1)$$

This result is consistent with the one found in S14 ($a = 2.3 \pm 0.1, b = 2.13 \pm 0.15$), obtained using the publicly available CvD12 models with only solar metallicity, as well as with previously published results (e.g. [6, 9]), and provide increasing support to the emerging notion of the non-universality of the IMF: a steeper IMF slope is required for systems with increasing galaxy mass. A Milky Way-like IMF is suitable for the least-massive galaxies, Salpeter IMF better matches the intermediate-mass galaxies and a possibly bottom-heavy IMF with $x < 3$ fits for the most massive SDSS ETGs bin.

Acknowledgements C.S thanks S.C. Trager and L.V.E. Koopmans for their valuable contribution to the project.

References

1. Arrigoni, M., Trager, S.C., Somerville, R.S., Gibson, B.K.: Galactic chemical evolution in hierarchical formation models - I. Early-type galaxies in the local Universe. MNRAS **402**, 173 (2010)
2. Auger, M.W., Treu, T., Bolton, A.S., et al.: The Sloan Lens ACS Survey. X. Stellar, dynamical, and total mass correlations of massive early-type galaxies. ApJ **724**, 511 (2010)
3. Barnabè, M., Czoske, O., Koopmans, L.V.E., Treu, T., Bolton, A.S.: Two-dimensional kinematics of SLACS lenses - III. Mass structure and dynamics of early-type lens galaxies beyond $z \simeq 0.1$. MNRAS **415**, 2215 (2011)
4. Cappellari, M., McDermid, R.M., Alatalo, K., et al.: Systematic variation of the stellar initial mass function in early-type galaxies. Nature **484**, 485 (2012)
5. Conroy, C.: Modeling the panchromatic spectral energy distributions of galaxies. ARA&A **51**, 393 (2013)
6. Conroy, C., van Dokkum, P.: Counting low-mass stars in integrated light. ApJ **747**, 69 (2012)
7. Davé, R.: The galaxy stellar mass-star formation rate relation: evidence for an evolving stellar initial mass function? MNRAS **385**, 147 (2008)
8. Dutton, A.A., Mendel, J.T., Simard, L.: Evidence for a non-universal stellar initial mass function in low-redshift high-density early-type galaxies. MNRAS L412 (2012)
9. La Barbera, F., Ferreras, I., Vazdekis, A., et al.: SPIDER VIII - constraints on the stellar initial mass function of early-type galaxies from a variety of spectral features. MNRAS **433**, 3017 (2013)
10. Napolitano, N.R., Romanowsky, A.J., Tortora, C.: The central dark matter content of early-type galaxies: scaling relations and connections with star formation histories. MNRAS **405**, 2351 (2010)
11. Spiniello, C., Trager, S.C., Koopmans, L.V.E., Chen, Y.P.: Evidence for a mild steepening and bottom-heavy initial mass function in massive galaxies from sodium and titanium-oxide Indicators. ApJl **753**, LL32 (2012)
12. Spiniello, C., Trager, S., Koopmans, L.V.E., Conroy, C.: The stellar IMF in early-type galaxies from a non-degenerate set of optical line indices. MNRAS **438**, 1483 (2014)

13. Spiniello, C., Trager, S.C., Koopmans, L.V.E.: The non-universality of the low-mass end of the IMF is Robust against the choice of SSP model. ApJ **803**, 87 (2015)
14. Trager, S.C., Faber, S.M., Worthey, G., González, J.J.: The stellar population histories of early-type galaxies. II. Controlling parameters of the stellar populations. AJ **120**, 165 (2000)
15. van Dokkum, P.G., Conroy, C.: A substantial population of low-mass stars in luminous elliptical galaxies. Nature **468**, 940 (2010)
16. Vazdekis, A., Ricciardelli, E., Cenarro, A.J., et al.: MIUSCAT: extended MILES spectral coverage - I. Stellar population synthesis models. MNRAS **424**, 157 (2012)
17. Worthey, G.: Comprehensive stellar population models and the disentanglement of age and metallicity effects. ApJs **95**, 107 (1994)
18. Zaritsky, D., Gonzalez, A.H., Zabludoff, A.I.: The fundamental manifold of spheroids. ApJ **638**, 725 (2006)

Chapter 35
Luminosity Functions in the CLASH-VLT Cluster MACS J1206.2-0847: The Importance of Tidal Interactions

A. Mercurio, M. Annunziatella, A. Biviano, M. Nonino, P. Rosati, I. Balestra, M. Brescia, M. Girardi, R. Gobat, C. Grillo, M. Lombardi, B. Sartoris, and the CLASH-VLT team

Abstract We present the optical luminosity functions (LFs) of galaxies for the CLASH-VLT cluster MACS J1206.2-0847 at $z = 0.439$, based on HST and SUBARU data, including ~600 spectroscopically confirmed member galaxies. The LFs on the wide SUBARU FoV are well described by a single Schechter function down to M~M*+3, whereas this fit is poor for HST data, due to a faint-end upturn visible down M~M*+7, suggesting a bimodal behaviour. We also investigate the effect of local environment by deriving the LFs in four different regions, according to the distance from the centre, finding an increase in the faint-end slope going from the core to the outer rings. Our results confirm and extend our previous findings

A. Mercurio (✉) • M. Brescia
INAF/Oss. Astronomico di Capodimonte, via Moiariello 16, 80131, Napoli, Italy
e-mail: mercurio@na.astro.it

M. Annunziatella • M. Girardi • B. Sartoris
Dip. di Fisica, Univ. di Trieste, via Tiepolo 11, 34143 Trieste, Italy

INAF/Oss. Astronomico di Trieste, via G. B. Tiepolo 11, 34143, Trieste, Italy

A. Biviano • M. Nonino • I. Balestra
INAF/Oss. Astronomico di Trieste, via G. B. Tiepolo 11, 34143, Trieste, Italy

P. Rosati
Dip. di Fisica e Scienze della Terra, Univ. di Ferrara, via Saragat 1, 44122, Ferrara, Italy

R. Gobat
Korea Institute for Advanced Study, KIAS, 85 Hoegiro, Dongdaemun-gu Seoul 130-722, Republic of Korea

C. Grillo
Dark Cosmology Centre, Niels Bohr Inst., Univ. of Copenhagen, Juliane Maries Vej 30, 2100 Copenhagen, Denmark

M. Lombardi
Dipartimento di Fisica, Universitàädegli Studi di Milano, via Celoria 16, I-20133 Milan, Italy

CLASH-VLT team: https://sites.google.com/site/vltclashpublic/clash-vlt-team

N.R. Napolitano et al. (eds.), *The Universe of Digital Sky Surveys*, Astrophysics and Space Science Proceedings 42, DOI 10.1007/978-3-319-19330-4_35

on the analysis of mass functions, which showed that the galaxies with stellar mass below $10^{10.5}\,M_{\odot}$ have been significantly affected by tidal interaction effects, thus contributing to the intra cluster light (ICL).

35.1 Introduction

The galaxy LF, which describes the number of galaxies per unit volume as function of luminosity, is a powerful tool to study the properties of galaxy populations and to constrain their evolution through comparisons with the dark matter halo mass function. The dependence of the observed galaxy LF on the environment provides a powerful discriminator among the environmental-related mechanisms that have been suggested to drive galaxy transformations, e.g. merging, ram-pressure stripping, tidal interactions. Interestingly, in dense environments, a single Schechter function was found to be a poor fit of the LF, due to the presence of an upturn at fainter magnitudes (e.g. Agulli et al. [1] and references therein). This feature is present in dynamically evolved regions (i.e. regions with a large fraction of elliptical galaxies, a high galaxy density, and a short crossing time) but absent in unevolved regions, such as the Ursa Major cluster and the Local Group (e.g. Trentham and Hodgkin [12]). The LFs can vary from cluster to cluster or/and also in different cluster regions, depending on the mixture of different galaxy types induced by cluster-related processes. In order to asses the relative importance of the processes that may be responsible for the galaxy transformations, we have performed a photometric study of the cluster MACS J1206.2-0847 at $z = 0.439$, by examining the effect of local environment through the comparison of the LFs in four cluster regions. This study has an unprecendeted combination of depth and area, taking advantage of both multiband HST data and Subaru panoramic imaging, as well as extensive spectroscopic coverage.

35.2 Data and Catalogues

Ground-based photometric observations were carried out with the SuprimeCam at Subaru, covering $30' \times 30'$, with total exposure times of 2,400, 2,160, 2,880, 3,600 and 1,620 s in B, V, R_c, I_c and z band respectively, and seeing values between 0.7 and 1.0 arcsec. The photometric catalogues were extracted using the software SExtractor [3] in conjunction with PSFEx [4], which performs PSF fitting photometry. Spectroscopic data were acquired with the VIMOS instrument at the ESO VLT, as part of the ESO Large Programme CLASH-VLT *Dark Matter Mass Distributions of Hubble Treasury Clusters and Foundations of ΛCDM Structure Formation Models* (P.I. Piero Rosati; see [11]).

The final photometric catalogue contains ∼34,000 objects down to $R_c = 24$ mag. The CLASH-VLT campaign provided 2,749 reliable spectroscopic redshifts. To complete the spectroscopic sample, we have used photometric redshifts, calibrated on the large redshift catalog, as described in Biviano et al. [5]. Thus, our final sample includes 2,468 cluster members (590 of which are spectroscopically confirmed). In order to derive the LFs, we also applied the completeness corrections to the observed galaxy counts reported in Annunziatella et al. [2].

MACS J1206.2-0847 was observed with HST in 16 broadband filters, from the UV to the near-IR as part of the CLASH multi-cycle treasury program (see Postman et al. [9]). As described in Grillo et al. [8], in the HST/WFC3 FoV we selected cluster members by measuring the probability for a galaxy to be cluster member according to its multi-dimensional color space distribution from 12 CLASH bands (excluding the F225W, F275W, F336W, and F390W bands, due to the low signal-to-noise), based on the color distribution of the 233 galaxies in the spectroscopic sample. With this method, we obtained a total sample of 1,177 spectroscopic or photometric members, down to F814W = 26.5 mag.

This dataset has allowed us to investigate the cluster galaxy population down to M$\sim$M*+7 (corresponding to stellar masses of $\mathscr{M} \sim 10^{8.2}\,M_{\odot}$) in the central region (R<0.4 r_{200}) and M$\sim$M*+3 ($\mathscr{M} \sim 10^{9.5}\,M_{\odot}$) out to 3.5 r_{200} ($r_{200} = 1.96$ Mpc; [5]).

In order to match the photometry of the HST data in the F814W band, and of the SUBARU data in the Rc band, we derived a linear relation between the magnitudes measured in these two bands

In order to match the photometry of the HST data in the F814W band and of the SUBARU data in the R_c band, we derived a linear relation between the magnitudes measured in these two bands on.

35.3 Results

We show in Fig. 35.1 (left panel) the R_c LFs obtained from SUBARU data in four different cluster environments. We differentiate the environment in the cluster with the clustercentric distance from the brightest cluster galaxy. As reported in Annunziatella et al. [2], this is equivalent to referring to the local galaxy number density in this dynamically relaxed cluster (but see also Girardi et al. [7]). The LFs are well described by a single Schechter function. Environmental effects are apparent in the slope of LFs, which is found to be steeper in the outskirts compared to the central region, by more than 1σ. We assess the statistical significance of the difference between LFs non parametrically, via a Kolmogorov-Smirnov test, which gives a very low probability (<0.04 % in the better case) to the hypothesis that the distribution of galaxies are drawn from the same population in the first region respect to the three adiacent areas. This steepening is also in agreement with the stellar mass function (MF) presented by Annunziatella et al. [2]. As shown in Fig. 35.1 (right panel), we find a deficit of low mass galaxies in the cluster core, which adds up to $\sim 6\times10^{11}\,M_{\odot}$, computed by extrapolating the slope of the MF in the external region. Interestingly, this stellar mass matches the ICL stellar mass estimated by Presotto et al. [10]. Thus, the deficit of these galaxies

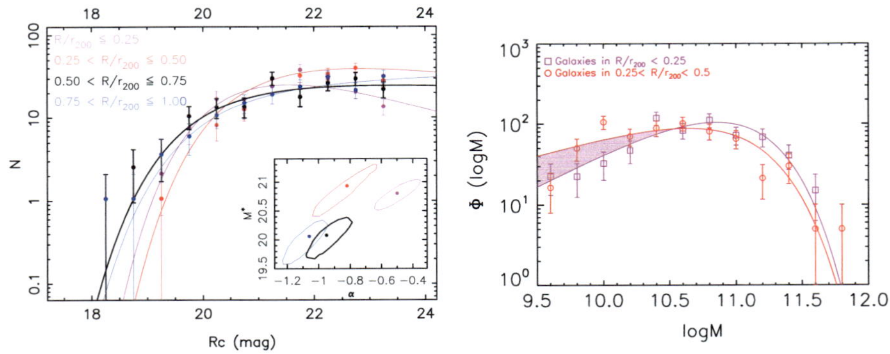

Fig. 35.1 *Left panel*: LFs of galaxies in MACS J1206.2-0847, covering regions with increasing distance from the centre. *Continuous lines* are the fits with the Schechter function. The best–fit Schechter parameters for α and M* with their 1σ contours for the corresponding LFs are also reported in the *small panel*. *Right panel*: MFs of galaxies in the core of MACS J1206.2-0847 and the outer adjacent ring. The *shaded area* highlights the deficit of low mass galaxies in the core, which adds up to $\sim 6 \times 10^{11} M_{\odot}$ (see Annunziatella et al. [2] for details)

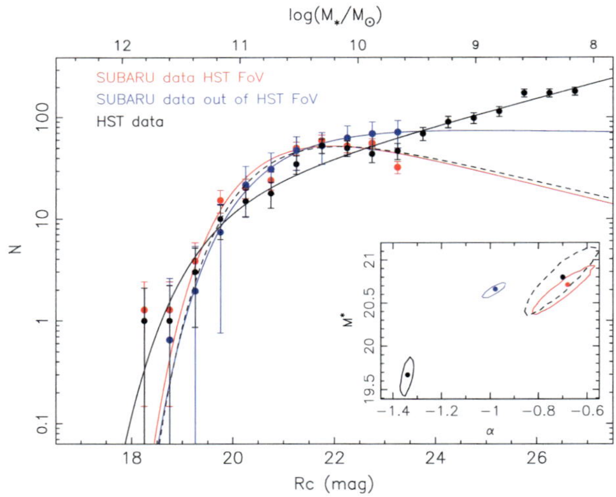

Fig. 35.2 LFs of galaxies in MACS J1206.2-0847, from HST data (*black*) and from SUBARU data in the same region covered by HST (*red*) and in the external region (*blue*). *Continuous lines* are the fits with the Schechter function. *Dashed line* is the fit to HST data down to M = M*+3. The best–fit Schechter parameters for α and M* with their 1σ contours for the corresponding LFs are also reported in the *small panel*

in the core can be interpreted as stripped stellar mass which went to populate the ICL component during the cluster assembly process. The deficit of faint galaxies could be also related to galaxy merging processes, however, we don't observe an excess of massive galaxies in cluster cores compared to more external regions (see Annunziatella et al. [2]). In Fig. 35.2, we present the LF of the cluster members in HST data (in black), compared with that obtained from SUBARU data in the same region (in red) and in the outer region (in blue). Ground-based photometric data are adequately fitted by a single Schechter function. Instead, as for the central ~0.4 r/r_{200} region mapped by HST, the fit with a single Schechter function is poor. The residuals show that the fit systematically under- and over-predicts the observed counts in the range $21<R_c<23.5$ ($10^{10.5} < \mathcal{M} < 10^{9.5}\,M_\odot$), and there is an upturn at $R_c > 23.5$, suggesting that a bimodal behaviour of the LF is more suitable to describe the LF down to $M = M^*+7$. This result confirms the scenario reported above for the ICL and supports the idea that the ICL is built-up by the tidal stripping of $\sim M^*+2$ mag galaxies (DeMaio et al. [6]). We are now extending the analysis of the LF and MF to other clusters from the CLASH-VLT dataset, three of which include ~1,000 spectroscopic members. This high-number statistics will allow us to also take into account structural and stellar population properties of the cluster members, to elucidate the impact of tidal effects in the mass assembly history of massive clusters.

Acknowledgements The authors acknowledge financial support from PRIN-INAF 2014: *Glittering Kaleidoscopes in the sky, the multifaceted nature and role of galaxy clusters* (PI M. Nonino).

References

1. Agulli, I., Aguerri, J.A.L., Sànchez-Janssen, R., et al.: Deep spectroscopic luminosity function of Abell 85: no evidence for a steep upturn of the faint-end slope. MNRAS **444**, 34 (2014)
2. Annunziatella, M., Biviano, A., Mercurio, A., et al.: CLASH-VLT: the stellar mass function and stellar mass density profile of the z = 0.44 cluster of galaxies MACS J1206.2-0847. A&A **571**, A80 (2014)
3. Bertìn, E., Arnouts, S.: SExtractor: software for source extraction. A&AS **117**, 393 (1996)
4. Bertìn, E.: Automated morphometry with SExtractor and PSFEx. ASP Conf. Ser. **442**, 393 (2011)
5. Biviano, A., Rosati, P., Balestra, I., et al.: CLASH-VLT: the mass, velocity-anisotropy, and pseudo-phase-space density profiles of the z = 0.44 galaxy cluster MACS J1206.2-0847. A&A **558**, 1 (2013)
6. DeMaio, T., Gonzalez, A., Zabludoff, A., Zaritsky, D., Bradac, M.: On the origin of the intracluster light in massive galaxy clusters. MNRAS **448**, 1162 (2015). arXiv:1501.02225
7. Girardi, M., Mercurio, A., Balestra, I., et al.: CLASH-VLT: substructure in the galaxy cluster MACS J1206.2-0847 from kinematics of galaxy populations. A&A **579**, 4 (2015). arXiv:1503.05607
8. Grillo, C., Suyu, S.H., Rosati, P., et al.: CLASH-VLT: insights on the mass substructures in the Frontier fields cluster MACS J0416.1-2403 through accurate strong lens modeling. ApJ **800**, 38 (2015)
9. Postmann, M., Coe, D., Benitez, N., Bradley, L., et al.: The cluster lensing and supernova survey with Hubble: an overview. ApJS **199**, 25 (2012)
10. Presotto, V., Girardi, M., Nonino, M., et al.: Intracluster light properties in the CLASH-VLT cluster MACS J1206.2-0847. A&A **565**, 126 (2014)
11. Rosati, P., Balestra, I., Grillo, C., et al.: CLASH-VLT: a VIMOS large programme to map the dark matter mass distribution in galaxy clusters and probe distant lensed galaxies. Msngr **158**, 48 (2014)
12. Trentham, N., Hodgkin, S.: The luminosity function of the Virgo Cluster from MB= −22 to −11. MNRAS **335**, 712 (2002)

Chapter 36
A Search for Giant Radio Galaxy Candidates and Their Radio-Optical Follow-up

I. del C. Santiago-Bautista, C.A. Rodríguez-Rico, H. Andernach, R. Coziol, J.P. Torres-Papaqui, E.F. Jiménez Andrade, I. Plauchu-Frayn, and E. Momjian

Abstract We present results of a search for giant radio galaxies (GRGs) larger than 1 Mpc in projected size. We designed a computer algorithm to identify contiguous emission regions, large and elongated enough to serve as GRG candidates, and applied it to the entire 1.4-GHz NRAO VLA Sky survey (NVSS) image atlas. Subsequent visual inspection of 1,000 such regions revealed 15 new GRGs, as well as many other candidate GRGs, some of them previously reported, for which no redshift was known. Our optical spectroscopy of 25 host galaxies with two 2.1-m telescopes in Mexico, and four others with the 10.4-m Gran Telescopio Canarias (GTC), yielded another 24 GRGs. We also obtained higher-resolution radio images with the Karl G. Jansky Very Large Array for some unconfirmed GRG candidates.

I. del C. Santiago-Bautista (✉) • C.A. Rodríguez-Rico • H. Andernach • R. Coziol • J.P. Torres-Papaqui
Deptartamento de Astronomía, DCNE, Universidad de Guanajuato, Guanajuato, Mexico
e-mail: isantiago@astro.ugto.mx; carlos@astro.ugto.mx; heinz@astro.ugto.mx; rcoziol@astro.ugto.mx; papaqui@astro.ugto.mx

E.F. Jiménez Andrade
Instituto Nacional de Astrofísica, Óptica y Electrónica, Tonantzintla, Pue., Mexico
e-mail: eric@inaoep.mx

I. Plauchu-Frayn
Instituto de Astronomía, UNAM, Ensenada, B.C., Mexico
e-mail: ilse@astrosen.unam.mx

E. Momjian
NRAO, Socorro, NM 87801, USA
e-mail: emomjian@nrao.edu

N.R. Napolitano et al. (eds.), *The Universe of Digital Sky Surveys*, Astrophysics and Space Science Proceedings 42, DOI 10.1007/978-3-319-19330-4_36

36.1 Introduction

GRGs are usually defined as having radio emission extending over a (projected) largest linear size (LLS) >1 Mpc for $H_0 = 50$–$100\,\mathrm{km\,s^{-1}\,Mpc^{-1}}$. GRGs are very rare. Indeed, in the literature prior to our study (e.g., [10, 12, 13, 16]), unifying the sizes with $H_0 = 75\,\mathrm{km\,s^{-1}\,Mpc^{-1}}$, we found only ~100 of these objects.

Statistical analyses of samples of GRGs [10, 11] suggest that their extreme sizes neither can be explained by a preferred orientation in the plane of the sky, nor by a location in less dense regions of the Universe, nor by more powerful jets feeding their lobes and thus reaching further out in intergalactic space. Instead, [11] argued that it is an exceptionally long-lasting radio activity in ~10 % of FR II sources [9] that allows GRGs to develop. On the other hand, [14] found evidence for the lobes of GRGs to be oriented normal to the major axes of galaxy overdensities near the hosts. Nevertheless, the reason why some radio galaxies become giants is still not fully understood. Moreover, only ~10 % of AGN are radio loud and a much smaller fraction still are GRGs. Determining how frequent GRGs really are can help to clarify why radio-loud AGNs are so rare in the first place. Thus, larger samples of GRGs are desirable to shed light on their origin.

Many new GRGs were recently discovered by us [2] from a visual inspection of large-area radio surveys and subsequent identification of the host, e.g., in NED [15] or the SDSS [1]. In order to increase our discovery rate we designed a computational algorithm that can be applied directly to radio survey images.

36.2 Search for GRGs in the NVSS Image Atlas

The NVSS at 1.4 GHz [7] currently provides the best combination of sensitivity to radio sources of large angular size ($\lesssim 16'$), low brightness ($1\text{-}\sigma \sim 0.45\,\mathrm{mJy\,beam^{-1}}$), angular resolution ($45''$), and coverage ($\delta_{2000} > -40°$ or 82 % of the sky). The NVSS image atlas contains 2,326 images of $4° \times 4°$ with pixels of $15'' \times 15''$. To detect new GRGs in the NVSS, we designed an algorithm to find contiguous emission regions, large and elongated enough to suggest the presence of a GRG.

First, the images were binarized by setting all pixels above 3 times the noise level to 1, and all others to 0. Then we applied the *closing* procedure, which consists of two steps: (1) the *erosion* operator sets a pixel to zero if any of its 8 neighbors is zero; (2) the *dilatation* operator sets a pixel to 1 if any of its 8 neighbor pixels has value 1. Thus, *closing* provides an image cleaned from noise pixels. After that, we perform *region growing*, which selects only contiguous regions of pixels of value 1 that are larger than a minimum number of pixels. To avoid spurious detections, we also excluded from our search those regions with noise levels $\gtrsim 0.6\,\mathrm{mJy\,beam^{-1}}$, e.g. near strong sources or close to the Galactic plane, as explained in [17].

Limiting the region size to at least 18 arcmin^2 we obtained 1,000 such regions. Since our regions were chosen to be contiguous, and many radio galaxies are known to show two or three separate, neighboring emission regions (core and lobes), we used visual inspection to detect these cases. To find the host galaxy, we overlaid NVSS contours with optical images from DSS [8] or SDSS [1]. When available, we used FIRST images [4] to look for faint ($\lesssim$2 mJy) radio cores between the radio lobes in NVSS. We found optical hosts for 160 candidates with redshifts in NED, of which 15 were previously unreported as GRGs. For many of the remaining candidates, plus several of those found in [2], we retrieved photometric redshifts from SDSS [1] or [5, 6]. This allowed us to estimate their linear sizes, and select the largest sources with sufficiently bright host galaxies for optical spectroscopy.

36.3 Follow-Up with Optical Spectroscopy and Radio Imaging

From the list of candidates with LLS $\gtrsim$ 0.7 Mpc, we obtained optical spectroscopy for hosts brighter than ~16.5 mag, using two 2.1-m telescopes in northern Mexico: at Obs. Astronómico Nacional (OAN, San Pedro Mártir) during 2013 and 2014, and at Obs. Astrofísico G. Haro (OAGH, Cananea) in April 2014. The resulting redshifts confirmed 18 GRGs with LLS >1 Mpc, and several more with smaller radio sizes. Spectra of four fainter hosts ($r \gtrsim$ 18 mag) were obtained in 2014 with the OSIRIS instrument on the 10.4-m Gran Telescopio Canarias (GTC) in Spain. Three of these were found to have LLS = 1.2–1.5 Mpc, and the other had LLS ~0.8 Mpc. The spectral activity types of the hosts observed here are as varied as we found them to be for a much larger sample we studied [3]. GRG hosts may be QSOs, Sy 1s or Sy 2s, LINERs, LLAGNs, or have no line emission, only that with increasing redshift the fraction of LLAGN decreases and the number of QSOs and Sy 1s increases.

For many of our GRG candidates the angular resolution of NVSS is not sufficient to reveal the optical host galaxy, often because the lobes are far apart and no central compact source is detected that would indicate the host. For 14 of these we obtained Karl G. Jansky Very Large Array (VLA) observations at higher angular resolution and/or frequency. Two of the most clear-cut results are shown in Fig. 36.1. Note that the core radio luminosity of the source with larger LLS is lower than that of the source with smaller LLS. These results agree with the absence of a correlation between core radio luminosity and LLS found in larger samples (see e.g. [10]).

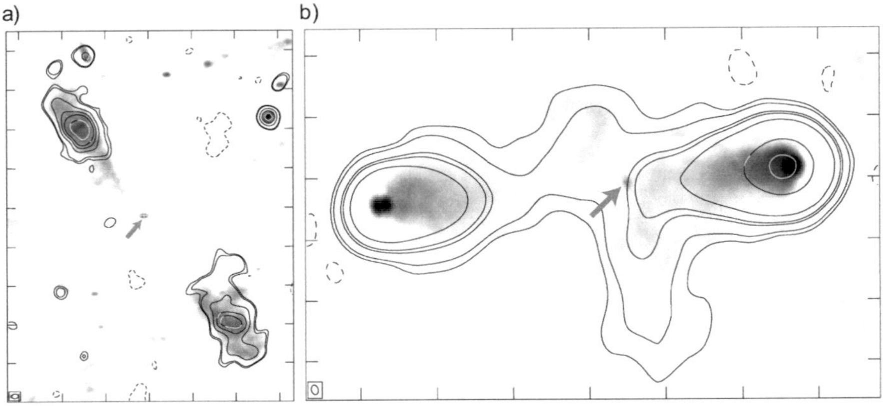

Fig. 36.1 Two of our VLA C-configuration images in grey-scale, with a noise of 0.1 mJy beam^{-1}. Contours are from NVSS, starting at 1.3 mJy beam^{-1} (3σ). (**a**) J0047+5339: NVSS shows two amorphous lobes, with no obvious radio core. Our new VLA L-band (1–2 GHz) image with $21'' \times 14''$ ($\alpha \times \delta$) resolution reveals collimated structures in the lobes for the first time. We also detect a 0.8-mJy core ($\log P_{1.4} = 3.8 \times 10^{22}$ W Hz^{-1}) of an R = 16.3 mag host galaxy, for which we obtained a redshift of 0.146 at OAN. The source's angular size of $\sim 16'$ implies an LLS of 2.3 Mpc. (**b**) J0157+0209: The NVSS image only shows diffuse emission and no radio core. Our new VLA S-band (2–4 GHz) image, with $7'' \times 10''$ ($\alpha \times \delta$) resolution, clearly detects a 1.8-mJy core ($\log P_{1.4} = 2.1 \times 10^{23}$ W Hz^{-1}), coincident with a host of $r' = 17.8$ mag at $z = 0.2217$ [1]. The source's angular size of $7'$ thus corresponds to an LLS of 1.4 Mpc

36.4 Conclusions

Our new method to find GRGs in the NVSS radio survey, together with radio-optical overlays, led us to find the host galaxy and redshift for 15 previously unreported GRGs. We obtained new redshifts for another 29 candidates using 2-m and 10-m class telescopes. Of these, we confirmed 24 new GRGs with sizes from 1.0 to 2.3 Mpc and several smaller ones. So far, our project has uncovered 39 new GRGs.

We also obtained new radio observations with the VLA of 14 GRG candidates, to better understand their radio structures, and to estimate the dynamical ages of the lobes. Analysis of these data is in progress.

With our GTC spectroscopy of GRG candidates with LLS $\gtrsim$ 2.5 Mpc at redshifts $z > 0.5$ we aim to probe the density and physical conditions of GRGs in the intermediate-redshift Universe. Among other goals, we wish to understand why the cosmological surface brightness dimming with a factor of $(1 + z)^4$ does not seem to prevent us from finding GRGs out to redshifts beyond unity.

Acknowledgements ISB, CARR, HA & JPTP were supported by DAIP-UG grant #318/13, and ISB by travel grants from CONACyT and Univ. de Guanajuato. EFJA is grateful to Academia Mexicana de Ciencias (AMC) for a summer student grant. We appreciate the help of R. F. Maldonado Sánchez in the inspection of radio survey images. This work is based upon observations acquired at Obs. Astron. Nacional, San Pedro Mártir (OAN-SPM), B.C., Mexico, at the Obs. Astrofísico G. Haro (OAGH), Cananea, Son., Mexico, and with the Gran Telescopio Canarias, La Palma, Spain. The National Radio Astronomy Observatory is a facility of the National Science Foundation operated under cooperative agreement by Associated Universities, Inc.

References

1. Alam, S., Albareti, F.D., Allende Prieto, C., et al.: The eleventh and twelfth data releases of the sloan digital sky survey: final data from SDSS-III. ApJS **219**, 12 (2015)
2. Andernach, H., Jiménez Andrade, E.F., Maldonado Sánchez, R.F., Vásquez Báez, I.R.: Conference poster, see http://adsabs.harvard.edu/abs/2012sngi.confP...1A (2012)
3. Andernach, H., Jiménez Andrade, E.F., Coziol, R.: Conference poster, see http://deep15.oal.ul.pt/wp-content/uploads/sites/4/2014/02/deep15_poster_HAndernach.pdf (2015)
4. Becker, R.H., White, R.L., Helfand, D.J.: The FIRST survey: faint images of the radio sky at twenty centimeters. ApJ **450**, 559 (1995)
5. Bilicki, M., Jarrett, T.H., Peacock, J.A., Cluver, M.E., Steward, L.: Two micron all sky survey photometric redshift catalog: a comprehensive three-dimensional census of the whole sky. Apjs **210**, 9 (2014)
6. Brescia, M., Cavuoti, S., Longo, G., De Stefano, V.: A catalogue of photometric redshifts for the SDSS-DR9 galaxies. A&A **558**, A126 (2014)
7. Condon, J.J., Cotton, W.D., Greisen, E.W., et al.: The NRAO VLA Sky Survey. AJ **115**, 1693 (1998)
8. Digitized Sky Survey, e.g., http://archive.stsci.edu/cgi-bin/dss_form.html
9. Fanaroff, B., Riley, J.: The morphology of extragalactic radio sources of high and low luminosity. MNRAS **167**, P31 (1974)
10. Ishwara-Chandra, C.H., Saikia, D.J.: Giant radio sources. MNRAS **309**, 100 (1999)
11. Komberg, B.V., Pashchenko, I.N.: Giant radio galaxies: old long-lived quasars? Astron. Rep. **53**, 1086 (2009)
12. Lara, L., Cotton, W.D., Feretti, L., et al.: A new sample of large angular size radio galaxies. I. The radio data. A&A **370**, 409 (2001)
13. Machalski, J., Jamrozy, M., Zola, S., Koziel, D.: The new sample of giant radio sources. II. Update of optical counterparts, further spectroscopy of identified faint host galaxies, high-frequency radio maps, and polarisation properties of the sources. A&A **545**, 85 (2006)
14. Malarecki, J.M., Jones, D.H., Saripalli, L., Staveley-Smith, L., Subrahmanyan, R.: Giant radio galaxies – II. Tracers of large–scale structure. MNRAS **449**, 955 (2015)
15. NASA/IPAC Extragalactic Database (NED; http://ned.ipac.caltech.edu)
16. Schoenmakers, A.P., de Bruyn, A.G., Röttgering, H.J.A., van der Laan, H.: A new sample of giant radio galaxies from the WENSS survey. I. Sample definition, selection effects and first results. A&A **374**, 861 (2001)
17. Silerio-Vázquez, M.: Herramientas para el reconocimiento de radiofuentes extendidas en grandes rastreos del Cielo. MSc thesis, CIMAT, Guanajuato, Mexico (2012)

Chapter 37
The PN Population in the M87 Halo and the Virgo Cluster Core

A. Longobardi, M. Arnaboldi, and O. Gerhard

Abstract Cosmological simulations allow us to make detailed predictions for the evolution of galaxy halos and intracluster light (ICL) in cluster environments, but the extremely low surface brightnesses of such contents make it difficult to gather observational constraints. Planetary nebulas (PNs) offer a unique tool to investigate these environments owing to their strong [OIII] emission line. Making use of a deep and extended PN sample we show that the halo of M87 and the Virgo ICL are dynamically distinct components with different parent stellar populations.

37.1 Introduction

Galaxy halos are faint stellar components made of stars gravitationally bound to the individual galaxies. In galaxy clusters these halos may be surrounded by IC stars [6]. The formation of ICL and extended halos around brightest cluster galaxies (BCGs) is closely related to the morphological transformation of galaxies in clusters [3] and, hence, their physical properties reflect the dynamical evolution of the galaxies and their hosting clusters. When studying cluster central galaxies, one main questions is where does the galaxy halo end and the IC component begin, or do we need to make a distinction at all? In observational literature the presence of an additional ICL component is usually inferred from a change of slope at large radii in the surface brightness profile [e.g., 10]. In some ΛCDM simulations [2] the system BCG+ICL is treated as a single entity consisting of all stars which are not bound to any sub-halos. Others adopt a more dynamical definition of the ICL, which identifies this component in terms of either binding energy or velocity distribution [1, 5]. With such definitions, the model BCG halo and ICL have distinct stellar populations in terms of kinematics, spatial distribution and physical properties.

A. Longobardi (✉) • O. Gerhard
MPI für extraterrestrische Physik, Giessenbachstrasse 1, D-85741 Garching, Germany
e-mail: alongobardi@mpe.mpg.de; gerhard@mpe.mpg.de

M. Arnaboldi
European Southern Observatory, Karl-Schwarzschild-Strasse 2, D-85748 Garching, Germany
e-mail: marnabol@eso.org

N.R. Napolitano et al. (eds.), *The Universe of Digital Sky Surveys*, Astrophysics and Space Science Proceedings 42, DOI 10.1007/978-3-319-19330-4_37

The Virgo cluster, the nearest large scale structure in the Universe, and its central galaxy M87, are prime targets to study the relation between BCG halo and IC component. PNs, make it possible to study M87 in its outermost regions where the halo blends into the ICL. In what follows we investigate the halo-ICL dichotomy making use of the largest PN sample in M87 acquired to date [9].

37.2 PN Photometric and Spectroscopic Surveys

We first carried out a photometric survey with Suprime-Cam@Subaru that covered a 0.43 deg^2 region centred on M87 [8], obtaining images with a narrow-band (on-band) filter centred on the redshifted [OIII]λ5,007 Å emission line (at the Virgo cluster redshift) and a broadband (off-band) V-filter. As a consequence of their bright [OIII], PNs can be identified as unresolved emission sources with positive flux on the on-band image and no detection on the off-band image. This led to a large and deep photometric sample of PNs, that we spectroscopically followed-up with FLAMES@VLT. The 287 sources classified as PNs were identified from their narrow and symmetric redshifted λ5,007 Å [OIII] emission line, the presence of the second λ4,959 Å [OIII] emission line, and the absence of significant continuum.

37.3 Kinematical Separation of M87 Halo and ICL

At a distance of $\sim$150 kpc from the centre, it is a region at which the light of M87 blends into the ICL. Analysis of our photometric sample showed that it was consistent with the superposition of different PN populations associated with M87's halo and the ICL [8]. To confirm this spectroscopically, we fit two histograms to the line-of-sight velocity distribution of the PNs (see Fig. 37.1, left). We found that the velocity distribution can be represented as the sum of a narrow Gaussian component centred on the systemic velocity of the galaxy, and a broad Gaussian component centred on $V = 999.5\,\mathrm{kms^{-1}}$, associated with the ICL. We implemented a robust technique to measure the halo velocity dispersion from the projected phase-space, and applied a sigma-clipping algorithm to identify PNs associated with the M87 halo and ICL (for further details see [9]). The two population of PNs can be seen in Fig. 37.1 (right).

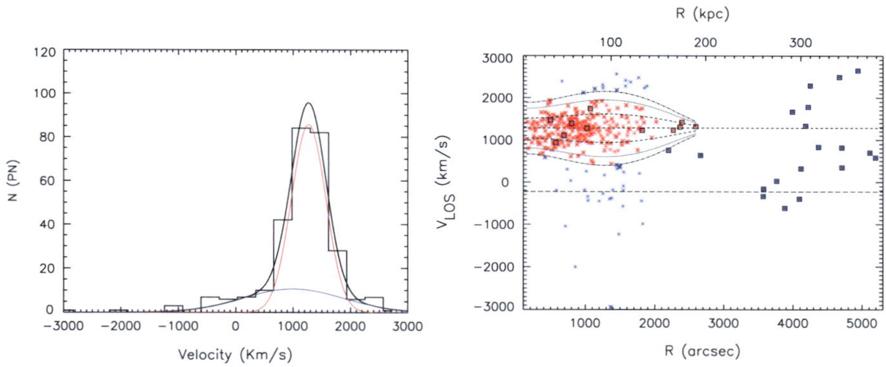

Fig. 37.1 *Left*: Line-of-site velocity distribution (LOSVD) of spectroscopically confirmed PNs, fit with a double Gaussian (*black line*). The M87 halo and ICL contribution to the total LOSVD are shown by the *red* and *blue lines*, respectively. *Right*: LOSV plotted against major axis distance from the centre of M87 for all spectroscopically confirmed PNs from this work [9] (*stars*), and [4] (*black-squares*). PNs bound to the halo are denoted in *red*, and the ICPNs are shown in *blue*. The *dashed*, and *long dashed lines* denote M87 and M86 systemic velocities, respectively

37.4 The M87 Halo and IC Component: Physical Properties

Number Density Profiles of M87 Halo and IC PNs

Using our kinematically separated PNs, we constructed density profiles for both populations and compared them to the surface brightness of M87 (see Fig. 37.2, left). We found that halo and IC PNs have different spatial distributions: the number density of halo PNs follows the galaxy's surface brightness profile, whereas the ICPNs are characterised by a shallower power-law profile, $I_{ICL} \propto R^{\gamma}$ with γ in the range $[-0.34, -0.04]$. These results are consistent with hydro-dynamical simulations that predict steeper profiles for the bound halo than for the diffuse IC component [5].

The α-Parameter

The study of the composite PN number density profile (Fig. 37.2, right) confirms the superposition of different PN populations associated with the M87 halo and the ICL, characterised by different PN specific numbers α. The observed flattening at large radii can be reproduced with a two-component photometric model [8, 9] when $\alpha_{\rm halo} = 1.06 \times 10^{-8}\ {\rm N_{PN}L^{-1}_{\odot,bol}}$ and $\alpha_{\rm ICL} = 2.72 \times 10^{-8}\ {\rm N_{PN}L^{-1}_{\odot,bol}}$, i.e. the ICL component is found to contribute three times more PNs than the bound halo population. This result is consistent with a gradient towards bluer colours at large radii and a gradient in metallicity with higher values associated to the halo [9].

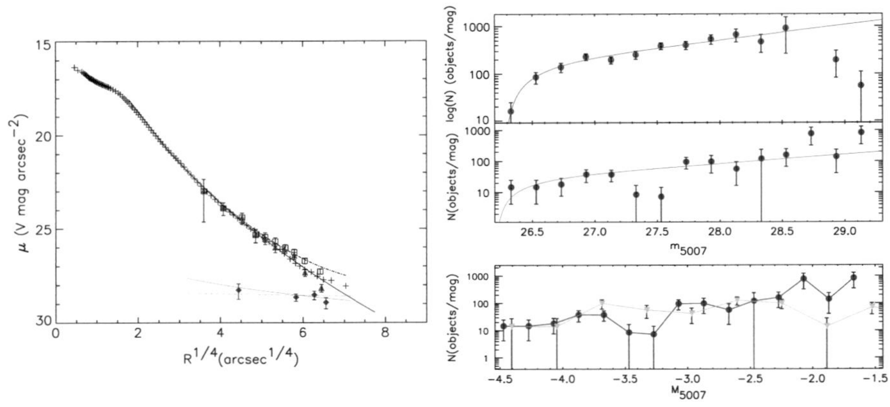

Fig. 37.2 *Left*: Logarithmic PN number density profiles for the M87 halo (*red triangles*) and IC PNs (*blue diamonds*), compared with the M87 surface brightness (*crosses* [7]). The *continuous black line* shows their Sersic fit to the halo light with $n = 11.8$. *Black squares* show the combined halo and IC PN number density profile which is well modelled by the a two-component model (see text). *Right*: Halo (*top panel*) and IC (*middle panel*) PNLFs. The *bottom panel* compares the ICPNLF (*blue circles*) with the PNLF of M33 (*cyan triangles*): both systems show dips in their LFs

PNLFs for Halo and ICL

We used our large and deep PN sample to study the PN luminosity function of halo and IC PNs down to $m_{5,007} = 28.8$ mag. This function depicts the number of PN expected for a given luminosity. We show the two PNLFs for the two PNs populations in Fig. 37.2 (right). Both PNLFs can be fit by a generalised PNLF law [9] with a cut-off magnitude at around $m_{5,007} \sim 26.3$ mag. However, the ICPNLF shows a dip at a magnitude of $m_{5,007} \sim 27.5$ mag, which is usually observed for star-forming galaxies (Fig. 37.2, right).

37.5 Conclusion

Using PNs as tracers we showed that the stellar halo of the M87 is distinct from the surrounding ICL in its kinematics, density profile, and parent stellar population, consistent with the halo of M87 being redder and more metal-rich than the ICL. Hence, in Virgo, BCG halo and ICL cannot be considered as a single component with a gradual transition in their kinematics. This supports results from analysis of galaxy cluster simulations, which suggest that the ICL component consists of unrelaxed accreted stars bound to the cluster potential, while the stellar halo of the BCG consists of a relaxed accreted component bound to the galaxy. Based on its PN population properties, we propose that the progenitors of the Virgo ICL were low-mass, star-forming galaxies.

References

1. Contini, E., De Lucia, G., Villalobos, Á., Borgani, S.: On the formation and physical properties of the intracluster light in hierarchical galaxy formation models. MNRAS **437**, 3787–3802 (2014)
2. Cooper, A.P., Gao, L., Guo, Q., Frenk, C.S., Jenkins, A., Springel, V., White, S.D.M.: Surface photometry of BCGs and intracluster stars in Lambda-CDM. ArXiv e-prints (2014)
3. De Lucia, G., Weinmann, S., Poggianti, B.M., Aragón-Salamanca, A., Zaritsky, D.: The environmental history of group and cluster galaxies in a Λ cold dark matter universe. MNRAS **423**, 1277–1292 (2012)
4. Doherty, M., Arnaboldi, M., Das, P., Gerhard, O., Aguerri, J.A.L., Ciardullo, R., Feldmeier, J.J., Freeman, K.C., Jacoby, G.H., Murante, G.: The edge of the M 87 halo and the kinematics of the diffuse light in the Virgo cluster core. Astr. Astrophys. **502**, 771–786 (2009)
5. Dolag, K., Murante, G., Borgani, S.: Dynamical difference between the cD galaxy and the diffuse, stellar component in simulated galaxy clusters. MNRAS **405**, 1544–1559 (2010)
6. Gonzalez, A.H., Zabludoff, A.I., Zaritsky, D.: Intracluster light in nearby galaxy clusters: relationship to the halos of brightest cluster galaxies. ApJ **618**, 195–213 (2005)
7. Kormendy, J., Fisher, D.B., Cornell, M.E., Bender, R.: Structure and formation of elliptical and spheroidal galaxies. ApJs **182**, 216–309 (2009)
8. Longobardi, A., Arnaboldi, M., Gerhard, O., Coccato, L., Okamura, S., Freeman, K.C.: The planetary nebula population in the halo of M 87. Astr. Astrophys. **558**, A42 (2013)
9. Longobardi, A., Arnaboldi, M., Gerhard, O., Hanuschik, R.: The outer regions of the giant Virgo galaxy M87 II. Kinematic separation of stellar halo and intracluster light. ArXiv e-prints (2015)
10. Zibetti, S., White, S.D.M., Schneider, D.P., Brinkmann, J.: Intergalactic stars in z˜ 0.25 galaxy clusters: systematic properties from stacking of sloan digital sky survey imaging data. MNRAS **358**, 949–967 (2005)

Chapter 38
The Stellar Mass Functions of the CLASH-VLT Clusters MACS J1206-0847 and Abell 209

M. Annunziatella, A. Biviano, A. Mercurio, M. Nonino, P. Rosati, I. Balestra, M. Girardi, C. Grillo, and the CLASH-VLT Team

Abstract The study of the galaxy stellar mass function, and in particular its dependence from the environment, represents a key observable to discriminate between different models of galaxy evolution. We determined the stellar mass function (SMF) of passive and star-forming (SF) galaxies in different regions of two clusters in the CLASH-VLT sample, MACS J1206.2-0847 and Abell 209. Since these two clusters are at different redshifts, the comparison between the results obtained in the two cases can inform us about the evolution of the SMF with cosmic time.

38.1 Introduction

It is well known that many galaxy properties have a bimodal distribution (e.g., [3]). Two main populations of galaxies can be identified, the bulge-dominated, high-mass, red galaxies with low star formation rate (SFR), and the disk-dominated, low-mass, blue, and high-SFR galaxies. Blue galaxies are the dominant population

M. Annunziatella (✉) • M. Girardi
Dip. di Fisica, Univ. di Trieste, via Tiepolo 11, 34143 Trieste, Italy

INAF – Osservatorio Astronomico di Trieste, via G. B. Tiepolo 11, 34131, Trieste, Italy
e-mail: annunziatella@oats.inaf.it

A. Biviano • M. Nonino • I. Balestra
INAF – Osservatorio Astronomico di Trieste, via G. B. Tiepolo 11, 34131, Trieste, Italy

A. Mercurio
INAF – Osservatorio Astronomico di Capodimonte, via Moiariello 16, 80131, Napoli, Italy

P. Rosati
Dip. di Fisica e Scienze della Terra, Univ. di Ferrara, via Saragat 1, 44122, Ferrara, Italy

C. Grillo
Dark Cosmology Centre, Niels Bohr Institute, University of Copenhagen, Juliane Maries Vej 30, 2100 Copenhagen, Denmark

CLASH-VLT Team: https://sites.google.com/site/vltclashpublic/clash-vlt-team

N.R. Napolitano et al. (eds.), *The Universe of Digital Sky Surveys*, Astrophysics and Space Science Proceedings 42, DOI 10.1007/978-3-319-19330-4_38

at high redshifts and in low-density environments, while red galaxies take over at lower redshifts and in high-density environments. This suggests a link between the redshift evolution of these two populations and the environments in which they reside. Studying the distribution of galaxy masses (i.e. the stellar mass function, SMF) in different environments and at different redshifts can help us constrain galaxy evolutionary processes. In this context, we studied in detail the SMF of passive and star-forming (SF) galaxies in different regions of the CLASH-VLT cluster of galaxies MACS J1206.2-0847 (M1206) at redshift $\overline{z} \sim 0.44$ [1] and are now comparing these results with those obtained for another CLASH cluster, Abell 209 (A209) at $\overline{z} \sim 0.21$.

38.2 Data Sample

Both clusters were observed with VIMOS at the ESO VLT as part of the CLASH-VLT ESO Large Programme Dark Matter Mass Distributions of Hubble Treasury Clusters and Foundations of ΛCDM Structure Formation Models [P.I. Piero Rosati; see 5]. Stellar masses of specroscopic and photometric member galaxies have been obtained by fitting their spectral energy distributions (SED) with the MAGPHYS [2] code, using five-band photometry data (B, V, R, I and z) obtained at the Subaru telescope.

We reach a mass completeness limit of $10^{9.5}\,M_\odot$ for the ∼1,300 members of M1206, and $10^{8.6}\,M_\odot$ for the ∼2,200 members of A209. The ratio between the number of spectroscopic and photometric members is ∼33 % and 50 % in M1206 and A209, respectively.

38.3 Results

We determined the SMF of SF and passive galaxies in different regions of the two clusters. The SMF is usually described by a [6] function:

$$\Phi(\log M) = \ln(10)\,\Phi^* \left(\frac{M}{M^*}\right)^{1+\alpha} \exp\left(-\frac{M}{M^*}\right) d(\log M), \tag{38.1}$$

where Φ^* is the normalization, α is the low-mass end slope, and M^* is the exponential cutoff of the SMF at high masses. A double Schechter function is a better fit to the SMF of the whole population of cluster members in both clusters. The SMFs of, separately, SF and passive cluster members are adequately fitted by a single Schechter function, although there is an indication that in A209 the passive galaxy SMF deviates somehow from a single Schechter. In both clusters the passive SMF dominates over the SF one, at all masses down to the completeness limit in

A209, and only down to $M_\star/M_\odot \sim 10^{10.1}$ in M1206. This could be related to the fact that A209 has a higher fraction of passive galaxies, being located at a smaller redshift than M1206.

We define the environment in the two clusters either as the clustercentric distance from the brightest cluster galaxy or in terms of local density (number of galaxies per arcmin^2). In the dynamically relaxed cluster M1206 the two definitions lead to almost coincident regions, especially close to the cluster center. This is not the case in A209, which is an elongated, dynamically not fully relaxed cluster. In Fig. 38.1 we show the SMF for passive galaxies in regions of different densities in the two clusters. There is a striking difference in the slopes of the two cluster SMF. Another difference is that we detect a clear environmental effect on the SMF of M1206, but not on the SMF of A209. In M1206 the slope of the SMF in the densest region is different from those of the other regions, in the sense that the SMF decreases more strongly at lower masses. In A209 there is a hint of a similar environmental effect, but it is not statistically very significant. We interpret the environmental effect

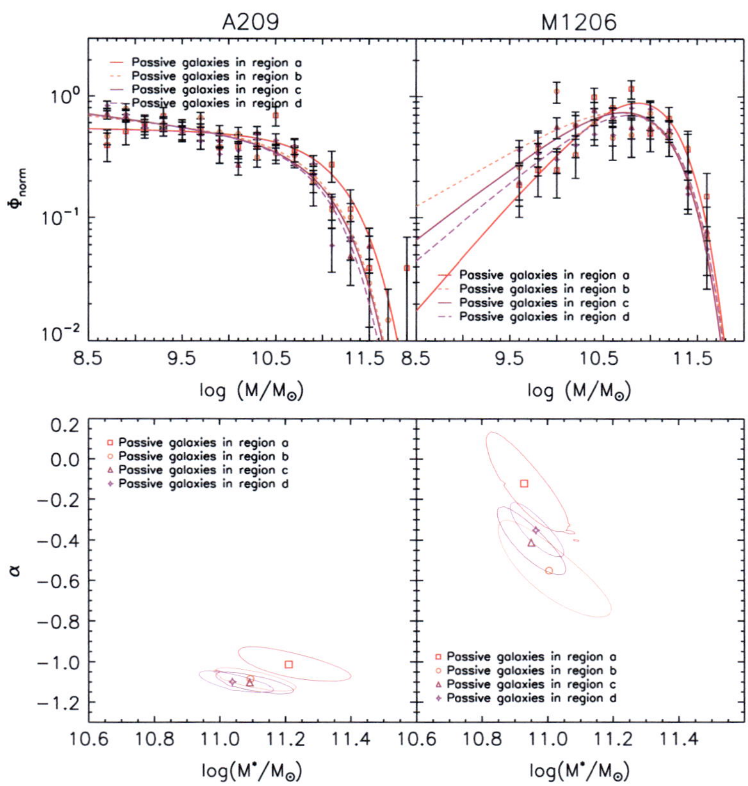

Fig. 38.1 *Upper panels*: SMFs of passive galaxies for M1206 (*left*) and A209 (*right*) in different cluster regions characterized by decreasing densities from a to d. *Lower panels*: Best-fit Schechter parameters with their 1 σ likelihood contours for the corresponding SMFs

observed for the M1206 SMFs as follows. In the most dense region of the cluster, subgiant galaxies ($M_\star \leq 10^{10.5} M_\odot$) have been destroyed by tidal interactions and have become part of the intra-cluster light (ICL). This interpretation is supported by the analysis of the cluster ICL [4]. We are now investigating the reason why such a process should be more effective in M1206 than in A209.

References

1. Annunziatella, M., Biviano, A., Mercurio, A., et al.: CLASH-VLT: the stellar mass function and stellar mass density profile of the z = 0.44 cluster of galaxies MACS J1206.2-0847. A&A **571**, A80 (2014)
2. da Cunha, E., Charlot, S., Elbaz, D.: A simple model to interpret the ultraviolet, optical and infrared emission from galaxies. MNRAS **388**, 1595 (2008)
3. Kauffmann, G., Heckman, T.M., White, S.D.M., et al.: Stellar masses and star formation histories for 105 galaxies from the Sloan Digital Sky Survey. MNRAS **341**, 33 (2003)
4. Presotto, V., Girardi, M., Nonino, M., et al.: Intracluster light properties in the CLASH-VLT cluster MACS J1206.2-0847. A&A **565**, 126 (2014)
5. Rosati, P., Balestra, I., Grillo, C., et al.: CLASH-VLT: a VIMOS large programme to map the dark matter mass distribution in galaxy clusters and probe distant lensed galaxies. Msngr **158**, 48 (2014)
6. Schechter, P.: An analytic expression for the luminosity function for galaxies. ApJ **203**, 297 (1976)

Part IV
The High-Energy Universe

Chapter 39
Supernovae as Cosmological Probes

E. Cappellaro

Abstract I review the use of SN Ia as distance indicators for measuring H_0, the Hubble constant, and the expansion history of the Universe. Most current estimates of H_0 are in the range 74–76 km s^{-1}Mpc^{-1}, in significant disagreement with the PLANCK's CMB estimate that is 10 % smaller. The main issues for SN Ia calibration, namely the luminosity vs. light curve shape relation and the correction for dust extinction are briefly addressed. SN Ia have been the key for the discovery of the acceleration of the cosmic expansion and in the near future they are expected to give a significant contribution to reveal the nature of dark energy.

39.1 Introduction

The link of supernovae with cosmology began with their distinction from ordinary novae in the context of the *Great Debate* on the nature of *nebulae* that, a century ago, marked the beginning of modern cosmology [43]. It was argued that if the *island nebulae* were extra-galactic, the novae that they were hosting would have an "impossibly great absolute magnitude", far more luminous than ordinary Galactic novae [11, 39]. On the other hand, there was evidence for a significant dispersion of the magnitudes of galactic novae that prompted for the separation into a lower and an upper class, the latter reaching an absolute magnitude similar to that of the system where they appear [21]. After the determination of the actual scale of the Universe, the upper class novae were labelled supernovae (SNe) with the suggestion that they were related to the formation of neutron stars [5].

Hereafter, SNe have become important as tracers of the galaxy stellar population and of the cosmic star formation rates (see Botticella et al., this volume) and as counterparts of enigmatic high energy events, i.e. cosmic rays, gamma-ray bursts

E. Cappellaro (✉)
INAF, Osservatorio Astronomico di Padova, vicolo dell'Osservatorio 5, 35122 Padova, Italy
e-mail: enrico.cappellaro@oapd.inaf.it

N.R. Napolitano et al. (eds.), *The Universe of Digital Sky Surveys*, Astrophysics and Space Science Proceedings 42, DOI 10.1007/978-3-319-19330-4_39

and gravitational waves (Grado et al., this volume). More important, SNe have consolidated their role of accurate cosmological probes for the measurement of the Hubble constant and of the cosmic expansion history. In the following I will briefly review the contribution of SNe in this field and the prospects for future experiments.

39.2 SNe and the Hubble Constant

The history of the Hubble constant determination is shown in an instructive plot by Huchra [19]. The early measurements were affected by severe underestimation of the nebulae distances that it took some time to sort out until, in the early 1970s, the estimates of H_0 appeared to converge. However, at this point a controversy began that lasted over two decades between different groups that, while progressively reducing their error-bars, were claiming incompatible values at the extremes of the range 50–100 km s^{-1} Mpc^{-1} (in the following H_0 units are omitted). The first attempts to use SN Ia to measure the Hubble constant did not solve the controversy [12, 35] as shown in Fig. 39.1. An exception was the result of Capaccioli et al. [9] that, by combining novae and SN Ia, derived the relative distance of Virgo and Coma clusters and a value of $H_0 = 70 \pm 15$. In fact, this value is very close to those obtained in the last decade using Cepheid's calibrated SN Ia [3, 15]. The most recent analysis appear to converge to H_0 values between 74–76 with errors <5 % [14, 16, 33, 40]. We note however that there is a group still arguing for a smaller value, $H_0 = 63.7 \pm 2.3$ [37, 42] that, given the small errors, is incompatible with the above estimates.

The apparent comforting convergence of (most) H_0 measurements was perturbed by the publication of the cosmological parameters derived from the CMB observations by Planck. The temperature and lensing data interpreted with a six-parameter inflationary ΛCDM cosmology give $H_0 = 67.8 \pm 0.9$ [31]. Although formally this value sits in between the SN Ia measurements listed above, most researchers

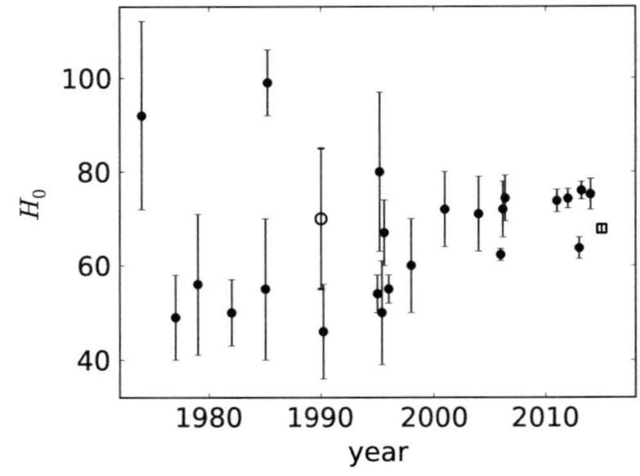

Fig. 39.1 The history of H_0 measurements through SNe. The *open circle* is the Capaccioli et al. [9] estimate discussed in the text. The *open square* is the estimate obtained from the analysis of the CMB map obtained by the Planck satellite [31]

would agree that there is a real tension between Planck and SN Ia estimates. This is an important issue because as stressed by the Planck collaboration [31] if this is confirmed it will be strong evidence for additional physics beyond the base ΛCDM model.

39.3 SN Ia Calibration Issues

The use of SN Ia as distance indicators requires accurate calibration and correction for dust extinction. Until the early 1990s SN Ia were considered standard candles with a very low dispersion of the absolute magnitude at maximum ($\sim$0.2 mag) mainly attributed to observational errors rather than to intrinsic differences [36]. Actually, it was soon obvious that there is some diversity in light curve evolutions with *fast* and *slow* events, although at the beginning no evidence was found for a difference in absolute magnitudes between the two sub-types [6]. Eventually, the improved photometry obtained with CCD detectors revealed the existence of a correlation between light curve shape and absolute magnitude, with slow declining objects being more luminous [18, 29]. This correlation, now a fundamental tool for the accurate calibration of SN Ia, has been related to variation of the mass of ^{56}Ni produced in the explosion. However, the residual scatter around the relation appears to be intrinsic and pointing to a second parameter [7] possibly related to asymmetries in the explosion along with varying viewing angles for different events [22].

A key issue for SN calibration is the correction for dust extinction that is typically derived from the color excess measured after assuming that SN Ia have all similar intrinsic color [30]. The color excess is converted into an absorption estimate by means of the relation $A_V = R_V \times E(B - V)$. A surprising finding was that the dispersion of the SN Ia absolute magnitudes is minimized by adopting a value of R_V that is significantly smaller than the standard reference value of 3.1 [9]. This result has been confirmed in all recent studies based on similar statistical arguments [3, 8, 20] but also on detailed analysis of single events [4, 13] that return $R_V \simeq 1.4$–1.8. Whether this effect is due to peculiar properties of the dust around SN Ia or is a consequence of multiple scattering in dense clouds is still unclear [24, 45].

39.4 SNe and the Acceleration of Cosmic Expansion

The first attempt, in the mid 1970s, to use SNe to test cosmic expansion was defeated by large errors of the photometry and poor SN calibration [34]. At that time, the paradigm was that by measuring the deceleration parameter it was possible to obtain an estimate of the matter density in the Universe. Measuring cosmological distances appeared to require a new generation of telescopes/instruments and in particular, a project under study at that time, a Space Telescope [10, 44]. Actually, the advancement of CCD detectors allowed to start paving the road already using ground based telescopes: the first distant SN, at $z = 0.31$, was found with the Danish 1.5 m telescope in 1989 [23] followed in 1992 by a SN Ia at $z = 0.46$ found with the INT 2.5 m telescope [26]. From here on the work of two groups, the SN Cosmology Project (SCP[1]) and the High-z team[2] and, finally, the overwhelming power of HST, made the progresses very rapid. Early reports, based on small SN samples, were cautious [17, 27] but eventually the two competing groups almost simultaneously announced the same surprising result: the Universe expansion was accelerated [28, 32, 38]. This results properly complemented by CMB measurements, that constrain the Universe to a flat curvature, required the presence of a dominant cosmological constant or of some sort of dark energy. The analysis of the latest SN data, a combination of local SN Ia, Sloan Digital Sky Survey for low redshift, Supernova Legacy Survey for intermediate redshifts and HST for high redshift, confirms that for a flat Universe $\Omega_M = 0.27$, hence $\Omega_\Lambda = 0.73$, with an error of only 5 % [41].

The discovery of the acceleration of the Universe, that led to the Nobel Prize in Physics in 2011, immediately raised a new question: what is the nature of the mysterious *dark energy* ?

Answering this question requires an accurate measurement of w, the ratio of pressure to energy density for this component, i.e. the equation of state of dark energy. In particular, considering that in general w may not be constant, the equation of state is parametrized as $w(a) = w_0 + (1-a)w_a$ where a is the scale factor, w_0 the present value and w_a its rate of change. A conventional estimate of the ability of a given experiment, or combination of experiments, to measure the equation of state is the figure of merite defined as $FoM = 1/\sigma(w_0)\sigma(w_a)$ [1].

Current estimates using combination of SNe and CMB measurements indicate that $w_0 \sim 1$ with no evidence for a dynamical dark energy, i.e. w_a consistent with 0, though the errors are still fairly large, with $FoM \sim 10$ [41].

The goal of future experiments is to measure all cosmological parameters with an accuracy that is one order of magnitude better than those available today. A detailed analysis has shown that this requires the combination of all four principal techniques, namely baryon acoustic oscillations, galaxy clustering, weak lensing and SNe and, for all probes, improving significantly statistical and systematic uncertainties [1, 25]. For SNe, this means to increase the number of high redshift

---

[1] http://supernova.lbl.gov

[2] https://www.cfa.harvard.edu/supernova/HighZ.html

SN Ia with good photometry from hundreds to thousands. For systematic errors the emphasis is in cross-band calibration because typically nearby and distant SNe are observed in different bands. We also need to constrain the possible evolution of SN Ia properties, that requires both empirical tests but also improvements of our understanding of SN Ia physics.

39.5 On-Going and Future SN Surveys

A running project that is aimed to make a significant progress in this field is the Dark Energy Survey[3] (DES). This project consists in a wide area survey using the 3 deg^2 DECam instrument installed at the 4 m Blanco telescope at CTIO. DES will exploit the different techniques mentioned above; for SNe it features a 30 deg^2 time-domain survey that is expected to produce few thousand SN Ia up to $z = 1$. The redshift limit is set by the fact that, at higher redshifts, the bulk of SN emission moves in the near-infrared (IR). In fact, to obtain a large sample of SN Ia at $z > 1$ requires an IR, wide field imager in space. A unique opportunity in this respect is offered by EUCLID,[4] the ESA medium size mission that is planned to flight in the early 2020s. The main tools of EUCLID are weak lensing and baryonic acoustic oscillations that will be addressed by mean of a combined optical/near-infrared survey of 15,000 deg^2. In addition, it was argued that by using 6 months of the EUCLID time we can obtain a sample of 1,700 SN Ia in the redshift range $0.75 < z < 1.55$ that, combined with a LSST sample (see next), can give a $FoM \sim 200$, from SNe only [2]. The decision about the actual implementation of a SN survey in EUCLID is still pending.

LSST[5] is a 8.4 m telescope that is expected to perform a wide-field astronomical survey of over 20,000 deg^2. Each patch of sky will be visited about 1,000 times in 10 years. In the standard operating mode, providing frequent all-sky coverage, LSST is expected to discover roughly 250,000 SNe per year, with SN Ia at a mean redshift of about 0.45. SNe will be discovered also from a "staring mode" search of a more limited area of sky. After 10 years and 10 min per night spent staring at a single field will yield 14,000 SNe with a mean redshift of 0.75.

Finally, the Wide-Field Infrared Survey Telescope (WFIRST[6]), a NASA space observatory, is planned to perform wide-field imaging and slitless spectroscopic surveys in the near-IR. The current design of the mission (named AFTA) makes use of an existing 2.4 m telescope. For SNe the baseline is to perform a multi-tier survey scanning different sky areas with progressively deeper limits that is expected to produce 2,500 SN Ia up to $z \sim 1.7$.

---

[3]http://www.darkenergysurvey.org

[4]http://sci.esa.int/euclid

[5]http://www.lsst.org/lsst/

[6]http://wfirst.gsfc.nasa.gov

Therefore, the prospects are that in the next two decades SN Ia will maintain a key role as cosmological probes and they will give a major contribution to address the nature of dark energy.

References

1. Albrecht, A., Bernstein, G., Cahn, R., et al.: Report of the dark energy task force. Astro (2006). arXiv:astro-ph/0609591
2. Astier, P., Balland, C., Brescia, M., et al.: Extending the supernova Hubble diagram to $z \sim 1.5$ with the Euclid space mission. A&A **572**, A80 (2014)
3. Altavilla, G., Fiorentino, G., Marconi, M., et al.: Cepheid calibration of type Ia supernovae and the Hubble constant. MNRAS **349**, 1344 (2004)
4. Amanullah, R., Goobar, A., Johansson, J., et al.: The peculiar extinction law of SN 2014J measured with the Hubble space telescope. ApJ **788**, L21 (2014)
5. Baade, W., Zwicky, F.: Cosmic rays from super-novae. PNAS **20**, 259 (1934)
6. Barbon, R., Capaccioli, M., Ciatti, F.: Studies of supernovae. A&A **44**, 267 (1975)
7. Benetti, S., Cappellaro, E., Mazzali, P.A., et al.: The diversity of type Ia supernovae: evidence for systematics?. ApJ **623**, 1011 (2005)
8. Burns, C.R., Stritzinger, M., Phillips, M.M., et al.: The Carnegie supernova project: intrinsic colors of type Ia supernovae. ApJ **789**, 32 (2014)
9. Capaccioli, M., della Valle, M., D'Onofrio, M., Rosino, L.: Distance of the Large Magellanic Cloud through the maximum magnitude versus rate of decline relation for novae. ApJ **360**, 63 (1990)
10. Colgate, S.A.: Supernovae as a standard candle for cosmology. ApJ **232**, 404 (1979)
11. Curtis H.: The scale of the universe. Bull. Nat. Res. Coun. **2**, 194 (1921)
12. de Vaucouleurs G., Corwin H.G., Jr.: The distance of the Hercules supercluster from supernovae and SBC spirals, and the Hubble constant. ApJ **297**, 23 (1985)
13. Elias-Rosa, N., Benetti, S., Cappellaro, E., et al.: Anomalous extinction behaviour towards the type Ia SN 2003cg. MNRAS **369**, 1880 (2006)
14. Fiorentino G., Musella I., Marconi, M.: MNRAS **434**, 2866 (2013)
15. Freedman, W.L., Madore, B.F., Gibson, B.K., et al.: Cepheid theoretical models and observations in HST/WFC3 filters: the effect on the Hubble constant H_0. ApJ **553**, 47 (2001)
16. Freedman, W.L., Madore, B.F., Scowcroft, V., et al.: Carnegie Hubble program: a mid-infrared calibration of the Hubble constant. ApJ **758**, 24 (2012)
17. Garnavich, P.M., Jha, S., Challis, P., et al.: Supernova limits on the cosmic equation of state. ApJ **509**, 74 (1998)
18. Hamuy, M., Phillips, M.M., Suntzeff, N.B., Schommer, R.A., Maza, J., Aviles, R.: The absolute luminosities of the Calan/Tololo type IA supernovae. AJ **112**, 2391 (1996)
19. Huchra, J.: The Hubble Constant (2008). https://www.cfa.harvard.edu/~dfabricant/huchra/hubble/index.htm
20. Kelly, P.L., Filippenko, A.V., Burke, D.L., Hicken, M., Ganeshalingam, M., Zheng, W.: Distances with $< 4\%$ precision from type Ia supernovae in young star-forming environments. Sci. **347**, 1459 (2015)
21. Lundmark, K.: The motions and the distances of spiral. MNRAS **85**, 865 (1925)
22. Maeda, K., Benetti, S., Stritzinger, M., et al.: An asymmetric explosion as the origin of spectral evolution diversity in type Ia supernovae. Nature **466**, 82 (2010)
23. Norgaard-Nielsen, H.U., Hansen, L., Jorgensen, H.E., Aragon Salamanca, A., Ellis, R.S.: The discovery of a type IA supernova at a redshift of 0.31. Nature **339**, 523 (1989)
24. Patat, F., Taubenberger, S., Cox, N.L.J., et al.: Properties of extragalactic dust inferred from linear polarimetry of type Ia supernovae. A&A **577**, 53 (2014). arXiv:1407.0136

25. Peacock, J.A., Schneider, P., Efstathiou, G., et al.: Report by the ESA-ESO working group on fundamental cosmology. ewg3.rept (2006). arXiv:astro-ph/0610906
26. Perlmutter, S., Pennypacker, C.R., Goldhaber, G., et al.: Cosmology from 7 high-Redshift supernovae: type Ia homogeneity at $Z \sim 0.4$ and the measurement of q_0. ApJ **440**, L41 (1995)
27. Perlmutter, S., Gabi, S., Goldhaber, G., et al.: Measurements of the cosmological parameters Ω and Λ from the first seven supernovae at $z \geq 0.35$. ApJ **483**, 565 (1997)
28. Perlmutter, S., Aldering, G., Goldhaber, G., et al.: Measurements of Ω and Λ from 42 high-redshift supernovae. ApJ **517**, 565 (1999)
29. Phillips M.M.: The absolute magnitudes of type ia supernovae. ApJ **413**, L105 (1993)
30. Phillips, M.M., Lira, P., Suntzeff, N.B., Schommer, R.A., Hamuy, M., Maza, J.: The Reddening-Free decline rate versus luminosity relationship for type ia supernovae. AJ **118**, 1766 (1999)
31. Planck Collaboration, et al.: Planck 2015 results. XIII. Cosmological parameters (2015). arXiv:1502.01589
32. Riess, A.G., Filippenko, A.V., Challis, P., et al.: Observational evidence from supernovae for an accelerating universe and a cosmological constant. AJ **116**, 1009 (1998)
33. Riess, A.G., Macri, L., Casertano, S., et al.: A 3% Solution: determination of the Hubble constant with the Hubble space telescope and wide field camera 3. ApJ **752**, 76 (2012)
34. Rust, B.W.: Use of supernovae light curves for testing the expansion hypothesis and other cosmological relations. PhDT (1974)
35. Sandage A., Tammann G.A.: Steps toward the Hubble constant. VIII – the global value. ApJ **256**, 339 (1982)
36. Sandage A., Tammann G.A.: The Hubble diagram in V for supernovae of type IA and the value of H(0) therefrom. ApJ **415**, 1 (1993)
37. Sandage, A., Tammann, G.A., Saha, A., Reindl, B., Macchetto, F.D., Panagia, N.: The Hubble constant: a summary of the Hubble space telescope program for the luminosity calibration of type Ia supernovae by means of Cepheids. ApJ **653**, 843 (2006)
38. Schmidt, B.P., Suntzeff, N.B., Phillips, M.M., et al.: The high-Z supernova search: measuring cosmic deceleration and global curvature of the universe using type IA supernovae. ApJ **507**, 46 (1998)
39. Shapley, H.: The scale of the universe. Bull. Nat. Res. Coun. **2**, 171 (1921)
40. Sorce, J.G., Tully, R.B., Courtois, H.M., Jarrett, T.H., Neill, J.D., Shaya, E.J.: From Spitzer Galaxy photometry to Tully-Fisher distances. MNRAS **444**, 527 (2014)
41. Sullivan, M., Guy, J., Conley, A., et al.: SNLS3: constraints on dark energy combining the supernova legacy survey three-year data with other probes. ApJ **737**, 102 (2011)
42. Tammann, G.A., Reindl B.: The luminosity of supernovae of type Ia from tip of the red-giant branch distances and the value of H_0. A&A **549**, AA136 (2013)
43. Trimble, V.: The 1920 Shapley-Curtis discussion: background, issues, and aftermath. PASP **107**, 1133 (1995)
44. Wagoner, R.V.: The deceleration parameter, supernovae and the space telescope. Grg. Conf. 390 (1977)
45. Wang, L.: Dust around type Ia supernovae. ApJ **635**, L33 (2005)

Chapter 40
Unveiling Accreting White Dwarf Binaries in Hard X-Ray Surveys

D. de Martino

Abstract Among hard X-ray sources detected in the *INTEGRAL/IBIS* and *Swift/BAT* surveys, those identified as accreting white dwarf binaries recently boosted in number, representing ~20 % of the Galactic sample. The majority are identified as Cataclysmic Variables with magnetic white dwarf primaries suggesting that this subclass could be an important costituent of the galactic population of X-ray sources. In this paper I present the results of an on-going follow-up progamme in the X-rays aiming at identifying the true nature of newly discovered sources.

40.1 Introduction

Since mid-1970s, accreting white dwarf (WD) binaries were known as sources of X-rays, belonging to the broader class of Low Mass X-ray Binaries (LMXBs) but a few tens were known. Thanks to a prominent soft X-ray emission (kT $\leq$ few eV), about few hundreds were detected in the *ROSAT* All Sky Survey (RASS) [1–3]. These include Cataclysmic Variables (CVs) (most non-magnetic Dwarf Novae (DNs) in quiescence and magnetic polars), AM CVn and Super Soft X-ray sources, the latter encompassing Novae in the Milky Way and in the Magellanic Clouds.

Our view of the X-ray sky has today dramatically changed thanks to the recent deep *INTEGRAL* and *Swift* hard X-ray surveys with more than 1,000 sources detected above 20 keV [4, 5]. The *INTEGRAL/IBIS* extensive survey of the galactic plane has shown that our knowledge of the Galactic X-ray binaries was poor, surprisingly detecting a large number of Catalysmic Variables (CVs), most (~70 %) of the magnetic Intermediate Polar (IP) type [4, 6]. The complementary *Swift/BAT* survey mainly covering high latitudes, still confirms the high (~60 %) incidence

D. de Martino (✉)
INAF–Capodimonte Astronomical Observatory, Via Moiariello 16, 80131 Naples, Italy
e-mail: demartino@oacn.inaf.it

N.R. Napolitano et al. (eds.), *The Universe of Digital Sky Surveys*, Astrophysics and Space Science Proceedings 42, DOI 10.1007/978-3-319-19330-4_40

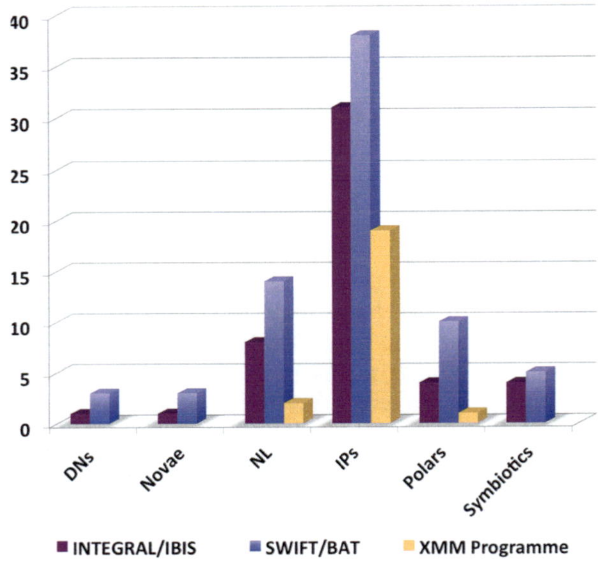

Fig. 40.1 The distribution of CV types detected by *INTEGRAL* and *Swift* using latest catalogue releases (misidentifications were corrected). Identified systems from our follow-up programme are also shown

of IPs. Figure 40.1 shows the CV types detected in both surveys that include of DNs, Nova-like (NLs) (many are disputed to be magnetic), old Novae, Symbiotics and the magnetic IPs and polars. IPs are believed to harbour weakly magnetized accreting white dwarfs (WDs) ($\leq$10 MG), whilst polars contain strongly magnetized (B $\sim$ 10–240 MG) primaries. Basic differences are the asynchronous WD rotation (P_{spin} $\sim$mins) and a hard optically thin ($kT_{brem.}$ $\sim$20–40 keV) emission in the IPs, whilst the polars have orbitally-locked ($P_{spin=orb}$ $\sim$hrs) WDs and possess a strong soft optically thick (kT_{bb} $\sim$30–50 eV) emission. The current roster of hard detected CVs totals to 72 systems representing $\sim$20 % of Galactic X-ray sources in these surveys.

The existence of an extended population of yet unrecognized low-luminosity X-ray sources of galactic origin contributing to the X-ray background (XRB) was already suggested at the time of *ROSAT* survey [7]. A decade later deep *Chandra* observations of the Galactic centre revealed thousands of dim sources and ascribed to CVs of the magnetic IP type [8]. This was corroborated by further pointings of Galactic bulge fields [9, 10]. Observations with *INTEGRAL*, *Rossi-XTE Suzaku* and *XMM-Newton* satellites also allowed to resolve the Galactic X-ray Ridge Emission (GRXE) into thousands of discrete low-luminosity sources mainly attributed to IPs [11–14]. However the true contribution to the X-ray luminosity function at faint levels is still disputed as well as the true space densities of these binaries [15, 16]. The knowledge of X-ray binary populations is crucial to understand close binary evolution. The high ($\sim$20 %) incidence of magnetism in CVs with respect to that in single WDs ($\sim$10 %) would either imply that CV formation is favoured by magnetism or CV production enhances magnetism [17].

The negligible absorption in the hard X-rays and the flux limits of the *INTEGRAL* and *Swift* surveys can allow one to detect magnetic CVs up to $\sim$1 kpc ($\sim 10^{33}$ erg s^{-1}), hence to obtain the first volume-limited sample of these systems.

40.2 Results from X-Ray Follow-Ups

The newly identified hard X-ray sources require follow-ups. While optical spectroscopy [18–20] provides first selection of suitable candidates, the true nature can be inferred through the detection of X-ray pulses at the WD spin period (that implies magnetically channeled accretion) and the study of broad-band X-ray spectra. To this purpose we are pursuing a follow-up programme for faint sources with *XMM-Newton* and for bright sources also with *NuSTAR*.

Among 24 new sources observed so far we could classify 20 IPs, 2 hard NLs and a hard polar [21–26]. We also disproved a CV classification for XSS J12270-4859 which we identified as a LMXB, unexpectedly associated with a high energy *Fermi/LAT* source [27, 28] and recently found to be one of the few transitional millisecond pulsar binaries [29, 30]. To date 56 IPs are confirmed of which 41 detected in the hard X-ray surveys. The polar group instead amounts to $\sim$100 systems with only 10 identified as hard sources.

Magnetic accretion produces a strong shock above the WD magnetic poles below which the flow cools by bremsstrahlung (hard X-rays) and cyclotron optical radiation that are partially thermalized and re-emitted in the soft X-rays and/or EUV/UV domains. The efficiency of these cooling mechanisms depends on the magnetic field strength: cyclotron is increasingly efficient in high field systems (polars) and is able to suppress high temperatures [31]. It is then likely that low-field systems (IPs) preferentially emit hard X-rays. Since 3 hard X-ray polars are also slightly desynchronised, asynchronism seems a common characteristics. Early attempts to relate the hard X-ray emission with the degree of asynchronism fail with the current enlarged sample which now includes short period and weakly desynchronised IPs (Fig. 40.2) and 10 polars with field strengths ranging from 7 to 40 MG. Other parameters such as the WD mass could play a role, but significantly massive primaries are not favoured [24]. The local mass accretion rate is also a key parameter under investigation.

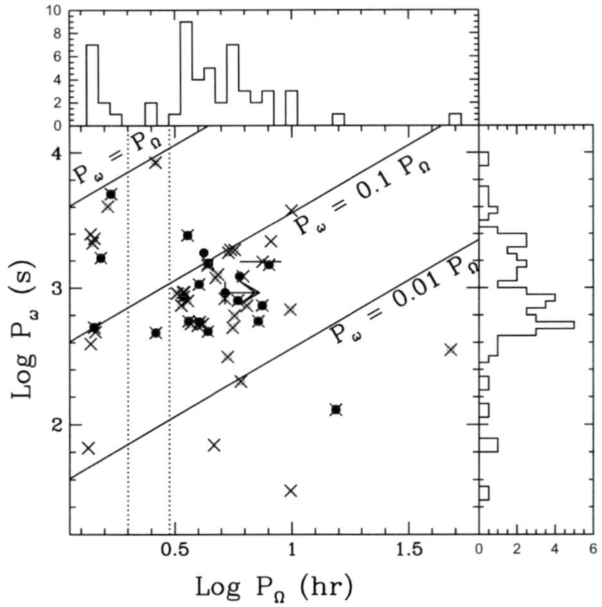

Fig. 40.2 The spin-orbit period plane of confirmed IPs (*crosses*) including our programme targets (*filled circles*). The *solid lines* mark synchronism and two levels of asynchronism (0.1 and 0.01). *Vertical lines* mark the orbital CV gap. The spin (*right panel*) and orbital (*upper panel*) period distributions are reported. Hard X-ray IPs are marked as *red boxes*

While previously known systems were concentrated in a limited range of the spin-orbit period plane, the new identifications enlarged it at long orbital periods ($\geq$5 h) and below the 2–3 h orbital period gap (Fig. 40.2), where mCVs are expected to have reached synchronism [32]. The discovery of these short orbital period IPs may hint to a low-luminosity population yet to be unveiled [16].

Acknowledgements DDM acknowledges collaboration with K. Mukai, F. Bernardini, M. Falanga and N. Masetti and financial support from ASI/INAF I/037/12/0.

References

1. Beuermann, K., Thomas, H.C.: The ROSAT view of the cataclysmic variable sky. Adv. Space Res. **13**, 115 (1993)
2. Verbunt, F., Bunk, W.H., Ritter, H., Pfeffermann, E.: Cataclysmic variables in the ROSAT PSPC all sky survey. Astr. Astrophys. **327**, 602 (1997)
3. Orio, M., Covington, J., Ögelman, H.: X-ray emission from classical and recurrent novae observed with ROSAT. Astr. Astrophys. **373**, 542 (2001)
4. Bird, A.J., Malizia, A., Bazzano, A., Barlow, E.J., Bassani, L., Hill, A.B., Bélanger, G., Capitanio, F., Clark, D.J., Dean, A.J., Fiocchi, M., Götz, D., Lebrun, F., Molina, M., Produit,

N., Renaud, M., Sguera, V., Stephen, J.B., Terrier, R., Ubertini, P., Walter, R., Winkler, C., Zurita, J.: The fourth IBIS/ISGRI soft Gamma-ray survey catalog. ApJs **186**, 1 (2010)
5. Baumgartner, W., Tueller, J., Markwardt, C., Skinner, G., Barthelmy, S., Mushotzky, R., Evans, P., Gehrels, N.: The 70 month Swift-BAT All-sky hard X-ray survey. ApJs **297**, 19 (2013)
6. Barlow, E.J., Knigge, C., Bird, A.J., J Dean, A., Clark, D.J., Hill, A.B., Molina, M., Sguera, V.: 20-100keV properties of cataclysmic variables detected in the INTEGRAL/IBIS survey. MNRAS **372**, 224 (2006)
7. Maoz, E., Grindlay, J.E.: An extended galactic population of low-luminosity X-ray sources (CVs?) and the diffuse X-ray background. ApJ **444**, 183 (1995)
8. Muno, M.P., Baganoff, F.K., Bautz, M.W., Feigelson, E.D., Garmire, G.P., Morris, M.R., Park, S., Ricker, G.R., Townsley, L.K.: Diffuse X-ray emission in a deep Chandra image of the Galactic Center. ApJ **613**, 326 (2004)
9. Hong, J.: Dominance of magnetic cataclysmic variables in the resolved Galactic Ridge X-ray emission of the limiting window. MNRAS **427**, 1633 (2012)
10. Morihana, J., Tsujimoto, M., Yoshida, T., Ebisawa, K.: X-ray point source populations constituting the Galactic Ridge X-ray emission. ApJ **766**, 14 (2013)
11. Revnivtsev, M., Sazonov, S., Churazov, E., Forman, W., Vikhlinin, A., Sunyaev, R.: Discrete sources as the origin of the Galactic X-ray ridge emission. Nature**458**, 1142 (2009)
12. Revnivtsev, M., Sazonov, S., Forman, W., Churazov, E., Sunyaev, R.: Luminosity function of faint Galactic sources in the Chandra bulge field. MNRAS **414**, 495 (2011)
13. Yuasa, T., Makishima, K., Nakazawa, K.: Suzaku studies of magnetic cataclysmic variables and the Galactic ridge X-ray emission. ApJ **753**, 129 (2012)
14. Warwick, R.S.: Low-luminosity X-ray sources and the Galactic ridge X-ray emission. MNRAS **445**, 66 (2014)
15. Reis, R., Wheatley, P., Gänsicke, B., Osborne, J.: X-ray luminosities of optically selected cataclysmic variables and application to the Galactic ridge X-ray emission. MNRAS **430**, 1994 (2013)
16. Pretorius, M.L., Mukai, K.: Constraints on the space density of intermediate polars from the Swift-BAT survey. MNRAS **442**, 2580 (2014)
17. Tout, C.A., Wickramasinghe, D.T., Liebert, J., Ferrario, L., Pringle, J.E.: Binary star origin of high field magnetic white dwarfs. MNRAS **387**, 897 (2008)
18. Masetti, N., Parisi, P., Palazzi, E., Jiménez-Bailón, E., Chavushyan, V., Bassani, L., Bazzano, A., Bird, A.J., Dean, A.J., Charles, P.A., Galaz, G., Landi, R., Malizia, A., Mason, E., McBride, V.A., Minniti, D., Morelli, L., Schiavone, F., Stephen, J.B., Ubertini, P.: Unveiling the nature of INTEGRAL objects through optical spectroscopy.VIII. Identification of 44 newly detected X-ray sources. Astr. Astrophys. **519**, A96 (2010)
19. Masetti, N., Parisi, P., Jiménez-Bailón, E., Palazzi, E., Chavushyan, V., Bassani, L., Bazzano, A., Bird, A.J., Dean, A.J., Galaz, G., Landi, R., Malizia, A., Minniti, A.D., Morelli, L., Schiavone, L., Stephen, J.B., Ubertini, P.: Unveiling the nature of INTEGRAL objects through optical spectroscopy. IX. Twenty two more identifications, and a glance into the far hard X-ray Universe. Astr. Astrophys. **538**, A123 (2012)
20. Parisi, P., Masetti, N., Rojas, A.F., Jiménez-Bailón, E., Chavushyan, V., Palazzi, E., Bassani, L., Bazzano, A., Bird, A.J., Galaz, G., Minniti, D., Morelli, L., Ubertini, P.: Accurate classification of 75 counterparts of objects detected in the 54-month Palermo Swift/BAT hard X-ray catalogue. Astr. Astrophys. **561**, A67 (2014)
21. de Martino, D., Matt, G., Mukai, K., Bonnet-Bidaud, J.M., Falanga, M., Gänsicke, B.T., Haberl, F., Marsh, T.R., Mouchet, M., Littlefair, S.P., Dhillon, V.: 1RXSJ173021.5-055933: a cataclysmic variable with a fast-spinnning magnetic white dwarf. Astr. Astrophys. **481**, 149 (2008)
22. Anzolin, G., de Martino, D., Bonnet-Bidaud, J.M., Mouchet, M., Gänsicke, B.T., Matt, G., Mukai, K.: Two new intermediate polars with a soft X-ray component. Astr. Astrophys. **489**, 1243 (2008)

23. Anzolin, G., de Martino, D., Falanga, M., Mukai, K., Bonnet-Bidaud, J.M., Mouchet, M., Terada, Y., Ishida, M.: Broad-band properties of the hard X-ray cataclysmic variables IGRJ00234+6141 and 1RXSJ213344.1+510725. Astr. Astrophys. **501**, 1047 (2009)
24. Bernardini, F., de Martino, D., Falanga, M., Mukai, K., Matt, G., Bonnet-Bidaud, J.M., Masetti, N., Mouchet, M.: Characterization of new hard X-ray cataclysmic variables. Astr. Astrophys. **542**, A22 (2012)
25. Bernardini, F., de Martino, D., Mukai, K., Falanga, M., Andruchow, I., Bonnet-Bidaud, J.M., Masetti, N., Buitrago, D.H.G., Mouchet, M., Tovmassian, M.: On the nature of the hard X-ray sources SWIFTJ1907.3-2050, IGRJ12123-5802 and IGRJ19552+0044. MNRAS **435**, 2822 (2013)
26. Bernardini, M., de Martino, D., Mukai, K., Falanga, M.: SwiftJ2218.4+1925: a new hard-X-ray selected polar observed with XMM-Newton. MNRAS **445**, 1403 (2014)
27. de Martino, D., Falanga, M., Bonnet-Bidaud, J.M., Belloni, T., Mouchet, M., Masetti, N., Andruchow, I., Cellone, S., Mukai, K., Matt, G.: The intriguing nature of the high-energy Gamma-ray source XSSJ12270-4859. Astr. Astrophys. **515**, A25 (2010)
28. de Martino, D., Belloni, T., Falanga, M., Papitto, A., Motta, S., Pellizzoni, A., Evangelista, Y., Piano, G., Masetti, N., Bonnet-Bidaud, J.M., Mouchet, M., Mukai, K., Possenti, A.: X-ray follow-ups of XSSJ12270-4859: a low-mass X-ray binary with a Gamma-ray FERMI-LAT association. Astr. Astrophys. **550**, A89 (2013)
29. de Martino, D., Casares, J., Mason, E., Buckley, D.A.H., Kotze, M.M., Bonnet-Bidaud, J.M., Mouchet, M., Coppejans, R., Gulbis, A.A.S.: Unveiling the redback nature of the low-mass X-ray binary XSSJ12270-4859 through optical observations. MNRAS **444**, 3004 (2014)
30. Papitto, A., de Martino, D., Belloni, T.M., Burgay, M., Pellizzoni, A., Possenti, A., Torres, D.F.: X-ray coherent pulsations during a sub-luminous accretion disc state of the transitional millisecond pulsar XSSJ12270-4859. MNRAS **449**, L26 (2015)
31. Fischer, A., Beuermann, K.: Accretion physics of AM Herculis binaries. I. Results from one-dimensional stationary radiation hydrodynamics. Astr. Astrophys. **373**, 211 (2001)
32. Norton, A.J., Butters, O., Parker, T., Wynn, G.A.: The accretion flows and evolution of magnetic cataclysmic variable. ApJ **672**, 524 (2008)

Chapter 41
The WiFeS S7 AGN Survey: Current Status and Recent Results on NGC 6300

J. Scharwächter, M.A. Dopita, P. Shastri, R. Davies, L. Kewley, E. Hampton, R. Sutherland, P. Kharb, J. Jose, H. Bhatt, S. Ramya, C. Jin, J. Banfield, I. Zaw, S. Juneau, B. James, and S. Srivastava

Abstract The Siding Spring Southern Seyfert Spectroscopic Snapshot Survey (S7) is a targeted survey probing the narrow-line regions (NLRs) of a representative sample of ∼140 nearby ($z < 0.02$) Seyfert galaxies by means of optical integral field spectroscopy. The survey is based on a homogeneous data set observed using the Wide Field Spectrograph WiFeS. The data provide a $25 \times 38\,\mathrm{arcsec}^2$ field-of-view around the galaxy centre at typically ∼1.5 arcsec spatial resolution and cover a wavelength range between ∼3,400–7,100 Å at spectral resolutions of $\sim 100\,\mathrm{km\,s^{-1}}$ and $\sim 50\,\mathrm{km\,s^{-1}}$ in the blue and red parts, respectively. The survey is primarily

J. Scharwächter (✉)
LERMA, Observatoire de Paris, PSL, CNRS, Sorbonne Universités, UPMC, F-75014 Paris, France
e-mail: julia.scharwaechter@obspm.fr

M.A. Dopita • R. Davies • L. Kewley • E. Hampton • R. Sutherland
Research School of Astronomy & Astrophysics, The Australian National University, Cotter Road, Weston Creek, ACT 2611, Australia

P. Shastri • P. Kharb • J. Jose • H. Bhatt • S. Ramya
Indian Institute of Astrophysics, Koramangala 2B Block, Bangalore 560034, India

C. Jin
Qian Xuesen Laboratory for Space Technology, Beijing, China

J. Banfield
CSIRO Astronomy & Space Science, P.O. Box 76, Epping, NSW 1710, Australia

I. Zaw
New York University (Abu Dhabi), 70 Washington Sq. S, New York, NY 10012, USA

S. Juneau
CEA-Saclay, DSM/IRFU/SAp, 91191 Gif-sur-Yvette, France

B. James
Institute of Astronomy, Cambridge University, Madingley Road, Cambridge CB3 0HA, UK

S. Srivastava
Astronomy and Astrophysics Division, Physical Research Laboratory, Ahmedabad 380009, India

N.R. Napolitano et al. (eds.), *The Universe of Digital Sky Surveys*, Astrophysics and Space Science Proceedings 42, DOI 10.1007/978-3-319-19330-4_41

designed to study gas excitation and star formation around AGN, with a special focus on the shape of the AGN ionising continuum, the interaction between radio jets and the NLR gas, and the nature of nuclear LINER emission. We provide an overview of the current status of S7-based results and present new results for NGC 6300.

41.1 Introduction

S7 is a targeted survey based on optical integral field spectroscopy, providing data for the central regions of ∼140 nearby Seyfert galaxies on ∼100 pc scales resolution [7–10]. S7 is designed to probe the AGN NLR by means of emission-line diagnostics and kinematics. The data set is homogeneous and gives access to a wide wavelength range covering a large number of optical emission lines. Line kinematic studies are facilitated by the comparatively high spectral resolution of ∼50 km s^{-1} in the red part of the spectra.

The S7 sample is a representative sample of nearby Seyfert galaxies, including LINERs and a number of H II galaxies, with an intentional bias toward radio-detected sources. Targets are selected from the "catalogue of quasars and active nuclei" [14]. The sources are typically chosen to have a declination of DEC $<$ $+10$ deg and to lie outside the Galactic plane by more than ∼20 deg. In order to guarantee a sufficient physical spatial resolution, the targets are limited to nearby sources ($z < 0.02$). For the S7 seeing conditions of ∼1.5–1.8 arcsec, this limit implies spatial resolutions of ∼100–700 pc. Those sources that have available radio data from the NRAO VLA Sky Survey (i.e. targets at DEC > -41 deg) are typically required to have a minimum flux of 20 mJy at 1.4 GHz.

The observations are carried out in a snapshot mode using the Wide Field Spectrograph (WiFeS) operated at the ANU 2.3 m telescope at Siding Spring in Australia [6]. WiFeS is an optical integral field spectrograph in image-slicer design with a field-of-view of 25 × 38 $arcsec^2$. The field-of-view is defined by 25 slitlets of 1 × 38 $arcsec^2$ size, for which the slit width has been adjusted to match the typical seeing conditions of >1 arcsec at Siding Spring. WiFeS is operated simultaneously in a blue and a red arm so that a wide wavelength coverage is achieved. For S7, the $R_S = 3{,}000$ and $R_S = 7{,}000$ gratings are used in the blue and red arm, respectively, which results in a continuous wavelength coverage between ∼3,400–7,100 Å. The data are reduced using standard procedures provided by the python-based WiFeS reduction pipeline PyWiFeS [4]. The reduction process yields calibrated sky-subtracted data cubes sampled on a 1 × 1 $arcsec^2$ pixel scale. For further details on the observations and data reduction for the first published data set of 64 galaxies, the reader is referred to the main survey paper [9]. Interferometric follow-up observations of S7 sources at radio frequencies using the Australia Telescope Compact Array (ATCA) and the Giant Metrewave Radio Telescope (GMRT) in India are in progress.

41.2 Overview of S7 Results

S7 is primarily designed to (i) study the AGN NLR and the mixing between photoionisation from the AGN and from star formation, (ii) analyse the spectral energy distribution of the AGN ionising continuum via detailed photoionisation models for individual sources, and (iii) derive constraints on the nature of the nuclear LINER phenomenon. Furthermore, the bias toward radio-detected sources has been introduced to add a special focus on the interaction between radio jets and the NLR gas. In this section, we provide a brief overview over the main science results from the S7 survey so far. These include the first sample paper on 64 S7 sources [9] and two case studies of S7 galaxies [7, 10], which showcase the above-mentioned science cases.

Three-colour [O III]-Hα-[N II] images and nuclear spectral properties for the first 64 S7 sources are compiled in [9]. This information can be used as a reference for source selection from S7 galaxies. A basic assessment of the occurrence of star forming rings and the relative orientation of ionisation cones is presented as well as a preliminary discussion of the nature of coronal emission lines.

The detailed analysis of the Seyfert 2 galaxy NGC 5427, discussed in [7], shows an example of S7-based photoionisation models for the NLR. Such models depend on the gas chemical abundance, the ionisation parameter, and the shape of the AGN ionising continuum. In [7], the NLR abundance is inferred from the chemical abundance measured for H II regions, resulting in improved constraints on the NLR model. The NLR model for NGC 5427 is used to derive limits for the shape of the AGN ionising continuum and to analyse the mixing between extended NLR emission and background H II regions.

The nature of nuclear LINER emission is addressed in a case study of the LINER NGC 1052 [10]. The optical-to-UV emission lines in NGC 1052 are interpreted in a double-shock scenario, in which gas already shocked in an accretion flow is subsequently affected by a cocoon shock driven by the radio jet. This scenario is supported by the WiFeS data for the stellar and Hα line-of-sight velocities and velocity dispersions in the inner $20 \times 20\,\mathrm{arcsec}^2$, suggesting two gas bubbles along the minor axis of the galaxy which are separated by a region of broad line widths.

41.3 Extended NLR of NGC 6300

NGC 6300 ($z = 0.0037$, [11]) is a ringed barred spiral galaxy hosting a Seyfert 2 nucleus [2, 3, 13]. The nuclear radio structure at 8.6 GHz has been reported to be slightly resolved, extending over 3.5 arcsec [12]. Fabry-Perot observations of the large-scale Hα velocity field show a blueshift of the nuclear Hα emission of $100\,\mathrm{km\,s^{-1}}$ with respect to the systemic velocity [3]. Recent high-spatial-resolution data for the H_2 1-0 S(1) line in the central $\sim 4 \times 4\,\mathrm{arcsec}^2$ ($\sim 320 \times 320\,\mathrm{pc}^2$) show evidence of a nuclear biconical outflow in approximately north-south direction with blueshifted velocities to the north and redshifted velocities to the south [5].

The WiFeS data for NGC 6300 include six fields covering the nucleus, bar, and parts of the ring (Fig. 41.1). The data reduction, fitting and subtraction of the stellar continuum, and the assignment of the World Coordinate System (WCS) is based on similar methods as described in [7]. The resulting six cubes were resampled into a single cube slice by slice based on the WCS solution using the software *Swarp* [1] with a bilinear interpolation to (binned) output pixels of $2 \times 2\,\mathrm{arcsec}^2$.

In Fig. 41.1, we present the first results from an analysis of Hα and [N II] λ6548 and λ6583, the strongest lines in the WiFeS spectra for the extended circum-nuclear gas in NGC 6300. The three lines were fitted simultaneously using a superposition of single Gaussian profiles per line forced to share the same kinematics. The gas and stellar line-of-sight velocities in the nucleus differ by about $\sim$100 km s^{-1} (upper left panel of Fig.41.1), in agreement with the Hα blueshift reported by [3]. Up to distances of $\sim$20 arcsec ($\sim$1.6 kpc) from the nucleus, the lines are found to have broad profiles ($\sim$60–90 km s^{-1}, compared to $\sim$20–30 km s^{-1} in the H II regions in

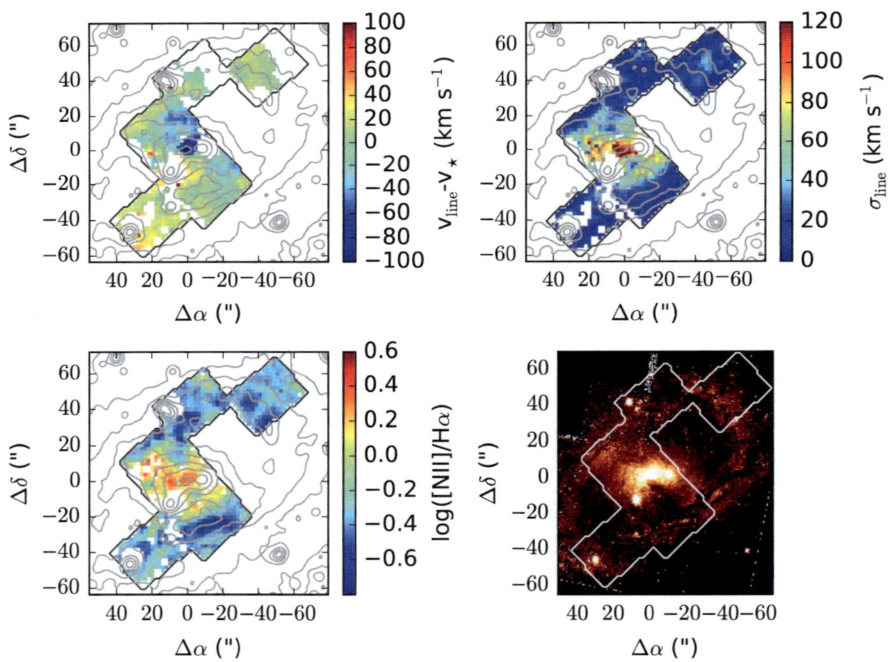

Fig. 41.1 Kinematic and line ratio maps for NGC 6300 derived from the Gaussian fits to [N II] λ6,548, Hα, and [N II] λ6583. *Upper left panel*: Difference between the emission-line and stellar line-of-sight velocities; *Upper right panel*: Intrinsic velocity dispersion of the emission lines; *Lower left panel*: [N II]λ6583/Hα ratios; *Lower right panel*: Overlay of the WiFeS field-of-view on a Hubble Space Telescope WFPC2 image in the *F450W* filter, taken from the Hubble Legacy Archive. A Digitized Sky Survey (DSS) image of NGC 6300 is overlaid on the WiFeS maps as *grey contours*. The maps have been clipped using a 3σ-limit for the line fluxes and a minimum line velocity dispersion of 10 km s^{-1}. In addition, a number of foreground stars (visible in the DSS contours) have been masked out

the ring) and enhanced [N II]/Hα ratios ($\sim$0.2, compared to log([N II]/Hα)$<$ 0 in the H II regions), (upper right and lower left panels of Fig. 41.1). The spectra for the lines east of the nucleus, which appear particularly broad in the single-Gaussian fits, show indications of double peaks that suggest more complex kinematics. The region of enhanced [N II]/Hα ratios indicates ionisation by the AGN continuum, shocks, or a mixture of both, on spatially extended scales around the active nucleus. The correlation of the line ratios with line width suggests that shocks are likely to play an important role in this region. Shocks could be associated with the nuclear outflow reported in [5], a radio jet, or with bar-driven perturbations.

Acknowledgements J. S. acknowledges the European Research Council for the Advanced Grant Program Number 267399-Momentum.

References

1. Bertin, E., Mellier, Y., Radovich, M., Missonnier, G., Didelon, P., Morin, B.: The TERAPIX pipeline. In: Bohlender, D.A., Durand, D., Handley, T.H. (eds.) Astronomical Data Analysis Software and Systems XI. ASP Conference Series, vol. 281, pp. 228. Astronomical Society of the Pacific, San Francisco (2002)
2. Buta, R.: The structure and dynamics of ringed galaxies. IV. Surface photometry and kinematics of the ringed barred spiral NGC 6300. ApJS **64**, 383 (1987)
3. Buta, R., Ryder, S.D., Madsen, G.J., Wesson, K., Crocker, D.A., Combes, F.: Dynamics of ringed barred spiral galaxies. I. Surface photometry and kinematics of NGC 1433 and NGC 6300. AJ **121**, 225 (2001)
4. Childress, M.J., Vogt, F.P.A., Nielsen, J., Sharp, R.G.: PyWiFeS: a rapid data reduction pipeline for the Wide Field Spectrograph (WiFeS). Ap&SS **349**, 617 (2014)
5. Davies, R.I., Maciejewski, W., Hicks, E.K.S., et al.: PyWiFeS: a rapid data reduction pipeline for the Wide Field Spectrograph (WiFeS). ApJ **792**, 101 (2014)
6. Dopita, M., Hart, J., McGregor, P., Oates, P., Bloxham, G., Jones, D.: The Wide Field Spectrograph (WiFeS). Ap&SS **310**, 255 (2007)
7. Dopita, M.A., Scharwächter, J., Shastri, P., et al.: Probing the physics of narrow-line regions of Seyfert galaxies. I. The case of NGC 5427. A&A **566**, A41 (2014)
8. Dopita, M.A., Shastri, P., Scharwächter, J., et al.: Probing the physics of Seyfert Galaxies through their ENLR & HII Regions. In: Ziegler, B.L., Combes, F., Dannerbauer, H., Verdugo, M. (eds.) Galaxies in 3D across the Universe. Proceedings of the International Astronomical Union, Cambridge University Press, vol. 10(309), pp. 200. Cambridge University Press, (2015)
9. Dopita, M.A., Shastri, P., Davies, R., et al.: Probing the physics of narrow line regions in active galaxies II: the siding spring southern seyfert spectroscopic snapshot survey (S7). ApJS **217**, 12 (2015)
10. Dopita, M.A., Ho, I.-T., Dressel, L.L., et al.: Probing the Physics of narrow line regions in active galaxies III: accretion and cocoon shocks in the LINER NGC1052. ApJ **801**, 42 (2015)
11. Mathewson, D.S., Ford, V.L.: Parameters of 2447 Southern spiral galaxies for use in the Tully-Fisher relation. ApJS **107**, 97 (1996)
12. Morganti, R., Tsvetanov, Z.I., Gallimore, J., Allen, M.G.: Radio continuum morphology of southern Seyfert galaxies. A&AS **137**, 457 (1999)
13. Ryder, S.D., Buta, R.J., Toledo, H., Shukla, H., Staveley-Smith, L., Walsh, W.: Neutral hydrogen in the ringed barred galaxies NGC 1433 and NGC 6300. ApJ **460**, 665 (1996)
14. Véron-Cetty, M.-P., Véron, P.: A catalogue of quasars and active nuclei: 12th edition. A&A **455**, 773 (2006)

Chapter 42
Variability-Selected AGNs in the VST-SUDARE Survey of the COSMOS Field

D. De Cicco, S. Falocco, M. Paolillo, G. Covone, G. Longo, A. Grado, L. Limatola, M.T. Botticella, G. Pignata, E. Cappellaro, M. Vaccari, D. Trevese, F. Vagnetti, M. Salvato, M. Radovich, W.N. Brandt, M. Capaccioli, N.R. Napolitano, and P. Schipani

D. De Cicco (✉) • M. Capaccioli
Department of Physics "Ettore Pancini", University "Federico II" of Naples, Italy
e-mail: demetra.decicco@unina.it

S. Falocco • G. Covone • G. Longo
Department of Physics "Ettore Pancini", University "Federico II" of Naples, Italy

INFN – Sezione di Napoli, via Cinthia 9, 80126 Napoli, Italy

M. Paolillo
Department of Physics "Ettore Pancini", University "Federico II" of Naples, Italy

INFN – Sezione di Napoli, via Cinthia 9, 80126 Napoli, Italy

ASI Science Data Center, via del Politecnico snc, 00133 Roma, Italy

A. Grado • L. Limatola • M.T. Botticella • N.R. Napolitano • P. Schipani
INAF – Osservatorio Astronomico di Capodimonte, via Moiariello 16, 80131 Napoli, Italy

G. Pignata
Universidad Andres Bello, Avda. Republica 252, Santiago, Chile

Millennium Institute of Astrophysics, Santiago, Chile

E. Cappellaro • M. Radovich
INAF – Osservatorio Astronomico di Padova, vicolo dell'Osservatorio 5, I-35122 Padova, Italy

M. Vaccari
University of the Western Cape, Private Bag X17, 7535 Bellville, Cape Town, South Africa

D. Trevese
University of Roma "La Sapienza", Piazzale Aldo Moro 2, 00185 Roma, Italy

F. Vagnetti
University of Roma "Tor Vergata", via della Ricerca Scientifica 1, 00133 Roma, Italy

M. Salvato
Max Planck Institut für Extraterrestrische Physik, Giessenbachstraße 1, D-85748 Garching bei München, Germany

W.N. Brandt
The Pennsylvania State University, University Park, PA 16802, USA

Institute for Gravitation and the Cosmos, The Pennsylvania State University, University Park, PA 16802, USA

N.R. Napolitano et al. (eds.), *The Universe of Digital Sky Surveys*, Astrophysics and Space Science Proceedings 42, DOI 10.1007/978-3-319-19330-4_42

Abstract Variability is characteristic of active galaxies; luminosity variations generally occur at all wavelengths, affecting both continuum and line emission, with timescales and amplitudes depending on the wavelength. We made use of optical variability to select AGNs in the VST-SUDARE survey of the COSMOS field, and confirmed 95 % of them by means of multiwavelength data from overlapping surveys. The analysis returns a sample of AGN candidates with 81 % purity and 15 % completeness with respect to the X-ray confirmed AGNs in the field with an optical counterpart. This study provides an estimate of the reliability of the AGN selection techniques based on variability, which will be used by future wide-field surveys, such as those by LSST.

42.1 Overview

Emission from active galactic nuclei (AGNs) is generally explained in terms of accretion onto a supermassive black hole in the center of the galaxy. It is common belief that all galaxies go through an active phase during their lives, and several empirical laws relate physical properties of the black hole to properties of the host galaxy. These empirical laws constitute evidence for the existence of a strong feedback between galaxy evolution and nuclear black hole formation. As a consequence, in order to understand galaxy evolution, it is crucial that we gain a deeper knowledge of black hole formation and evolution.

The search for AGNs has been based on a wide range of selection techniques over the last decades (e.g., [1, 3, 9, 13]), but none of them is completely bias-free. Provided the availability of multiepoch data, a way to identify AGNs is based on the detection of their variability, which is usually attributed to instabilities in the accretion disk and changes in the accretion rate (e.g., [12]). In the work that we illustrate here, we select AGNs in the COSMOS field on the basis of their optical variability; for details, see [7]. A complementary analysis of optically variable AGNs in the Chandra Deep Field South is presented in the work by Falocco et al. in this same volume.

Our analysis is based on data from the VLT Survey Telescope (VST; see [5]), located at Cerro Paranal Observatory, Chile. The analysis of optical variability in the COSMOS field is based on images from the Supernova Diversity And Rate Evolution (SUDARE) survey (see [2] for details): the dataset consists of multiple $1 \times 1\,\text{deg}^2$ images of the cosmos field spread over a time baseline of $\approx$5 months.

42.2 The Sample of AGN Candidates

Following the approach described in [14], we derived from the light curve of each source i an average magnitude $\overline{\mathrm{mag}}_i$ and the corresponding r.m.s. deviation σ_i; hence, in order to obtain a sample of optically variable sources, we computed the running average of the r.m.s. deviation $\langle\sigma_i^{ltc}\rangle$ and its own r.m.s. deviation $\mathrm{r.m.s.}_{\langle\sigma_i^{ltc}\rangle}$ over a 0.5 mag wide bin, and then defined the variability threshold as $\sigma_i^{ltc} \geq \langle\sigma_i^{ltc}\rangle + 3 \times \mathrm{r.m.s.}_{\langle\sigma_i^{ltc}\rangle}$. Our analysis is restricted to the objects detected in at least 20 % of the epochs and with $r(\mathrm{AB}) < 23$ mag. We visually inspected each of the optically variable sources and the corresponding light curves; after removing all the candidates whose variability is likely spurious (because of close neighbors, defective pixels, etc.), the final sample consists of 83 sources.

In order to investigate the nature and properties of our AGN candidates, we made use of data from multiwavelength COSMOS catalogs available from other observatories.

Spectroscopy is widely used to identify AGNs, since their spectra are characterized by prominent emission lines and are generally broadband, with a continuum emission of approximately the same energy per decade in the frequency range 10^{13}–10^{20} Hz [8]. Another reliable indicator of the active nature of a galaxy is its X-ray emission, especially when it is characterized by variability. 76 % of our sources have an X-ray counterpart in the catalogs from *Chandra* X-ray Observatory [6] and XMM-Newton [4], and can be classified as AGNs on the sole basis of their X-ray properties. All of them have an X-ray luminosity (both soft and hard) $L_X > 10^{42}\,\mathrm{erg\,s^{-1}}$: since this is generally assumed to be an upper limit for the X-ray luminosity of non-active galaxies (e.g., [3]), it constitutes evidence for AGN activity.

A widely used diagnostic for AGN identification compares their X-ray and optical luminosities (e.g., [10]). On a diagram like the one shown in Fig. 42.1, AGNs generally place themselves in the region defined by $-1 \leq \mathrm{X/O} \leq 1$ (e.g., [11]); all the sources in our sample of AGN candidates with an X-ray counterpart do lie in that area if we take into account hard X-ray fluxes, while only two of them lie below the defined region if we refer to soft X-ray fluxes. It is worth mentioning that all but one of the VST sources with an X-ray counterpart are also classified as AGNs after their spectra ([4, 6]).

Another class of diagnostics makes use of colors to select AGNs: on a color vs. color plane, objects define distinct regions depending on their nature, because of their different spectral energy distributions. In Fig. 42.2 we show such a diagram in the optical/NIR: it is apparent that, apart from the stellar sequence and the area defined by galaxies, there is a different population of sources (65 % of the sample) which are quasar-like, i.e., point-like in shape but extended as for their colors; they are also optically variable, and are confirmed AGNs on the basis of their X-ray properties.

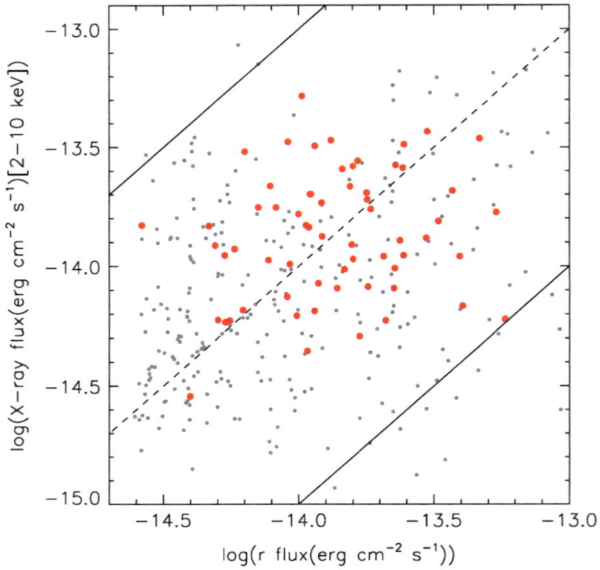

Fig. 42.1 Hard (2–10 keV) X-ray vs. optical flux. *Larger red dots*: AGN candidates with an X-ray counterpart from X-ray catalogs (*Chandra* or XMM); *smaller grey dots*: reference population of the X-ray sources in the *Chandra* catalog having a VST counterpart and r(AB) < 23 mag; *dashed line*: X/O = 0; *lower* and *upper solid lines*: X/O = −1 and X/O = 1, respectively

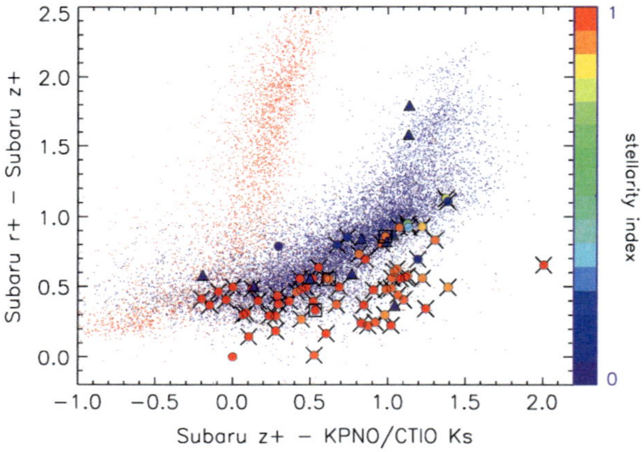

Fig. 42.2 *r-z vs z-k* diagram. *Larger dots*: AGN candidates; *smaller dots*: reference population of the VST-COSMOS sources for which stellarity index and color information are available; *crosses*: AGNs confirmed by X-ray properties; *triangles*: SNe; *boxes*: *ex-novo* confirmed quasar-like AGNs. Stellarity index: 0 = extended, 1 = point-like

42.3 Results and Conclusions

We confirm as AGNs 67 of the sources in our sample of candidates, while 12 are classified as SNe (see the work by Botticella et al. in this same volume). The purity of the sample is 81 %, while the completeness with respect to the X-ray confirmed AGNs with a VST counterpart is 15 %. We estimate that, if we had a longer (e.g., 2 year) baseline, the intrinsic variability of our sources would increase up to 50 %, returning a higher completeness. We also show (Fig. 42.3) that the subsample of VST sources with an X-ray counterpart is characterized by an average optical variability which is higher than the average optical variability of the whole VST sample; this means that, although 85 % of the X-Ray selected AGNs lie below our variability threshold, on average they are more variable than the rest of the optical sample, and thus could be detected using a lower threshold/longer baseline.

The selection techniques based on optical variability allow us to identify AGNs in wide sky areas with a ground-based telescope; this will prove of great importance in the scenario of time-domain astronomy, whose relevance has been growing exponentially in the last years and is certainly going to grow even more in the near future.

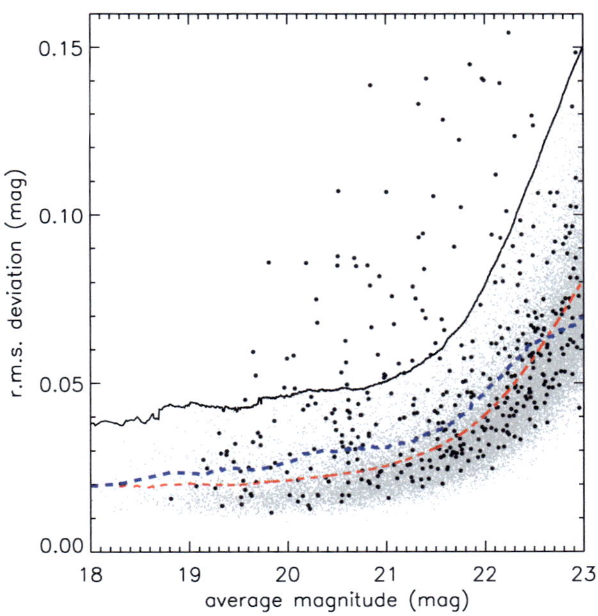

Fig. 42.3 Light curve r.m.s. as a function of magnitude. *Black symbols*: X-ray emitters that are confirmed AGNs; *grey dots*: non-variable sources in the VST complete sample. *Thin dashed red line* and *thick dashed blue line*: running average of the r.m.s. deviation of the complete sample and of the subsample of X-ray emitters, respectively; *solid line*: variability threshold

References

1. Baldwin, J.A., Phillips, M.M., Terlevich, R.: Classification parameters for the emission-line spectra of extragalactic objects. PASP **93**, 5 (1981)
2. Botticella, M.T., et al.: SUDARE at the VST. Messenger **151**, 29 (2013)
3. Brandt, W.N., Hasinger, G.: Deep extragalactic X-ray surveys. ARA&A **43**, 827 (2005)
4. Brusa, M., et al.: The XMM-newton wide-field survey in the COSMOS field (XMM-COSMOS): demography and multiwavelength properties of obscured ans unobscured luminous active galactic nuclei. ApJ **716**, 348 (2010)
5. Capaccioli, M., Schipani, P.: The VLT survey telescope opens to the sky: history of a commissioning. Messenger **146**, 2 (2011)
6. Civano, F., et al.: The Chandra COSMOS survey. III. Optical and infrared identification of X-ray point sources. ApJS **201**, 30 (2012)
7. De Cicco, D., et al.: Variability-selected active galactic nuclei in the VST-SUDARE/VOICE survey of the COSMOS field. A&A **574**, A112 (2015)
8. Elvis, M., et al.: Atlas of quasar energy distributions. ApJS **95**, 1 (1994)
9. Fan, X.: Simulation of stellar objects in SDSS color space. AJ **117**, 2528 (1999)
10. Maccacaro, T., et al.: The X-ray spectra of the extragalactic sources in the EINSTEIN extended medium sensitivity survey. ApJ **326**, 680 (1988)
11. Mainieri, V., et al.: XMM-Newton observations of the Lockman Hole. II. Spectral analysis. A&A **393**, 425 (2002)
12. Pereyra, N.A., et al.: Colors of 2625 quasars at $0 < z < 5$ measured in the Sloan digital sky survey photometric system. ApJ **642**, 87 (2006)
13. Richards, G.T., et al.: Variability-selected active galactic nuclei from supernova search in the Chandra deep field south. AJ **121**, 2308 (2001)
14. Trevese, D., et al.: A&A **488**, 73 (2008)

Chapter 43
A New Search for Variability-Selected Active Galaxies Within the VST SUDARE-VOICE Survey: The Chandra Deep Field South and the SERVS-SWIRE Area

S. Falocco, D. De Cicco, M. Paolillo, G. Covone, G. Longo, A. Grado, L. Limatola, M. Vaccari, M.T. Botticella, G. Pignata, E. Cappellaro, D. Trevese, F. Vagnetti, M. Salvato, M. Radovich, L. Hsu, W.N. Brandt, M. Capaccioli, N. Napolitano, A. Baruffolo, E. Cascone, and P. Schipani

---

S. Falocco • D. De Cicco • G. Covone • G. Longo • M. Capaccioli
Department of Physics "Ettore Pancini", University "Federico II" of Naples, Italy

INFN – Sezione di Napoli, via Cinthia 9, 80126 Napoli, Italy
e-mail: falocco@fisica.unina.it

M. Paolillo (✉)
Department of Physics "Ettore Pancini", University "Federico II" of Naples, Italy

INFN – Sezione di Napoli, via Cinthia 9, 80126 Napoli, Italy

ASI Science Data Center, via del Politecnico snc, 00133 Roma, Italy
e-mail: paolillo@na.infn.it

A. Grado • L. Limatola • M.T. Botticella • N. Napolitano • E. Cascone • P. Schipani
INAF Osservatorio Di Capodimonte Naples, Napoli, Italy

M. Vaccari
Astrophysics Group, Physics Department, University of the Western Cape, Cape Town, South Africa

M. Radovich • E. Cappellaro • A. Baruffolo
INAF- Osservatorio di Padova, Asiago, Italy

L. Hsu • M. Salvato
Max Plank Institute fur Extraterrestrische Physik, Garching, Germany

G. Pignata
Departamento de Ciencias Fisicas, Universidad Andres Bello, Santiago, Chile

D. Trevese
Department of Physics, University La Sapienza Roma, Rome, Italy

F. Vagnetti
Department of Physics, University Tor Vergata Roma, Rome, Italy

W.N. Brandt
Department of Astronomy and Astrophysics, The Pennsylvania State University, University Park, PA 16802, USA

N.R. Napolitano et al. (eds.), *The Universe of Digital Sky Surveys*, Astrophysics and Space Science Proceedings 42, DOI 10.1007/978-3-319-19330-4_43

Abstract This work makes use of the VST observations to select variable sources. We use also the IR photometry, SED fitting and X-ray information where available to confirm the nature of the AGN candidates. The IR data, available over the full survey area, allow to confirm the consistency of the variability selection with the IR color selection method, while the detection of variability may prove useful to detect the presence of an AGN in IR selected starburst galaxies.

43.1 Aims and Method

The luminosity of virtually all AGN varies at every wavelength (see e.g. [6, 8, 12, 16] and references therein), thus making variability one of the most distinctive properties of these sources. The variability selection method assumes that all AGN vary intrinsically in the observed band, without requiring assumptions on the spectral shape, colours, and/or spectral line ratios.

In the work described in this proceeding we aim at constructing a new variability-selected AGN sample exploiting the data from the ongoing survey performed with the VST (VLT Survey Telescope), see [5] for details. We make use of data in the r band from the SUDARE-VOICE survey performed with the VST telescope [1, 2]. In a companion contribution by De Cicco et al. in this volume (also see [3], hereafter Paper I) we focused on the COSMOS region. Here we examine two fields around the CDFS region, that we label CDFS1 and CDFS2. We examined a total of 27 VST epochs for the CDFS1 and 22 epochs for the CDFS2 spanning 5 and 4 months respectively and covering an area of 2 deg^2. The data reduction and the analysis were performed using the procedure explained in Paper I (and in De Cicco et al., this volume) consisting in identifying as candidate AGNs all sources whose lightcurve showed an excess variability of 3 r.m.s. from the average variability of all sources with similar magnitude. To validate our catalogue of variable objects we exploited SWIRE by [10] and SERVS by [11]. We also used SED (Spectral Energy Distribution) classification given in [7] and [13].

43.2 Results and Discussion

We obtained a sample of 175 sources that we investigated further in detail analysing the diagnostics described below. Twelve percent of the selected sample are classified as SN, based on both visual inspection of the light-curves and template fitting by the SUDARE-I collaboration (Botticella et al., this volume and Cappellaro et al. in prep.).

We used the information contained in [7], to extract X-ray and SED data for our variable candidates located within the ECDFS area. There are only 15 sources in common between the sample presented in [7] and our selected sample. The 15 common sources belong to the CDFS1 which encloses the ECDFS. Twelve of the

15 common sources are detected in the X-rays and their SEDs require a strong AGN contribution (in particular in the NIR part of the spectrum). All these sources have also been identified as non-SN on the basis of the inspection of their lightcurves. The remaining three sources are non-detected in the X-rays and their best-fit SED template shows no evidence for a significant AGN contribution. These three sources were identified as SN according to their lightcurves. Therefore, we conclude that they are SN explosions in normal galaxies.

We validate our catalogue of variable objects with the overlapping surveys SWIRE [10] and SERVS [11] which provide data in the 3.6, 4.5, 5.6, 8, 24, 70, 160 μm bands, and in the U, g, r, i, z filters. In Fig. 43.1 we compare the $r-i$ versus the 3.6 μm to r band flux ratio of our variable candidates with the SERVS+SWIRE source catalog. This diagram has been proposed by [13] to separate stars from galaxies. The populations represented in the plot are segregated into two regions: stars and extragalactic objects. Figure 43.1 shows that six of 57 variable objects are along the stellar sequence, likely variable stars.

We further make use of the mid-IR colors in order to confirm the identification of our AGN candidates. Figure 43.2 shows the diagnostic developed by [9]. Due to the different dust content and temperature, normal galaxies, starforming galaxies and AGNs occupy different regions of this diagram. This allows us, as shown in [9], to define an empirical wedge (solid line in Fig. 43.2) which encloses a large fraction of the AGN population. Out of the 115 sources of the selected sample represented in the plot, 103 lie within the Lacy wedge, supporting their AGN nature. We also note from Fig. 43.2 that the average stellarity index of the variable candidates inside the Lacy region decreases towards the left side of the diagram, where the contamination is more severe. According to [15], many of such sources are low ionisation narrow

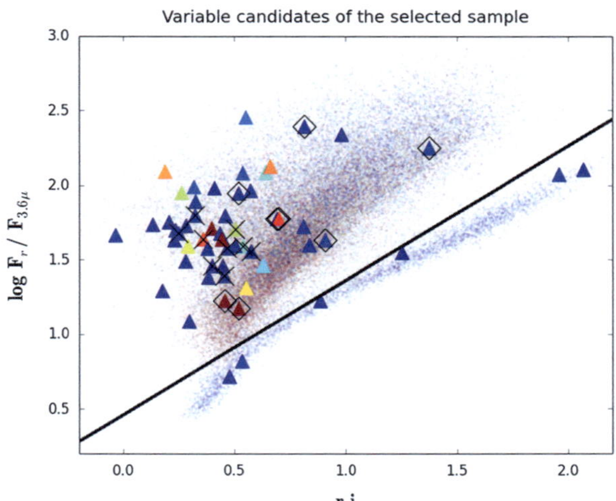

Fig. 43.1 Flux (F_λ) ratios between the r band and 3.6 μm versus $r-i$ colour. *Small points*: SERVS+SWIRE 82254 sources. *Triangles*: 57 sources in common with the selected sample. *Diamonds*: SN. *Crosses*: X-ray detected sources. The colours indicate the increasing stellarity, from *red* (extended sources) to *blue* (pointlike). The *solid line* separates the stellar sequence and the non-stellar region

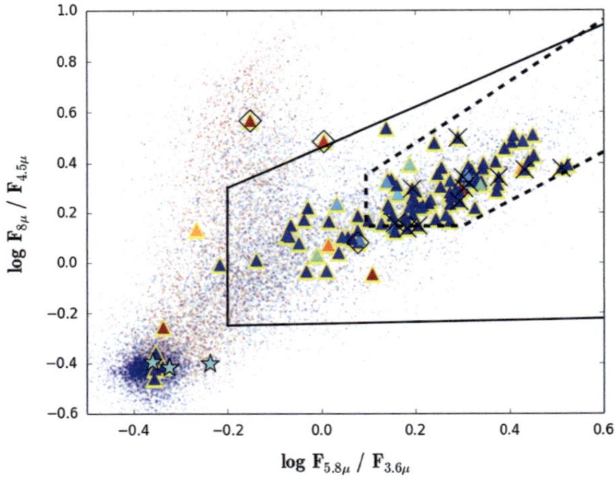

Fig. 43.2 Flux (F_λ) ratio (logarithmic) at 5.8 and 3.6 μm versus flux ratio at 8 and 4.5 μm. *Small points*: SERVS+SWIRE 18436 sources; *Triangles* (enclosed in *yellow* edges): 115 sources in common with the selected sample. *Cyan stars*: stars; *Diamonds*: SN. *Crosses*: X-ray detected sources. Colour code as in Fig. 43.1. The *solid line* is the Lacy region and the *dashed line* the Donley region (see text)

emission regions (LINERs). To improve the purity[1] of the IR selected AGN sample and to reduce the starburst contamination to IR-selected AGN samples, [4] defined a more restrictive criterion, which is shown in Fig. 43.2 as a dashed line. The majority of pointlike sources lie within the Donley wedge, strengthening the view that the Donley region is occupied prevalently by AGN-dominated galaxies.

43.3 Conclusions

We identified 175 candidates selected through variability using VST observations in the CDFS. To validate the sample, we used information available both within the VST-SUDARE consortium and in the literature. The total number of candidates for which we could employ the diagnostics discussed in the previous section is 137 out of a total 175 candidates in the selected sample. We found 103 confirmed AGN (by at least one diagnostic of those explored in the previous section), that is 75 % of the 137 candidates with ancillary data and 59 % of the selected sample. As expected, contaminants are mainly stars and SN: the stars constitutes 3 % of the selected sample of 175 candidates, while the SN constitute 12 %.

In conclusion the purity of our sample of optically variable sources is 75 %, close to the 80 % obtained for the COSMOS field in Paper I. The completeness of the variability-selected survey presented in this work is 22 % (computed with respect to the IR selection of [4]). In Paper I, the completeness (computed with respect to X-ray samples) has been estimated to be 15 % for a 5 months baseline; the two results are thus in broad agreement considering that they are estimated with respect

[1]We define the purity as the number of confirmed AGN divided by the number of AGN candidates

to different reference populations. These completeness levels can improve extending the monitoring baseline, as pointed out in previous papers (e.g. [14] and Paper I). The new observations which are currently being acquired for the COSMOS field (P.I: G. Pignata) with VST will allow us to directly compare these results using a 3-year long monitoring baseline.

References

1. Botticella, M.T., Cappellaro, E., Pignata, G.: SUDARE at the VST. The Messenger **151**, 29–32 (2013)
2. Cappellaro, E., et al: Supernova rates from the SUDARE VST-Omegacam search. I. A&A **584**, A62 (2015)
3. De Cicco, D., Paolillo, M., Covone, G., et al.: Variability-selected active galactic nuclei in the VST SUDARE/VOICE survey of the COSMOS field. A&A **574**, A112 (2015)
4. Donley, J.L., Koekemoer, A.M., Brusa, M., et al.: Identifying luminous active galactic nuclei in deep surveys: revised IRAC selection criteria. APJ **748**, 142 (2012)
5. Falocco S., Paolillo M., Covone G., et al.: SUDARE-VOICE variability-selection of active galaxies in the Chandra Deep Field South and the SERVS/SWIRE region. A&A **579**, A115 (2015)
6. García-González, J., Alonso-Herrero, A., Pérez-González, P.G., et al.: Selection of AGN candidates in the GOODS-South Field through Spitzer/MIPS 24 micron variability. MNRAS **446**(I 3), 3199–3223 (2015)
7. Hsu, L.-T., Salvato, M., Nandra, K., et al.: CANDELS/GOODS-S, CDFS, and ECDFS: photometric redshifts for normal and X-ray detected galaxies. ApJ **796**, 60 (2014)
8. Kawaguchi, T., Mineshige, S., Umemura, M.: Optical variability in active galactic nuclei: starburst or disk instabilities? APJ **504**, 671–679 (1998)
9. Lacy, M., Storrie-Lombardi, L.J., Sajina, A.: Obscured and unobscured active galactic nuclei in the Spitzer Space Telescope first look survey. APJs **154**, 166–169 (2004)
10. Lonsdale, C., Polletta, M.d.C., Surace, J.: First insights into the Spitzer wide-area infrared extragalactic legacy survey (SWIRE) galaxy populations. APJs **154**, 54–59 (2004)
11. Mauduit, J.-C., Lacy, M., Farrah, D.: The Spitzer extragalactic representative survey (SERVS): survey definition and goals. PASP **124**, 1135–1136 (2012)
12. Paolillo, M., Schreier, E.J., Giacconi, R., et al.: Prevalence of X-ray variability in the Chandra Deep Field-South. APJ **611**, 93–106 (2004)
13. Rowan-Robinson, M., Gonzalez-Solares, E., Vaccari, M., et al.: Revised SWIRE photometric redshifts. MNRAS **428**, 1958–1967 (2013)
14. Sesar, B., et al.: Exploring the variable sky with the sloan Digital Sky survey. AJ **134**, 2236–2251 (2007)
15. Trevese, D., Boutsia, K., Vagnetti, F., et al.: Variability-selected active galactic nuclei from the supernova search in the Chandra Deep Field South. A&A **488**, 73–81 (2008)
16. Ulrich, M.-H., Maraschi, L., Urry, C.M.: Variability of active galactic nuclei. ARAA **35**, 445–502 (1997)

Chapter 44
Fermi Continuous Survey of the High-Energy Sky and Its Serendipitous Results

P.A. Caraveo

Abstract Since more than 6 years, the Fermi gamma-ray telescope operates in scanning mode yielding a new image of the gamma-ray sky every 3 h. Such wealth of high-energy data (all immediately publicly available) has unveiled thousands of gamma-ray sources, steady as well as variable, 1/3 of which lacks even a tentative associations.

A number of approaches have been developed exploiting optical surveys, as well as radio observations and readily available X-ray data.

The gamma-ray sky provides also plenty of surprises which call for rapid multiwavelength response.

44.1 Introduction

Since Aug. 1st 2008, at the completion of the in orbit calibration and verification phase, the Fermi Large area Telescope (LAT-Atwood et al. [12]) is providing continuous coverage of the sky at energies >20 MeV. Each photon is accurately time tagged while its arrival direction and energy are reconstructed using both the LAT tracker and calorimeter. After data quality verification, Fermi LAT photons data are immediately made publicly available.

44.2 Gamma-Ray Sources: Variable and Steady

Exploiting its very large field of view, coupled with a carefully planned scanning sequence, Fermi LAT produces a new image of the entire sky every 3 h, making it possible to study gamma-ray source variability and to rapidly alert the community if multiwavelength observations are deemed necessary to address the appearance of a new source or an unusual behavior of a known one.

P.A. Caraveo (✉)
Istituto di Astrofisica Spaziale e Fisica Cosmica – INAF, Via Bassini 15, 20133 Milano, Italy
e-mail: pat@iasf-milano.inaf.it

N.R. Napolitano et al. (eds.), *The Universe of Digital Sky Surveys*, Astrophysics and Space Science Proceedings 42, DOI 10.1007/978-3-319-19330-4_44

In parallel, data are integrated over the, ever growing, mission lifetime to reach an, ever better, sensitivity for source detection. The results of such continuous effort are summarized in the Fermi LAT source catalogues which are one the most important legacy of the mission. For each source, the catalogue provides position, flux, spectral shape as well as an assessment of its variability computed on a monthly timescale. While the second Fermi LAT Catalogue (in short 2FGL, Nolan et al. [15]) encompasses 2 years of data and contains 1,873 sources, the recently released 3FGL (Acero et al. [7]) covers 4 years of data and lists 3,033 sources. The increase in the source number, however, did not change the relative contribution of different source classes. AGNs account for 58 % of the source listed in 3FGL while pulsars, which are the second most abundant class of gamma-ray sources, represent 6 % to which one should add a 2 % contribution from other galactic classes, such as SNR, PWN, binary systems. The remaining 33 %, namely 992 source, are still without an association and are the target of radio, optical and X-ray coverage which aim at single out possible counterparts.

Overall about 20 % of the LAT sources are flagged as variable: these are mainly high-latitude sources associated with AGNs but a small fraction of galactic sources is also contributing to the diversity of the variable gamma-ray source panorama.

44.3 The Extragalactic Sky: Variable AGNs

Among the >1,700 AGNs detected in gamma-rays, variability is rather common and, indeed, it turns out to be a very valuable tool to identify or, at least, to associate LAT sources with AGNs seen to vary at radio, optical, X-ray and VHE gamma-ray wavelengths. Apart from allowing the source identification, variability provides insight into the source emitting region as well as on the emission mechanisms at work. The diversity of the different AGN families is addressed also through their different spectral energy distribution.

In addition to the sources associated or identified with AGNs, one should remember that a fair fraction of the 658 unassociated sources found at galactic latitude $|b| > 5°$ are probably AGNs in sky area poorly covered by the catalogues currently available.

44.4 The Galactic Sky: Variable Sources

The galactic plane is the brightest gamma-ray source in the sky. Thus, the study of galactic gamma-ray sources is a challenging task requiring a careful modelling of the diffuse background. Among the 270 sources identified or associated to galactic objects such as pulsars, Pulsar Wind Nebulae, Supernova Remnant, binary systems, variability is far less common than in the extragalactic sky.

44.4.1 Expected Variability

Few binary systems are expected to be periodically variable with the gamma-ray flux varying as a function of their orbital cycle. LSI 61 303 [10] and LS 5039 [1] have been followed closely. The same is true for PSR B1259-63 which was scrutinized during two perigee passages, in 2010 and 2014, which yielded similar results with the maximum of the gamma-ray emission detected only during the second passage of the neutron star through the disk of the Be companion star [5]. The colliding wind binary ETA Car has also been studied [16] unveiling orbital variability. Cyg X3 was also found to exhibit sporadic emission [2, 17].

Blind search for long period variability unveiled 1FGL J1018.6-5856, the first (and so far the only) HMXB discovered by Fermi-LAT. It has an orbital period of 16 days [9].

In 2014, the scanning mode was modified to increase the coverage of the galactic center also in view of the perigee passage of the G2 cloud which was expected to be destroyed, and later accreted, by the supermassive black hole at the center of our galaxy. Copious gamma-ray emission was expected, but nothing happened.

44.4.2 Serendipitous Discoveries

44.4.2.1 Crab

On September 2010, a sudden enhancement of the overall Crab flux was reported first by AGILE and immediately confirmed by Fermi [4, 18]. A quick sequence of radio, X-ray, and optical observations made it clear that the pulsar was behaving normally. Indeed, in gamma rays the pulsed flux from the Crab was also unchanged, leaving the nebula as the only suspect for the flux increase. Moreover, the short time scale of the variability pointed to a quite small emitting region, possibly next to the pulsar. To achieve the highest angular resolution to study the interior of the Crab Nebula, next to the pulsar, the Chandra X-ray observatory and the Hubble Space Telescope (HST) were immediately triggered for Target of Opportunity (ToO) observations which were performed a few days after the event. The high-resolution optical and X-ray images failed to show any dramatic change in the notoriously active Crab nebula inner region. A search in both AGILE and Fermi data did prove that enhancements had been detected previously, making it clear that the Crab's Sept. 2010 flaring episode was not a unique event but rather a recurring one. A massive ToO campaign was organized, waiting for the next flares which were recorded in April 2011 and in March 2013 (Fig. 44.1). Unfortunately, comparing optical and X-ray images taken before, during and after the flare, nothing obvious was seen to change. The Crab flares shine only in gamma rays, and such events are possibly linked to sudden particle acceleration, driven perhaps by magnetic reconnection.

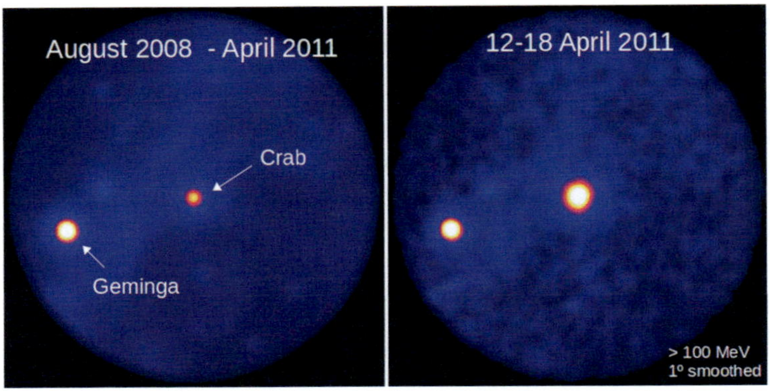

Fig. 44.1 Crab Variability. The *left panel* shows the usual behavior of the Crab in gamma-rays while the *right panel* shows its increased brightness during a flare

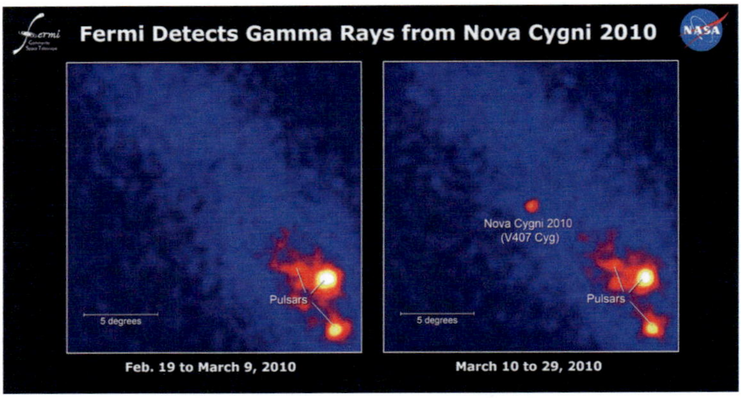

Fig. 44.2 Fermi LAT detection of Nova Cyg as a variable source detected from March 10 to 28, 2010 (*right panel*). The usual picture is given in the *left panel*

44.4.2.2 Novae as a New Class of Gamma-Ray Emitters

In March 2010 a source lit up in Cygnus. Within its error box, Nova V407 Cyg had just been reported, prompting the association between the variable gamma-ray source and the Nova [3] (Fig. 44.2). Indeed Novae were not expected to be able to accelerate particle so to produce gamma-rays, but such idea turned out to be most definitely wrong since, in the following years, few more Novae were detected by Fermi LAT [8]. Indeed, in the case of Nova Mon, the sudden appearence on June 12 2012 of a gamma-ray source, dubbed J0639+0548, in a region of the sky out of reach of the optical telescopes, proved to be a real Nova discovery which was confirmed in August, when optical observations became possible and unveiled V595 Mon. Novae are indeed a new class of galactic, variable gamma-ray sources.

44.5 Serendipity in the Pulsar Family

With the notable exception of PSR J2021+4026 [11], pulsars are steady sources [6]. The current grand total is 161 (//confluence.slac.standford.edu/display/GLAMCOG/Public+List+of+LAT-Detected+Gamma-Ray+Pulsars) but the number is constantly growing (Fig. 44.3). While the number of detections is well above the most optimistic pre-launch expectations, which were considering radio loud as well as radio quiet, Geminga like [13] neutron stars, the real surprise has been the detection of an ever growing number of millisecond pulsars (MSPs). Whereas gamma-ray emission from young, energetic pulsars (both radio-loud and radio-quiet) was widely expected, the detection of the millisecond pulsars came as a real surprise, mainly because such old, recycled pulsars were not supposed to be ideal gamma-ray emitters as their surface B fields are 10^4 times weaker than those of young neutron stars.

Thanks to the extremely successful synergy between radio and gamma-ray astronomy to find, characterize and phase-fold MSPs, the number of MSP discoveries currently accounts for two thirds of the new pulsars, making MSPs the dominant class among the gamma-ray pulsar types.

This is the unexpected, surprising and really revolutionary result of the Fermi mission that I have dubbed Gamma-ray pulsar revolution [14].

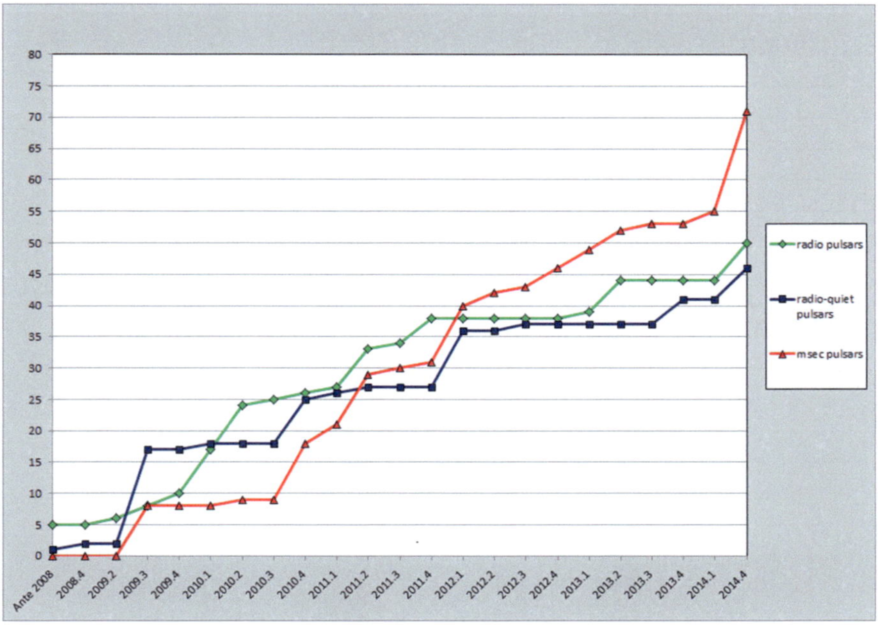

Fig. 44.3 Pulsar counting since the beginning of the Fermi mission. Radio pulsar detections are shown in *green*, radio-quiet in *blue* end MSP in *red* (See [14] for details)

References

1. Abdo, A.A., Ackermann, M., Ajello, M., et al.: Astrophys. J. **706**, L 56 (2009)
2. Abdo, A.A., Ackermann, M., Ajello, M., et al.: Science **326**, 1512– (2009)
3. Abdo, A.A., Ackermann, M., Ajello, M., et al.: Science **329**, 817–821 (2010)
4. Abdo, A.A., Ackermann, M., Ajello, M., et al.: (Fermi LAT Collab.) Science **331**, 739– (2011)
5. Abdo, A.A., Ackermann, M., Ajello, M., et al.: Astrophys. J. **736**, L 11 (2011)
6. Abdo, A.A., Ajello, M., Allafort, A., et al.: (Fermi LAT Collab.) Astrophys. J. Suppl. **208**, 17– (2013)
7. Acero, F., Ackermann, M., Ajello, M., Albert, A., Atwood, W. B., Axelsson, M., Baldini, L., Ballet, J., Barbiellini, G., Bastieri, D., et al.: Fermi large area telescope third source catalog. ApJS **218**, 23 (2015)
8. Ackermann, M., Ajello, M., Albert, A., et al.: Science **345**, 554–558 (2014)
9. Ackermann, M., Ajello, M., Ballet, J., et al.: Science **335**, 189– (2012)
10. Ackermann, M., Ajello, M., Ballet, J., et al.: Astrophys. J. Lett. **773**, L 35 (2013)
11. Allafort, A., Baldini, L., Ballet, J., et al.: (Fermi LAT Collab.) Astrophys. J. Lett. **777**, L 2 (2013)
12. Atwood, W.B., Abdo, A.A., Ackermann, M., et al.: (Fermi LAT Collab.) Astrophys. J. **69**, 1071– (2009)
13. Bignami, G.F., Caraveo, P.A.: Ann. Rev. Astron. Astrophys. **34**, 331– (1996)
14. Caraveo, P.A.: Ann. Rev. Astron. Astrophys. **52**, 211– (2014)
15. Nolan, P.L., Abdo, A.A., Ackermann, M., et al.: (Fermi LAT Collab.) Astrophys. J. Suppl. **199**, 31– (2012)
16. Reitberger, K., Reimer, O., Reimer, A., et al.: Astron. Astrophys. **544**, A 98 (2012)
17. Tavani, M., Bulgarelli, A., Piano, G., et al.: Nature **462**, 7273 (2009)
18. Tavani, M., Bulgarelli, A., Vittorini, V., et al.: Science **331**, 736– (2011)

Chapter 45
Selection of High-z Radio-Loud Quasars, and Their Luminosity Function

D. Tuccillo, J.I. González-Serrano, and C.R. Benn

Abstract We present the selection techniques based on the use of Neural Networks and on the cross match between FIRST and SDSS that lead to the spectroscopic identification of 15 new Radio Loud QSOs in the redshift range $3.6 \leq z \leq 4.4$. These QSOs did not have previous spectroscopical identification in SDSS or other works. Our selection method is highly complete (97 %) and it allows the estimation of the binned luminosity function of radio-loud quasar at $z \sim 4$ with unprecedented accuracy. Our luminosity function is compared with the results of other samples of RL QSOs in similar ranges of redshift and with the whole population of QSOs (RL+RQ). The evolution of the luminosity function with redshift was for many years interpreted as a flattening of the bright end slope, but has recently been re-interpreted as strong evolution of the break luminosity for high-z QSOs and our results, for the radio-loud population, are consistent with this. We also find indications of a constant radio-loud fraction for QSOs at high z. Our next investigation will select RL QSOs candidates in the range of redshift $4.4 \leq z \leq 5.7$, and will make use of data in the radio (FIRST), in the optical (SDSS DR10) and in the infrared (UKIDSS Large Area Survey DR10, and WISE).

45.1 Introduction

Nowadays the largest homogeneous sample of quasars is the SDSS Quasar Catalog V [6]. This is based on the SDSS Seventh Data Release (DR7) and contains 105,783 spectroscopically confirmed quasar. In spite of this remarkable success, the development of new techniques for targeting quasar candidates are still of great interest. Using different selection strategies it is possible to discover quasars that are missed by the SDSS selection, to reduce and better quantify the selection biases.

D. Tuccillo (✉) • J.I. González-Serrano
Instituto de Fisica de Cantabria (CSIC – Universidad de Cantabria), E-39005, Av. de los Castros s/n Santander, Cantabria, Spain
e-mail: tuccillo@ifca.unican.es; gserrano@ifca.unican.es

C.R. Benn
Isaac Newton Group, Apartado 321, E-38700, Santa Cruz de la Palma, Spain
e-mail: crb@ing.iac.es

N.R. Napolitano et al. (eds.), *The Universe of Digital Sky Surveys*, Astrophysics and Space Science Proceedings 42, DOI 10.1007/978-3-319-19330-4_45

In this context, the selection of Radio-Loud (RL) quasars is particularly interesting to check the completeness of samples selected on the basis of optical data only, since the use of radio-data reduces the biases due to reddening and/or dust obscuration. Also, increasing their population (representing $\sim$10 % of the whole population of quasars) and studying the evolution of their demography with more complete and unbiased samples, it is possible to gain some insight into the relationship between Radio-Loud and Radio-Quiet phenomena.

We present a resume of the results discussed in [9], for a selection of RL QSOs in the redshift range $3.6 \leq z \leq 4.4$ and our estimation of the RL luminosity function for $z \sim 4$. We discuss the implications of these results and the prospects of our investigation.

45.2 Selection of RL QSOs $3.6 \leq z \leq 4.4$

We selected a sample of 87 radio-loud QSOs in the redshift range $3.6 \leq z \leq 4.4$, detected in the FIRST radio survey ($S_{1.4\text{GHz}} > 1$ mJy) and having $r < 20.2$ in SDSS data release 7. Of these 87 QSOs, 15 of them are not included in the SDSS QSOs catalog and their spectra were obtained for the first time in our investigations. They were selected as QSOs candidates using our selection strategy based on the use of a Neural-Network. This selection method is proved to be 97 % complete and with an efficiency of $\sim$60 %. In Fig. 45.1 we show the spectra of the 7 QSOs within $3.6 \leq z \leq 4.4$, identified in [9]. Selection strategy, spectra of the candidates, assessment of the completeness and efficiency of our methodology are discussed in sections 2, 3 and 4 of [9].

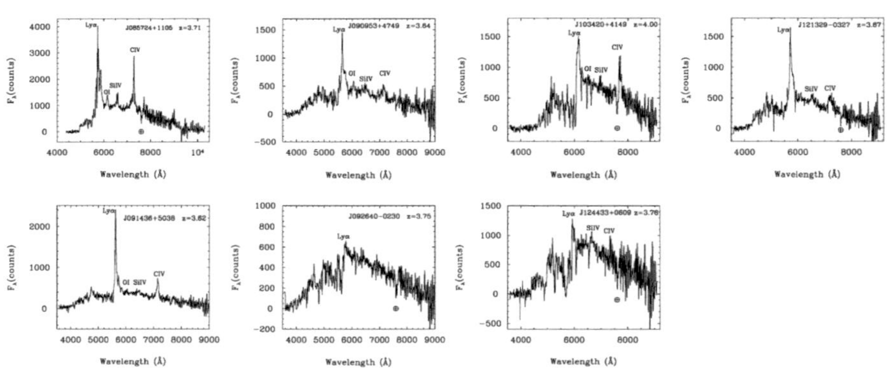

Fig. 45.1 The NOT/ALFOSC spectra of 7 Neural Network-selected in [9] as high-z candidates and identified as QSOs with $3.6 \leq z \leq 4.4$

45.3 RL QSOs Luminosity Function at $z \sim 4$

We use our sample of 87 RL QSOs to estimate the binned optical luminosity function of radio-loud QSOs at $z \sim 4$. From this, we estimate the luminosity function of the total QSOs population and their comoving density. For the radio-loud population at $M_{1,450} < -25.8$, we obtain values of the space densities: $\rho(z \approx 3.8, M_{1,450} < -25.8)_{\rm RL} = 4.51 \pm 0.61\,{\rm Gpc}^{-3}$ and $\rho(z \approx 4.2, M_{1,450} < -25.8)_{\rm RL} = 1.54 \pm 0.63\,{\rm Gpc}^{-3}$. In section 8.2 of [9], we argue that this determination of the space density suggests that the Radio Loud Fraction (RLF) at high-z is likely to be the same as at low-z, and that other authors may be underestimating the RLF at high-z.

In Sect. 9.3 we determine the slope of the optical luminosity function at $z = 3.8$ and at $z = 4.2$, finding respectively $\beta_{z=3.8} = -2.3 \pm 0.2$, and $\beta_{z=4.2} = -2.0 \pm 0.4$. Our results for the Radio-Loud population are consistent with the recent determinations of a strong evolution of the break luminosity for the whole population of high-z QSOs [4]. On the other hand our results can be also interpreted as suggestive of a flattening of the bright-end slope for the radio-loud population only. If confirmed, this implies an evolution of the density of super-massive black holes associated with radio-loud QSOs, in the sense that they were more abundant at $z \sim 4$ (Fig. 45.2).

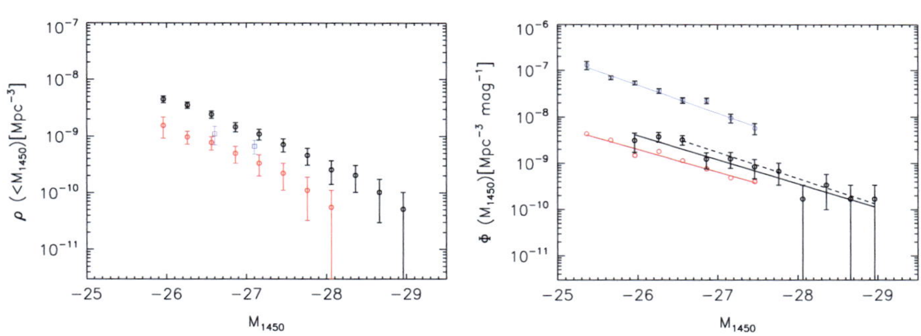

Fig. 45.2 Our determination of the space density and optical luminosity function for RL QSOs at $z \sim 4$. *On the left*: cumulative luminosity functions at $z = 3.8$ (*black circles*) and $z = 4.2$ (*red circles*). *Squares* represent densities derived by [7] (*blue*) and [1] (*purple*) at $z \sim 4.1$ and $z \sim 4$, respectively. *On the right*: *Black points* show the luminosity function we derived for $z \sim 3.8$. For comparison, *red points* show the luminosity function calculated by [3], for RL QSOs with radio-loudness $R > 70$; *blue points* the LF calculated by [5] for the entire population of QSOs

45.4 Conclusions and Ongoing Research

In [9] we show that our neural-network selection is highly complete, detecting ~97 % of the high-z RL QSOs, while SDSS only detects ~85 % of them. Our determinations of the luminosity function for the RL QSOs population at $z \sim 4$ are generally in agreement with other recent determinations. However, to clarify the evolution of the RL population, more observational constraints are needed, especially at $z > 4$.

Recent studies based on the SDSS QSOs catalog, find a trend for RL QSOs to be stronger radio sources [2] and with higher R-parameter [8] with increasing redshift. The average values of radio luminosity and of the radio-loudness (R parameter) of our 15 new discoveries do not contradict those results, because in agreement with the SDSS QSOs catalog.

Our ongoing investigations are focused on: (i) the selection of RL QSOs in the redshift range $4.4 \leq z \leq 5.7$ using data from FIRST, SDSS, UKIDSS and WISE; (ii) on the quantification of the selection biases affecting the statistical studies of the radio properties of high-z RL QSOs.

References

1. Carballo, R., Gonzalez-Serrano, J.I., Montenegro-Montes, et al.: A FIRST-APM-SDSS survey for high-redshift radio QSOs. MNRAS **370**, 1034 (2006)
2. Kratzer, R.M.: Mean and extreme radio properties of quasars and the origin of radio emission. PhD thesis, Drexel University (2014)
3. McGreer, I.D., Helfand, D.J., White, R.L.: Radio-selected quasars in the Sloan Digital Sky Survey. AJ **138**, 1925 (2009)
4. McGreer, I.D., Jiang, L., Fan, X.: The z = 5 quasar luminosity function from SDSS Stripe 82. ApJ **768**, 105 (2013)
5. Richards, G.T., Strauss, M.A., Fan, X., et al.: The Sloan Digital Sky Survey quasar survey: quasar luminosity function from data release 3. AJ **131**, 2766 (2006)
6. Schneider, D.P., Richards, G.T., Hall, P.B., et al.: The Sloan Digital Sky Survey quasar catalog. V. Seventh data release. AJ **139**, 2360 (2010)
7. Vigotti, M., Carballo, R., Benn, C.R., et al.: Decline of the space density of quasars between z = 2 and z = 4. ApJ **591**, 43 (2003)
8. Singal, J., Petrosian, V., Stawarz, L., Lawrence, A.: The radio and optical luminosity evolution of quasars. II. The SDSS sample. ApJ **734**, 43 (2013)
9. Tuccillo, D., Gonzalez-Serrano, J.I., Benn, C.R.: Neural-network selection of high-redshift radio quasars, and the luminosity function at z ~ 4. MNRAS **449**, 2818 (2015)

Chapter 46
Science with the EXTraS Project: Exploring the X-Ray Transient and Variable Sky

A. De Luca, R. Salvaterra, A. Tiengo, D. D'Agostino, M.G. Watson, F. Haberl, and J. Wilms on behalf of the EXTraS collaboration

Abstract The EXTraS project ("Exploring the X-ray Transient and variable Sky") will characterise the temporal behaviour of the largest ever sample of objects in the soft X-ray range (0.1–12 keV) with a complex, systematic and consistent analysis of all data collected by the European Photon Imaging Camera (EPIC) instrument onboard the ESA XMM-Newton X-ray observatory since its launch. We will search for, and characterize variability (both periodic and aperiodic) in hundreds of thousands of sources spanning more than nine orders of magnitude in time scale and six orders of magnitude in flux. We will also search for fast transients, missed by standard image analysis. Our analysis will be completed by multiwavelength characterization of new discoveries and phenomenological classification of variable sources. All results and products will be made available to the community in a public archive, serving as a reference for a broad range of astrophysical investigations.

A. De Luca (✉) • R. Salvaterra
INAF-IASF Milano, Via Bassini 15, I-20133 Milano, Italy
e-mail: deluca@iasf-milano.inaf.it; ruben@iasf-milano.inaf.it

A. Tiengo
IUSS Pavia, Piazza della Vittoria 15, I-27100 Pavia, Italy
e-mail: andrea.tiengo@iusspavia.it

D. D'Agostino
CNR-IMATI, Via de Marini 6, I-16149 Genova, Italy
e-mail: dagostino@ge.imati.cnr.it

M.G. Watson
Department of Physics and Astronomy, University of Leicester, Leicester, LE1 7RH, UK
e-mail: mgw@leicester.ac.uk

F. Haberl
MPG-MPE, Giessenbachstrasse, D-85748 Garching, Germany
e-mail: fwh@mpe.mpg.de

J. Wilms on behalf of the EXTraS collaboration
ECAP, Sternwartstrasse 7, D-96049 Bamberg, Germany
e-mail: joern.wilms@sternwarte.uni-erlangen.de

N.R. Napolitano et al. (eds.), *The Universe of Digital Sky Surveys*, Astrophysics and Space Science Proceedings 42, DOI 10.1007/978-3-319-19330-4_46

46.1 The EXTraS Project: Aim, Implementation

The EXTraS project ("Exploring the X-ray Transient and variable Sky") will systematically explore the temporal domain information stored in the database collected by the European Photon Imaging Camera (EPIC, [13, 15]) onboard the ESA XMM-Newton observatory. EPIC is the most powerful tool to study faint X-ray sources in the 0.1–12 keV range. Indeed, the Serendipitous Source Catalogue based on EPIC data, listing more than 560,000 detections in its most recent release (3XMM, see [9]), is the largest catalogue of X-ray sources ever compiled. However, time-domain information in such data, although very rich, remained mostly unused.

EXTraS will release to the community a full temporal characterization (both aperiodic and periodic variability) of hundreds of thousand of sources with flux spanning from 10^{-9} to 10^{-15} erg cm^{-2} s^{-1} (0.2–10 keV), on time scales ranging from $\sim$0.1 s to $\sim$10 years. EXTraS will also search for short-duration transients, not included in 3XMM, and will also perform a phenomenological classification of all sources based on their temporal and spectral properties. The different lines of temporal analysis implemented within the project are described in more detail in [3].

The project has received funding within the EU-FP7 framework (grant agreement n. 607452) and is carried out by a collaboration including INAF (Italy), IUSS (Italy), CNR/IMATI (Italy), University of Leicester (UK), MPE (Germany) and ECAP (Germany). More information, updates on the project as well as a full list of contacts can be found by visiting the project web site at www.extras-fp7.eu.

46.2 An (Incomplete) Overview of Science with EXTraS

The extremely broad range of variability timescales and luminosities investigated by EXTraS is shown in Fig. 46.1. The scientific discovery space is very large and we trust that EXTraS results will have a great impact in many areas of astrophysics and cosmology. The blind nature of our search will allow astronomers to measure (or give strong limits on) the intrinsic occurrence rate of different transient events and to perform population studies. Without demanding completeness, we include below a list of science cases that will benefit of our results (see also caption to Fig. 46.1).

Flaring stars: constraining, on a statistical basis, the duration, duty cycle and amount of energy released in flares of different stellar groups. Indeed, as an interesting first result of EXTraS, we may mention the unexpected detection of a flare from a *very young protostar* [6].

Cataclysmic variables and Novae: unveiling periodicities and bursts in about 100 known sources; searching for new systems.

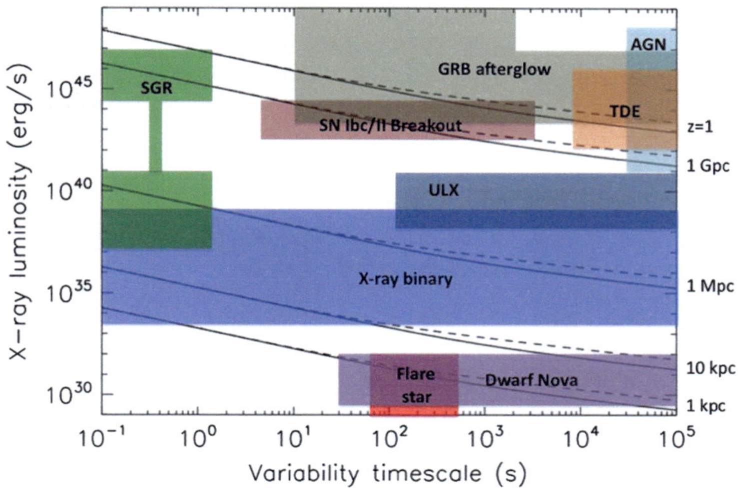

Fig. 46.1 Peak X-ray luminosity of different source populations as a function of their characteristic variability time scale (Adapted from [8]). Diagonal lines mark the approximate EXTraS 0.5–2 keV sensitivities (at 4σ confidence, EPIC/pn camera only) for various source distances. *Solid/dashed lines* correspond to min/max particle background levels. Even assuming a very conservative 6.5σ sensitivity level (a factor 2–5 higher flux than in the figure), implying an essentially null contamination from spurious detections, we can expect a very rich harvest of variable sources, based on the equivalent sky coverage of EPIC data and on previous observational results and theoretical models. Among such sources, several hundreds of flares from a broad population of stars and protostars; 3–300 and 0.3–30 bursts from persistent and transient LMXBs, respectively, in the Galactic Center region; 3–5 SFXTs active in bright flaring and 40–75 sources in an intermediate state of flaring; several new magnetars; a few TDEs up to $z \sim 5$–6; up to a dozen low-luminosity GRBs up to $z \sim 2$; about ten SN shock break-out events up to $z \sim 0.1$; moreover, it can be estimated that about 100 "Fast Radio Burst" should have occurred within the field of an ongoing EPIC pointed observation

Low-mass X-ray binaries (LMXBs): probing e.g. the properties of bursting LMXBs [12] – in particular, for the poorly known "burst only sources". Bursts from M31, the closest spiral galaxy, can also be detected [5].

High-Mass X-ray Binaries (HMXBs): constraining e.g. the census of Supergiant Fast X-ray Transients (SFXT) [10], which is crucial to unveil the evolutionary path and formation rate of massive stars.

Isolated Neutron Stars: constraining e.g. the population of magnetars (Anomalous X-ray Pulsars, AXPs, and Soft Gamma Repeaters, SGRs) [7] and its relation to the overall population of Neutron Stars (NS), which will also impact on our comprehension of the short GRBs. The case of GRB150301C/3XMM J004514.7+415035, with properties consistent with those of an active magnetar in M31, are a clear demonstration of the potentialities of EXTraS in this field (see GCN 17548 and ATel #7181).

ULtraluminous X-ray Sources (ULXs): assessing accretion physics in these poorly understood sources [4] by e.g. searching for, and unveiling, source "states"; systematically searching for bursts and periodicities (orbital, or rotational)

Tidal Disruption Events (TDEs): constraining their poorly known statistics (even the case of no detections would be interesting) and their physics [1].

Gamma-Ray Bursts (GRBs): constraining the rate of Low-Luminosity GRBs (e.g. [2]), possibly up to high redshift, and their progenitors; no detections would set the strongest available constraint on their population.

Supernovae (SNe): inferring a measure of the SN rate independent from optical surveys by detecting SN shock breakout events [11]; no detections would set important constraints on the poorly known underlying physics.

Active Galactic Nuclei (AGN): measuring the mass of $\sim$100 AGNs, based on their variability properties; calibrating the variability-luminosity relation locally; characterizing the variability of Blazars, their duty-cycle on short timescales and deriving constraints on particle injection/acceleration, magnetic field and BH mass.

Rare events: ranging from galactic events (e.g. TDEs of minor bodies falling onto NSs or BHs) to cosmic ones (e.g. orphan afterglows of GRB seen off-axis, GRBs from the first very massive stars); constraining the high energy properties of the puzzling "Fast Radio Bursts" [14].

Totally unexpected discoveries can be also foreseen, as has always been the case when a new region in parameter space has been explored.

References

1. Burrows, D.N., Kennea, J.A., Ghisellini, G., et al.: Relativistic jet activity from the tidal disruption of a star by a massive black hole. Nature **476**, 421 (2011)
2. Campana, S., Mangano, V., Blustin, A.J., et al.: The association of GRB 060218 with a supernova and the evolution of the shock wave. Nature **442**, 1008 (2006)
3. De Luca, A., Salvaterra, R., Tiengo, A., et al.: The EXTraS project: exploring the X-ray transient and variable sky. In: Swift: 10 yr of Discovery. To appear in PoS (2015)
4. Feng, H., Soria, R.: Ultraluminous X-ray sources in the Chandra and XMM-Newton era. New Astron. Rev. **55**, 166 (2011)
5. Pietsch, W., Haberl, F.: XMM-Newton detection of type I X-ray bursts in M31. A&A **430**, 25 (2005)
6. Pizzocaro, D., Stelzer, B., Paladini, R., et al.: Results from DROXO. IV. EXTraS discovery of an X-ray flare from the Class I protostar candidate ISO-Oph 85. A&A **587**, A36 (2016)
7. Mereghetti, S.: The strongest cosmic magnets: soft gamma-ray repeaters and anomalous X-ray pulsars. A&ARv **15**, 225 (2008)
8. Merloni, A., Predehl, P., Becker, W., et al.: eROSITA Science Book: Mapping the Structure of the Energetic Universe (2012). arXiv:1209.3114
9. Rosen, S., Webb, N.A., Watson, M.G., et al.: The XMM-Newton serendipitous survey. VII. The third XMM-Newton serendipitous source catalogue. Submitted to A&A (2015). arXiv:1504.07051

10. Sidoli, L.: Latest progress on supergiant fast X-ray transients and future direction. Adv. Spec. Res. **48**, 88 (2011)
11. Soderberg, A.M., Berger, E., Page, K.L., et al.: An extremely luminous X-ray outburst at the birth of a supernova. Nature **453**, 69 (2008)
12. Strohmayer, T., Bildsten, L.: New views of thermonuclear bursts. In: Lewin, W.H.G., van der Klis, M. (eds.) Compact Stellar X-Ray Sources, pp. 113–156. CUP, Cambridge (2010)
13. Strüder, L., Briel, U., Dennerl, K., et al.: The European photon imaging camera on XMM-Newton: the pn-CCD camera. A&A **365**, L18 (2001)
14. Thornton, D., Stappers, B., Bailes, M., et al.: A population of fast radio bursts at cosmological distances. Science **341**, 53 (2013)
15. Turner, M.J.L., Abbey, A., Arnaud, M., et al.: The European photon imaging camera on XMM-Newton: the MOS cameras. A&A **365**, L27 (2001)

Chapter 47
What's Next for VST: Electromagnetic Follow-Up of Gravitational Waves Events

A. Grado, E. Cappellaro, S. Piranomonte, E. Brocato, M. Branchesi, S. Covino, S. Campana, F. Getman, G. Greco, L. Nicastro, E. Pian, E. Palazzi, L. Stella, and G. Stratta on behalf of a larger collaboration

Abstract A big step forward in the long-standing quest for gravitational waves (GWs) will be made next year when the LIGO and VIRGO collaborations will start regular operations of their sensitive, upgraded interferometers. It is crucial that the electromagnetic counterparts of GW events are securely identified, a difficult task because of the large size of error box expected to be returned by the interferometers (dozens to hundreds of square degrees). Our group is tackling the challenge by organizing a follow-up campaign covering the widest possible range of the electromagnetic spectrum. The optical counterpart will be covered by the VST thanks to its characteristics. The sensitivity and optical quality of the telescope will allow us to probe faint transients (e.g. kilonovae and short GRBs) that are among the most promising GW source candidates.

A. Grado (✉)
INAF-Osservatorio Astronomico di Capodimonte, Via Moiariello 16, 80131, Naples, Italy
e-mail: agrado@na.astro.it

E. Cappellaro
INAF–Osservatorio Astronomico di Padova, vicolo dell'Osservatorio 5, 35122, Padova, Italy

S. Piranomonte • E. Brocato • L. Stella
INAF – Osservatorio Astronomico di Roma, Monte Porzio Catone (RM), Italy

M. Branchesi • G. Greco • G. Stratta
Università degli studi di Urbino, Urbino, Italy

S. Covino • S. Campana
INAF – Osservatorio Astronomico di Brera, Milan, Italy

F. Getman
INAF-Osservatorio Astronomico di Capodimonte, Via Moiariello 16, 80131, Naples, Italy

L. Nicastro • E. Palazzi
INAF-Istituto di Astrofisica Spaziale e Fisica Cosmica, Bologna, Italy

E. Pian
Scuola Normale Superiore, Pisa, Italy

N.R. Napolitano et al. (eds.), *The Universe of Digital Sky Surveys*, Astrophysics and Space Science Proceedings 42, DOI 10.1007/978-3-319-19330-4_47

47.1 Introduction

Direct detection of gravitational waves may become a reality in the coming years when a worldwide network of advanced versions of ground-based GW interferometers will start to operate [6, 11]. Besides representing a direct test of general relativity, the detection of GWs allows insight into radiation processes, explosion mechanisms and stellar structure in a way that is independent of absorption and propagation effects that often plague electromagnetic (EM) observations. GWs make possible to investigate the inner regions of astrophysical objects not accessible with photons at all the wavelength (optical, infrared, X, γ).

There is no doubt that to really exploit the huge efforts devoted to the detection of GWs, it is crucial that their electromagnetic counterparts are securely identified. The Italian National Institute of Astrophysics (INAF) has signed the MoU with LIGO-Virgo collaboration (LVC) thus recognising the need of a joint institutional effort to maximize the scientific return of these campaigns. The Italian observing facilities will include national telescopes directly controlled by INAF (e.g. CITE, TNG, REM, SRT, ASTRI, etc.) and large telescopes accessible through consortia (VST, VLT, LBT, CTA). In particular the VST equipped with the OmegaCam camera [8, 17] will play a major role due to his wide field-of-view (FOV), optical quality and angular resolution. We plan to use part of the INAF-VST and INAF-OmegaCam guaranteed observation time to activate a transient search immediately after an alert from the GWs Observatories is issued.

47.2 Gravitational Waves Sources

The Einstein's theory of The General Relativity predicts that mass distributions with time-varying mass quadrupole moments produce GW. These waves are oscillating perturbations of space-time travelling at the speed of light. GWs represent a unique way to investigate the interior regions of extremely dense astrophysical objects. Up to now, no direct GW detection has been obtained, even if indirect evidence has been found to explain the orbital evolution of the binary pulsar PSR B1913+16, [25]. The direct detection of GWs will open a new window in astronomy and the expectations of the scientific community is obviously extremely high.

To this end, a large effort is currently under way by Virgo [6] and LIGO [5] experiments in order to improve the detector sensitivity by a factor of 10 in the 40–1,000 Hz range and even by a larger factor at 10–40 Hz. In terms of detectable sources this means to increase their number by a factor of 1,000.

In the recent years a number of theoretical studies predicted that the most promising GW sources are expected to be emitted by extremely dense binary systems such like neutron stars (NS) and black hole (BH) systems during their coalescent phase (NS-NS or NS-BH). With the expected sensitivity of advanced LIGO and Virgo network is reasonable to foresee that this kind of sources can be detected within 200 Mpc (NS-NS) and 400 Mpc (NS-BH) [2]. According to the predicted rate of 1 event/week for these binary systems coalescence [1, 2] the unexplored GW astronomy is about to begin soon.

Other sources are expected to produce transient GW signals as well. Among them perturbed NS and BH [15, 16] and the core collapse of massive stars [21]. The event rate of core collapse supernovae (CC-SNe) is of the order of 100–1,000 events per year within a distance of 100 Mpc [10]. However, the number of CC-SNe, as well as for perturbed NS and BH, detectable by the advanced GW detectors is highly uncertain [3, 4, 20] owing to uncertainties in the amount of energy emitted in GWs.

It is worth noticing that CC-SNe explosions emit both gravitational wave and MeV neutrinos (e.g. SN 1987A [27, 28]). Therefore the contemporaneous detection of their GW, MeV neutrinos and EM signals will provide a unique tool to study the physics of such explosive events and open the era of the multi-messenger astronomy.

47.3 The GW Sources in the EM Window

How likely is it to observe the EM counterparts of GW events? In the current understanding the massive binary systems are thought to be the strongest GW emitters and these sources are expected to have an EM counterpart. The progenitors of long gamma-ray bursts (LGRBs) are thought to be due to the collapse of massive stars ("collapsars", e.g. [12, 29]). On the other hand and more promising for the GW detection, the short GRBs (SGRB) seems to be generated by the coalescence of NS-NS or NS-BH systems [7]. In spite of the extensive studies of GRBs from ground and space, a clear identification of the progenitors is still missing. A coincident detection of a GW signal from a SGRB will represent a strong evidence that the coalescing process of NS-NS or NS-BH system are the real origin of these high energy phenomena.

In this scenario the source orientation will play a fundamental role. SGRB observed in gamma-rays and hard X-rays will be in a face-on configuration. When the relativistic jets slows-down due to the interaction with the interstellar matter, GRBs will be visible as an afterglow radiating also at soft X-rays, optical and radio frequencies with a progressively larger beaming angle [18, 30]. When a GRB is observed off-axis, the emission at lower-energies (i.e. from radio to X with no emission in gamma-rays) is indeed expected to be the only signal from the jet that will reach the observer, generating the so-called "orphan afterglows". Mergers of compact objects likely to generate GWs, are believed to produce also the so-called kilo nova radiation, i.e. the isotropic thermal emission expected in the optical/IR bands powered by the radioactive decay of heavy elements [22]. The kilonova emission appears days to weeks [19, 22] after the merger and remains bright for a similar time (Fig. 47.1). The kilonova emission can provide information of the merger processes, as the light curves depend on the mass, velocity and geometry of the ejecta, while their opacity and spectral feature probe the ejecta composition. As discussed in the previous section, other possible sources of GWs exist, but the energy released as gravitational radiation is highly uncertain and are of minor interest for the EM follow-up.

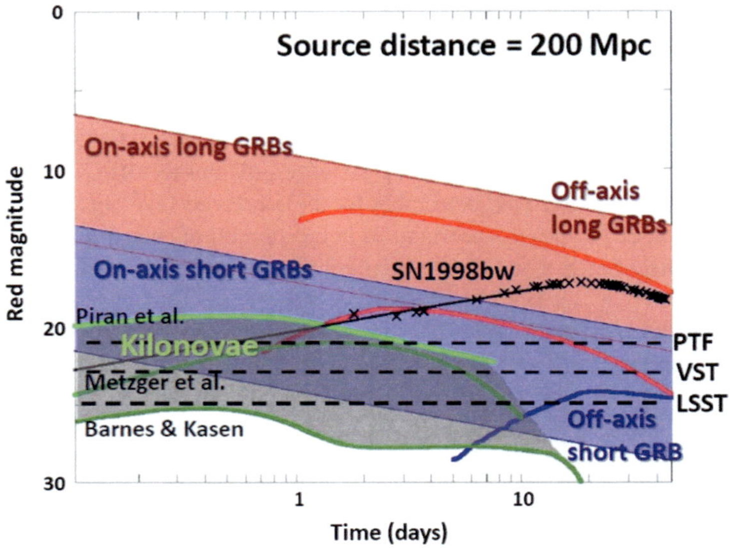

Fig. 47.1 GRB afterglow, kilonova and supernova R-band emissions as function of time from the GW emission. The *blue* (*red*) *band* indicates the region occupied by the observed on-axis short (long) GRB afterglows [13, 14]. The *blue (red) lines* indicate examples of off-axis afterglows of short (long) GRBs (modeled by [26]). The *grey band* is the region of the kilonova model [19, 22]. An example of GRB-associated SN emission is shown by *solid black line* (*black asterisks*; observation of SN 1998bw by [9]). The VST, PTF and LSST limiting magnitude for 80 s in r band for point sources with S/N 10 are indicated by *black dashed-lines*

47.4 Observations of Electromagnetic Counterpart of Gravitational Wave Events

To identify the EM counterpart of GW events at least two critical issues have to be addressed: (i) the determination of the GW source position and (ii) the short time before sources fade out. The GW trigger will be passed to the EM telescopes with a large positional uncertainties ranging from tens to hundreds square degrees. At the same time the high sensitivity of the advanced GWs detectors will access a large volume of the Universe. These two characteristics of GWs detection require, for the EM follow up, the use of both wide field of view and medium to large effective area telescopes. For what concern the time constraints the Fig. 47.1 shows the expected on- and off- axis afterglow emission from both long and short GRBs in the R-band as a function of time as well as all the quoted possible sources discussed in Sect. 47.3. One may appreciate that these sources will be observable only for few days by using medium to large sized telescopes. The VST meets all the requirements [8] needed for an effective follow-up of GWs.

The LIGO and Virgo collaborations (LVC) have engaged in a program of low-latency gravitational wave data analysis to generate triggers for significant GW candidates and share them with astronomical institutes which signed a MoUs with the LVC. INAF has signed the MoU and we are in the phase of set-up the hardware, software and observational facilities to react to such triggers and begin the observing campaigns aimed at identifying their possible optical counterpart. In particular the area surveyed by VST and its tiling strategy will be dictated by the GW signal direction (sky-position probability maps) and the telescope pointing constraints. Our initial strategy is to survey up to 100 deg^2 chosen as those sky map pixels with the highest probability to contain the GW source. We plan to perform the survey in two filters (g,r bands) and in three epochs, that is $t_0 + (0.1 \div 1)d$, $t_0 + (1 \div 3)d$, $t_0 + (20 \div 40)d$ where t_0 is the time of GW detection and d stays for day. The temporal sampling is chosen allowing for the rapid evolution of the expected GWs counterpart. Whenever suitable past archival images of the same field will be available, the transients can be identified already from the first epoch of the survey. Otherwise we will need a comparison with the second epoch for objects with rapid luminosity evolution (e.g. GRBs) or with the third epoch for slower evolving transients (e.g. SNe). Our immediate objective is to identify as early as possible all transients in the sky map area and select among them the best candidate for GWs counterparts [23, 24]. These will be further scrutinized by mean of dedicated photometric and spectroscopic observations at all wavelengths exploiting all the facilities controlled and/or participated by INAF. The VST needs approximatively one night of observation to cover 100 deg^2 in two filters with a limiting magnitude of 22.5 for point-like sources with S/N 10. This is what is required for the detection of faint transients in the local Universe.

47.5 Conclusions

There is no doubt that the detection of EM counterpart of GW signals will be one of the most challenging projects for space and ground based observatories. The INAF astronomical community is working to collect the human and financial resources to contribute to this effort. We plan to take advantage of the availability of the VST observational facilities and of the wide experience in the Time Domain Astronomy of the Italian community.

References

1. Aasi, J., et al.: (2013). Prospects for observing and localizing gravitational-wave transients with advanced LIGO and advanced virgo. arXiv:1304.0670
2. Abadie, J., et al.: TOPICAL REVIEW: predictions for the rates of compact binary coalescences observable by ground-based gravitational-wave detectors. CQG **27**(17) (2010). Art. No. 173001
3. Abadie, J., et al.: Search for gravitational wave bursts from six magnetars. ApJ **734**, id. L35, 9 (2011)
4. Abadie, J., et al.: Search for gravitational waves associated with the August 2006 timing glitch of the Vela pulsar. Phys. Rev. D **83**(4), id. 042001 (2011)

5. Abbott, B.P., et al.: LIGO: the laser interferometer gravitational-wave observatory. Rep. Prog. Phys. **72**, 076901 (2009)
6. Accadia, T., et al.: Virgo: a laser interferometer to detect gravitational waves. JINST **7**(03), 3012 (2012)
7. Berger, E.: The environments of short-duration gamma-ray bursts and implications for their progenitors. NewAR **55**(1), 1 (2011)
8. Capaccioli, M., Schipani, P.: The VLT survey telescope opens to the sky: history of a commissioning. Msngr **146**, 2 (2011)
9. Clocchiatti, A., Suntzeff, N.B., Covarrubias, R., Candia, P.: The ultimate light curve of SN 1998bw/GRB 980425. AJ **141**, 163 (2011)
10. Guetta, D., Della Valle, M.: On the rates of gamma-ray bursts and type Ib/c supernovae. ApJ **657**, L73 (2007)
11. Harry, G.M.: Advanced LIGO: the next generation of gravitational wave detectors. CQG **27**, 084006 (2010)
12. Hjorth, J., Bloom, J.S.: The gamma-ray burst - supernova connection. Camb. Astrophys. Ser. **51**, 169 (2012)
13. Kann, D.A., Klose, S., Zhang, B., et al.: The afterglows of swift-era gamma-ray bursts. I. Comparing pre-swift and swift-era long/soft (type II) GRB optical afterglows. ApJ **720**(2), 1513 (2010)
14. Kann, D.A., Klose, S., Zhang, B., et al.: The afterglows of swift-era gamma-ray bursts. II. Type I GRB versus type II GRB optical afterglows. ApJ **734**(2), 47 (2011)
15. Kokkotas, K., Schmidt, B.: Quasi-normal modes of stars and black holes. LRR **2**, 2 (1999)
16. Kokkotas, K.D., Stergioulas, N.: Gravitational waves from compact sources. In: Mourào, A.M. (ed.) New Worlds in Astroparticle Physics, Proceedings of the Fifth International Workshop. Held 8–10 January 2005 in Faro, Portugal, pp. 25–46. World Scientific (2006)
17. Kuijken, K., Bender, R., Cappellaro, E., et al.: OmegaCAM: the 16k×16k CCD camera for the VLT survey telescope. Msngr **110**, 15 (2002)
18. Meszaros, P.: Gamma-ray bursts. Rep. Prog. Phys. **69**, 2259 (2006)
19. Metzger, B.D., Martínez-Pinedo, G., Darbha, S., et al.: Electromagnetic counterparts of compact object mergers powered by the radioactive decay of r-process nuclei. MNRAS **406**, 2650 (2010)
20. Ott, C.D.: Probing the core-collapse supernova mechanism with gravitational waves. CQG **26**, id. 204015 (2009)
21. Ott, C.D., Abdikamalov, E., Mösta, P., et al.: General-relativistic simulations of three-dimensional core-collapse supernovae. ApJ **768**, 115 (2013)
22. Piran, T., Nakar, E., Rosswog, S.: The electromagnetic signals of compact binary mergers. MNRAS **430**(3), 2121 (2013)
23. Singer, L.P., Cenko, S.B., Kasliwal, M.M., et al.: Discovery and redshift of an optical afterglow in 71 deg2: iPTF13bxl and GRB 130702A. ApJ **776**, L34 (2013)
24. Singer, L.P., Kasliwal, M.M., Cenko, S.B.: The needle in the 100 deg^2 Haystack: uncovering afterglows of Fermi GRBs with the palomar transient factory. ApJ **806**(1), 21 (2015)
25. Taylor, J.H., Weisberg, J.M.: A new test of general relativity – gravitational radiation and the binary pulsar PSR 1913+16. ApJ **253**, 908 (1982)
26. van Eerten, H.J., MacFadyen, A.I.: Synthetic off-axis light curves for low-energy gamma-ray bursts. ApJ **733**, 2, L37, 7 (2011)
27. van Putten, M.H.P.M.: Gravitational waves from core-collapse supernovae and long GRBs-Long GRBs and massive stellar explosions from frame dragging around black holes. ASPC **482**, 177 (2014)
28. Wallace, J., Burrows, A., Dolence, J.: Detecting the supernova breakout burst in terrestrial neutrino detectors. ApJ **817**(2), 182, 24 (2016)
29. Woosley, S.E., Bloom, J.S.: The supernova gamma-ray burst connection. ARA&A **44**, 507 (2006)
30. Zhang, B., Meszaros, P.: Gamma-ray bursts: progress, problems & prospects. IJMPA **19**, 2385 (2004)

MIX
Papier aus verantwortungsvollen Quellen
Paper from responsible sources
FSC® C105338

If you have any concerns about our products,
you can contact us on
ProductSafety@springernature.com

In case Publisher is established outside the EU,
the EU authorized representative is:
Springer Nature Customer Service Center GmbH
Europaplatz 3, 69115 Heidelberg, Germany

Printed by Libri Plureos GmbH
in Hamburg, Germany